A HANDBOOK OF AGRICULTURAL SCIENCES

Other Books by Dr. K. Vanangamudi

ISBN 9789389907001

ISBN 9789389547986

K. VANANGAMUDI

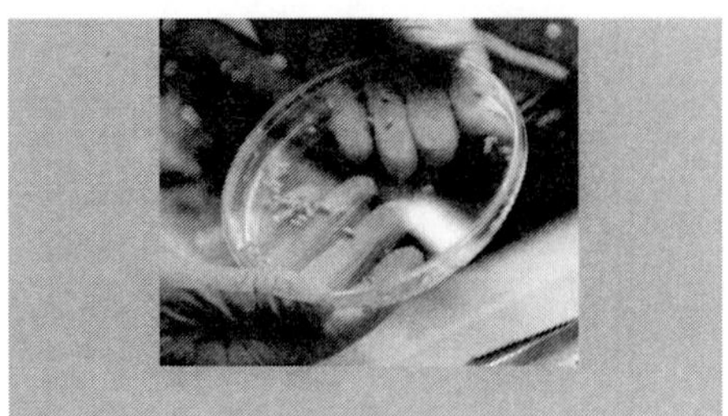

ISBN 9789389547900

ISBN 9788194849506

VALUED ADDED BENEFITS (VAB)
www.nipaers.com

VAB is new value added benefit service provided by NIPA to its customers in terms of e products being offered by and at **www.nipaers.com**, along side the print versions.

VAB marked books are being provided with a scrathable code inside the book, which can be claimed for a limited period of time by **registering on www.nipaers.com**

STEP 01: LOGON TO WWW.NIPAERS.COM

↓

STEP 02: CLICK ON EBOOKS/ONLINE TEST SERIES

↓

STEP 03: REGISTER YOURSELF BY ENTERING ALL REQUIRED DETAILS

↓

STEP 04: ENTER THE CODE (VISIBLE AFTER SCRATCHING THE SLIVER PATCH)

↓

STEP 05: VAB ARE NOW ACTIVATED

01. **e book:** User can claim an entire e book of the same print version book (already purchased) or a new book and can access it from any device for limited period from the date of claiming the VAB code. The user in this option, cannot download, save and print the book. This is only for reading.

02. **Online Test Series:** Users can claim a particular Online Test Series and can access it from any device for limited days from the date of claiming an Online Test Series. The user in this option, cannot download and print the test.

General Benefits

03. **Get in touch:** Always dreamt of being in touch with the likes of Professor M.S. Swaminathan, Professor K.V. Peter Professor V.K. Joshi, Dr. N.S. Rathore etc, Well your dream is now coming true as NIPA is now offering you this facility, just click on the Feedback Option of our book details page, mark, send it to author and the loop is started.

04. **Feedback to publisher:** Your valuable feedbacks help us to improve our contents, production and over all services we offer you, just click on the Feedback Option of our book details page, mark, send it to publisher and its done.

05. **Recommend it to a friend:** Just found a useful stuff for your friend, inform them about it, just click on the Feedback Option of our book details page, mark, recommend it to a friend, and its done.

Vol. 1

A HANDBOOK OF AGRICULTURAL SCIENCES

Importance of Agriculture / Crop Production / Natural Resource Management / Allied Agricultural Activities / Crop Improvement / Seed Science

K. Vanangamudi
Former Dean (Agriculture)
Tamil Nadu Agricultural University
Coimbatore - 641 003

M. Kokila
Junior Research Fellow
Trinity Cultural Academy
Coimbatore - 641 041

N. Chezhiyan
Former Professor and Head (Spices and Plantation Crops)
Tamil Nadu Agricultural University
Coimbatore - 641 003

New Delhi – 110 034

A Paperback Division of

NEW INDIA PUBLISHING AGENCY

101, Vikas Surya Plaza, CU Block, LSC Market
Pitam Pura, New Delhi – 110 034, India
Email: info@nipabooks.com
Web: www.nipabooks.com

For customer assistance, please contact
Phone: + 91-11-27 34 17 17 Fax: + 91-11- 27 34 16
16E-Mail: feedbacks@nipabooks.com

ISBN: 978-81-94766-87-2

Composed and Designed by NIPA

About the Authors

Dr. K. Vanangamudi, Former Dean (Agriculture) and Seed Scientist, graduated B.Sc., (Ag.) in 1975 from Agricultural College and Research Institute, Tamil Nadu Agricultural University, Coimbatore; M.Sc., (Ag.) and Ph.D. in Seed Science and Technology, respectively in 1977 and 1981 from TNAU; and Post Doctoral Fellow (1988-1990) at Mississippi State University, USA. He joined Department of Seed Science and Technology, TNAU as Assistant Professor in 1982 and served the University for 32 years as Associate Professor (1986-1988) at Agricultural Research Station, Bhavanisagar; Professor (1992-2001) at Forest College & Research Institute, Mettupalayam; Professor and Head (2001-2003) of Department of Seed Science and Technology; Dean, Adhiparasakthi Agricultural College (2003-2006), Kalavai; and Dean (Agriculture), Agricultural College and Research Institute, TNAU, Coimbatore (2006-2009). He retired from service on August 31, 2013.

He has undergone Post Doctoral Programme in Tree Seed Technology at Mississippi State University, USA from 1988-1990 and visited USA, Japan and Canada in connection with Forestry education and research under USAID-Winrock International program. He underwent training on e-learning and web based teaching at UC, Davis, California and Cornell University, Ithaca, New York, USA during 2017.

Dr. Vanangamudi's significant contributions in Agricultural Education are 1. starting B.Tech (Bioinformatics), B. Tech (Agricultural Information Technology) and BS (Agribusiness Management) degrees at AC&RI, Coimbatore, and B.Sc (Horti.) degree at AP Agricultural College; 2. He introduced e-education and online examination; 3. Under NAIP project on "Development of e-courses for B.Sc., (Ag) degree program as Co.PI, he developed e-learning materials for all the 50 courses and hosted in TNAU and ICAR websites; and 4. Video streaming of lecture presentation by the teachers in the classroom (video filming, editing and publishing).

He has prepared course materials for 90 courses for Industrial Training Institute (ITI) program in Agriculture, Horticulture, Agricultural Engineering, Animal husbandry, Poultry, Food processing under "Modular Employment Skills" and "Centre of Excellence" funded by National Instructional Media Institute, Ministry of Labour & Employment, GOI, Chennai, for the students and instructors of ITI all over India.

As a researcher, he was the Principal Investigator of 11 research projects funded by ICAR-NIAP, ICAR-World bank, ICFRE-World bank, Tamil Nadu Forest department, Ministry of Labour and Employment (GOI), CIDA-Gulpeh, Canada and Private agencies.

Dr. Vanangamudi has published 2 review papers, 35 research papers in International journals, 112 research papers in National journals, 29 popular articles in English and 53 in Tamil. He has written 46 books in English and 10 in Tamil. He has also written 11 technical bulletins in English and 29 in Tamil.

His recent books are "Objective Seed Science and Technology-2nd Edition, Revised & Enlarged", "MCQ's in Plant Breeding, Biotechnology & Seed Science", "Model test paper of Seed Science and Technology", "Seed Science and Technology: 2nd Enlarged and Fully Revised Edition", "Competitive Seed Science and Technology" and "Competitive Agriculture".

In recognition of his contribution in Agriculture Education, Research and Extension, he was awarded with Best PG teacher award in 1998 and Best scientist award in 1999 by TNAU; Best PG Teacher award 2000 by Madras Agricultural Students Union (MASU), Coimbatore; Best Seed Scientist award in 2003 by Hisar Agricultural University (HAU), Hisar; Tamil Nadu Scientist award in 2006 by Tamil Nadu Council for Science and Technology (TNCST), Chennai; Best Research Paper award in 2005 by Indian Society of Seed Technology (ISST), New Delhi and Bharat Ratna Dr. C. Subramaniam Best Teacher award in 2006 by Indian Council of Agriculture Research (ICAR), New Delhi.

He is a recognized guide of Bharathiar University, Coimbatore; Bharathidasan University, Trichy and Forest Research Institute, Dehradun.

He was a Research and Technical Consultant to Little's Orientals Balm and Pharmaceuticals Ltd., Chennai for seed priming, seed film coating and setting up of Seed Research Lab; to a farmer for 200 acres of farm land development; to a EU-KKID-Peace trust project for end term evaluation; and to a Kerala Forest Department project (World Bank) for developing training course syllabi and establishment of seed processing and testing units. He served as Senior Advisor and Consultant to Dr. Mohan's Health care Products Ltd., Chennai.

Ms. M. Kokila is now working as Junior Research Fellow with Faculty, Trinity Cultural Academy, Coimbatore-641 041. She graduated B.Sc., (Agri.) from Agricultural College and Research Institute (TNAU), Kudumiyanmalai, Pudukkottai in 2019.

She has published 2 books. She has participated in Personality Development Workshop held on 21st & 22nd January 2019 in AC & RI, Kudumiyanmalai and undergone a Industrial training at an Industrial Tie-Up Programme with J. M. Industry, at Pudukkottai, Tamil Nadu. She has undergone ten days internship with NGO (Vrutti). She has conducted training to the farmers on seedling treatment with biofertilizer, sugarcane trash composting during her UG program at KVK, Vamban. She also served as a National Service Scheme volunteer (2015-17) and committee leader in various college functions (2018-2019).

Dr. N. Chezhiyan worked as a Professor and Head, Department of Spices and Plantation crops, Horticultural College and Research Institute, Tamil Nadu Agricultural University, Coimbatore; and Professor and Head of Vegetable crops at Adiparasakthi Agricultural College, Kalavai. He graduated B.Sc., (Agri.) in 1972 and Ph.D (Floriculture) in 1991 from TNAU, Coimbatore; M.Sc., (Horticulture) from Allahabad Agricultural Institute, Allahabad.

Dr. N. Chezhiyan served in TNAU in various capacities including Professor and Head, Horticultural Reasearch Station, Thadiyankudisai for 30 years since 1979 to 2008. He also served as an Agricultural officer in the Department of Agriculture, Government of Tamil Nadu from 1972 to 1976.

He has rich experience in teaching, research and extension; and guided 18 M.Sc., (Horticulture) students and 16 Ph. D scholars in Horticulture. Two of his Ph.D scholars were awarded with ICAR Jawaharlal Nehru award for best thesis in Pomology. He released 13 varieties in fruits, vegetables, spices and flower crops. He has published 175 research papers in International and National journals, 9 books and 150 popular articles.

Dr. Chezhiyan attended 20 trainings and 20 workshops/seminars; and organized 35 trainings and 20 workshops/seminars. He served as a member of Academic Council and Board of Studies (Horticulture), TNAU, Coimbatore.

Preface

Agriculture is our wisest pursuit, because it will, in the end, contribute most to real wealth, good morals and happiness.

– **Thomas Jefferson**

Agricultural science is a broad multidisciplinary field of biology that encompasses the parts of natural, economic and social sciences that are used in practice and understanding of Agriculture.

The country has made significant advances in many of farm sectors such as service sector, industrial production etc, but agriculture continues to be the lifeline of the nation, especially for the Indians living in rural areas. Food is the most basic human need.

During the past 60 years, agricultural education has expanded rapidly in India to meet nation's demand for human resources and Agricultural technology.

By means of Research and Development, the farmer has been enabled to increase yield per acre, reduce losses from diseases, pests, and spoilage and augment net production by improved processing methods.

Students have variety of options to pursue courses in Agriculture. In India, need of Agricultural graduates is more to serve our nation for reducing the constraints in Agriculture.

This compendium provides an ample information for competitive exams point of view for aspiring students and will benefit maximum number of ambitious graduates to get into services of public and private sectors.

Hope this book "A Handbook of Agricultural Sciences Vol. 1" would be supportive to the Agriculture graduates in this competitive world.

Coimbatore
October 2, 2020

K. Vanangamudi

Contents

UNIT-III
NATURAL RESOURCE MANAGEMENT

Part 1: Soil Science

Part 2: Dryland Agriculture

Part 3: Watershed and Wasteland Management

Part 4: Water Management

UNIT-IV
CROP MANAGEMENT & ALLIED AGRICULTURAL ACTIVITIES

Part 1: Crop Management

Part 2: Agricultural Microbiology

Part 3: Mushroom

Part 4: Bee Keeping (Apiculture)

Part 5: Silkworm Rearing

Part 6: Agricultural Engineering

Part 7: Animal Husbandry

UNIT-V
CROP IMPROVEMENT

Part 1: Plant Breeding

Part 2: Plant Biotechnology

UNIT-VI
SEED SCIENCE AND TECHNOLOGY

UNIT-I
IMPORTANCE OF AGRICULTURE

1

Agriculture

1.1. Agriculture

- Derived from Latin word; Ager-soil and culture-Cultivation.
- **An art:** Governs the knowledge to perform the operations of the farm in a skillful ways.
- **A science:** Uses technologies including breeding, production technologies, crop protection, economics etc., to maximize the yield and profitability of the farmer.
- **A business:** Way of life for a rural population.
- Involves management of labour, water and capital, and employing technologies for production of food, fibre, feed and fuel through mechanization to achieve maximum returns.
- As per Agricultural act 1947, agriculture includes horticulture, fruit growing, seed growing, dairy farming, livestock breeding, grazing land, meadow land, market gardens and nursery gardens.

1.2. Scope

- Agriculture contribution to GDP = 17.9% (2018).
- Supports 2/3rd of total populations for their livelihood.
- 58% of countries population works in agriculture (employed).
- Single largest private sector employment (occupation).
- Accounts for 15% of the total export earnings.
- Provides inputs for agro-based industries like textiles, silk, sugar, flour mills, milk products.
- Maintains the food security and safety (Food Security Act-2016).
- Biggest market in rural areas for consumer goods and durables.
- Helps in resource mobilization in rural areas.

- Allied sectors like horticulture, animal husbandry, dairy and fisheries helps for the health and nutrition for rural mass.
- Agriculture is a dynamic change from brown (soil) to green (growing crop) and then to gold (matured crop) for bumper harvest.

1.3. Revolutions in agriculture

- White revolution (milk production): Quadrupled from 17 MT at independence to 108.5 MT (2016)
- Blue revolution (fish production): Increased to 7.6 MT from 0.75 MT during the last five decades independence
- Yellow revolution (Oil seeds): Increased 5 times from 5 MT to 25 MT since independence
- Brown revolution (Leather and cocoa production)
- Golden fibre revolution (Jute production)
- Golden revolution (Horticulture overall, honey and fruit production)
- Grey revolution (Fertilizers)
- Red revolution (Meat and tomato)
- Round revolution (Potato)
- Silver fibre revolution (Cotton)
- Pink revolution (Onion, pharmaceuticals and prawn)
- Black revolution (Petroleum products)
- Silver revolution (Egg): Increased from 2 MT to 28 MT
- Sugarcane tonnes (57 MT to 282 MT)
- Cotton production-3 million bales to 32 million bales (1 bale-170 kgs of lint)
- India is the second largest producer of fruits and vegetables in the world
- Largest producer of milk
- Green revolution (Agriculture includes cereals, millets, pulses and oilseeds)

1.3.1. Food grains production scenario

- 2012-13 – 257.13 MT (Million Tonnes)
- 2013-14 – 265.04 MT
- 2014-15 – 252.02 MT
- 2015-16 – 251.57 MT

- 2016-17 – 275.11 MT
- 2017-18 – 284.83 MT

1.4. Branches of agriculture

- **Agronomy-**Deals with production of field crops
- **Horticulture-**Deals with production of all horticultural crops
- **Forestry-**Deals with the cultivation of perennial trees for wood, timber, fibre, rubber, raw materials for industries and MFP (Minor Forest Produce) like gum, lac, wax, honey, tannins etc.
- **Animal husbandry-**Breeding and rearing of livestock includes cattle, goat, sheep, poultry, pigs and buffalo for human food, farm power, manures etc.,
- **Fishery-**breeding and rearing of fishes including marine and fresh water such as shrimps, prawn to provide food, feed and manures.
- **Agricultural engineering-**Deals with farm machineries for field preparation, sowing, weeding and harvesting and post harvest processes, soil and water conservation, bio-energy etc.,
- **Home science-**Deals with value added products (Value addition).

1.5. Importance of agriculture in Indian economy

- Contributes for national income: First 2 decades of independence, the contribution of agriculture to GDP ranged from 48-60%. In 2001-2002, declined to 26%.
- Plays a vital role in employment generation: 2/3rd of population are employed in agriculture
- Maintains food security
- Contributes for capital formulation
- Resource mobilization
- Supply raw materials to agro-based industries
- Market for industrial products in rural areas
- Internal and external trade and commerce
- Contribution in union government budget: First five year plan gave more importance to agriculture, government obtained huge revenue in agriculture and allied fields.
- Need of labour force.
- Greater competitive advantages: Export of agricultural products because of cheap labour and also self sufficiency in input supplies.

1.6. Facts and Figures (2018)

- 2017-18, the total work force in agriculture 50% ; In 2050, the work force will reduce to 25.7%
- GDP contribution is 17-18%
- Agriculture sector contribution to Indian economy is higher than world average (6.4%).
- Gross value added by agriculture, forestry and fishing is estimated at Rs.17.67 million rupees (US$ 274.23 billion) in FY18.
- Indian food and grocery market is the world's sixth largest and contributing 70% of the sales.
- Indian food processing industry accounts for 32% of the countries total food market
- In India, food industries is the largest and ranked 5th position in production, consumption and export; contributes for 8.8% of GVA (Gross Value Added) and 8.39% in agriculture
- India's food export is 13% and total industrial investment is 6%.

1.6.1. Market size (2017-18)

- Food production is 284.83 MT (million tonnes); targeted food production is 285.2 MT
- Milk production-165.4 MT (2017-18)
- Meat production-7.4 MT
- As on September 2018, total area sown in kharif crops is 105.78 million hectares
- Horticultural production is 311.70 MT
- Second largest fruit producer
- Total agricultural export is 16.45% over 2010-2018.
- In December 2016, the agricultural products export was 208.55 INR billion which reduced to 187.91 INR billion in January 2017; December 2013-276.15 billion INR; October 1991-4.95 billion INR.
- Tea export-248.68 million kgs in CY 2017
- Coffee export-3,50,000 tonnes

1.6.2. Investments

- DIPP (Department of Industrial Policy and Promotions) reported that, Indian food industry attracted Foreign Direct Investment (FDI) of US$ 8.57 billion from April 2000-June 2018.
- By 2019, India will export sugar to China.
- **First Mega food park:** Launched in Rajasthan in March 2018
- Agrifood start ups received funding of 1.66 billion US$ between 2013-2017 in 558 deals

1.6.3. Government initiatives

- First agricultural export policy 2018 approved in December 2018 to increase India's is agricultural exports to US$ 60 billion by 2022 and US$ 100 billion in the next few years.
- In September 2018, government announced Rs.15,053 crore procurement policy named PM – AASHA. (Pradhan Mantri Annadata Aay Sanraks Han Abhiyan): Compensation schemes and partnering with private agencies to ensure fair price for farmers.
- In September 2018, CCEA (Cabinet Committee on Economic Affairs) approved Rs.5,500 crores as package assistance to sugar industries.
- Primary Agricultural Co-operative Society (PACS) was given with 2,000 crores to implement digital technology.
- To boost innovation and entrepreneurship in agriculture, introduced AGRI-UDAAN programme to mentor Startup Company to connect with potential investors.
- Sanctioned PMKSY (Pradhan Mantri Krishi Sinchai Yojana) with an investment of 50,000 crores providing irrigation sources to alleviate the drought.
- Sanctioned 6,000 crores for investment for mega food park under the scheme for Agro-Marine.
- SAMPADA-Scheme for Agro Marine Processing and Development of Agro Processing, 100% FDI in marketing the food products and e-commerce.

1.6.4. Achievements in agriculture sector

- E-NAM-Electronic National Agricultural Market launched in April 2016 to create uniform National market for agricultural commodities by networking existing APMC; up to May 2018, 9.87 million farmers ,1,09,725 traders were registered under eNAM; 585 mandies were linked.

- Agricultural storage capacity increased at 4% CAGR between 2014-17 to store 131.8 million metric tonnes.
- 2014-18, 10,000 clusters were approved under PKVY-Paramparagat Krishi Vikas Yojana. 2.3 million metric tonnes were added in godowns.
- Steel silos with capacity of 6, 25,000 metric tonnes also created.
- During 2015-17, around 100 million Soil Health Cards (SHC) were distributed and launched soil health mobile app.

2

Agricultural Development Through Five Year Plan

2.1. Green revolution 1967-1968 to 1977-1978

- Father of green revolution-M.S.Swaminathan
- Started green revolution work since 1960 with Norman Borlaug (Noble prize winner in 1970).

2.1.1. Main events

- Boost for agriculture sector
- Creation of irrigation facility
- Introduction of improved varieties and technologies
- Provision for subsidies and credits
- Agricultural products are the fourth largest exported commodities.

2.2. First plan (1951-1956) (Harrod-Domar model)

- Two major problems: 1. Food crisis, 2. Shortage of industrial raw materials like raw jute and raw cotton
- Highest priority given to agriculture including irrigation power.
- Agriculture sector got 3% of the total plan outlay.
- Food production was successful 66 million tonnes against the target of 62 million tonnes due to favourable weather conditions and implementation of Community Development Programme involving every village.
- A new land policy was also adopted

2.3. Second plan (1956-1961) (Nehru-Mahalanobis model)

- Major priority given to industry instead of agriculture, to get foreign aid particularly food aid.
- Considered that heavy industrial launching based on the Soviet model of industrialization helps in economic development.

- Agriculture got 20% total plan outlay
- Food production not achieved due to adoption of wrong strategy, miss calculation and poor implementation of programmes.

2.4. Third plan (1961-1966) (Gadgil Yojna)

- Greater allocation for agriculture of Rs.1, 745 crores compared to 950 crores in the second plan.
- Agriculture got 20% of total plan outlay.
- Food production was targeted for consumption, industry and export.
- First four years record 89 million tonnes.
- At fifth year food production was not satisfactory due to disastrous drought and two wars (China war 1962, Pakistan 1965).
- However, in 1966-1967, introduced HYVP (High yielding varieties) and multiple cropping programmes.
- Due to poor performance in fifth year of third plan, the fourth plan delayed by three years.

2.5. Fourth plan (1969-1974)

- High priority-Agriculture technology as major input.
- Agriculture got 21% total plan outlay.
- Targeted to attain self sufficiency in food production, building up of sizeable buffer stock and stoppage of concessional import of food grains.
- Food production reached 108 million tonnes in 1970-71.
- But in last two years, there was a crop failure due to poor monsoons and created problems for price inflation.

2.6. Fifth plan (1974-79)

- More emphasis gave to agriculture; growth target in agriculture and allied sectors fixed at 3.94%.
- Total plan outlay was 21% for agriculture .
- But in the first year, agriculture was not satisfactory due to fall in food production.
- Achieved food production of 232.5 million tonnes in subsequent years.
- Plan period was cut short by one year by the change of government i.e., JANATA government

2.7. Sixth plan (1980-1985)

- Target of agriculture growth is fixed for domestic consumption and export.
- Growth rate was fixed at 3.8%.
- Due to severe drought during 1982-1983, food production was 138.1 million tonnes.

2.8. Seventh plan (1985-1990)

- Agriculture growth rate fixed as 4% and target for food production was 3.7%
- Agriculture got 22% of total plan outlay.
- Sanctioned 48,100 crores for agriculture, irrigation and rural development.
- Major programmes implemented were:
  - Special rice production programme in the eastern region.
  - National Oil Seeds Development Project.
  - National Watershed Development Program for rainfed agriculture.
  - Social forestry.
- Achieved 3.2% growth rate with food production of 176.92 MT.
- Area under irrigation proposed to be increased at the rate 2.5 million hectares per annum.

2.9. Eighth year plan (1992-1997)

- Priority to growth and diversification of agriculture to achieve self sufficiency and for export.
- Sanctioned 22% of the total plan outlay.
- Targeted agriculture growth rate was 4% and more efforts gave to rice, pulses and oil seed production.
- Total food production was 199.3 million tonnes against the target of 210 million tonnes.

2.10. Ninth plan (1997-2002)

- Thrust to achieve agriculture-led growth.
- Main focus was given to agriculture instead of industry.
- Targeted growth rate was 4.5%.
- Total food production was 212 million tonnes.

- Planning commission recommended 4 strategies
  - Minimum Support Price (MSP)
  - Input subsidy policy
  - Food security
  - Alleviation of hunger

2.11. The tenth plan (2002-2007)

- Agriculture did not receive high priority.
- Plan allocation for agriculture and irrigation was 3, 05,055 crores, an increase of 51.4% over ninth plan.
- Growth rate fixed as 4%.
- But achieved only 2.4% growth rate due to poor monsoon during 2002, 2004 and 2006.
- This resulted in poor agriculture growth, reduction in share of agriculture to GDP from 23.8% (2002, 2003) to 22.5% (2007).
- Total food production was 179.4 million tonnes (2002-2003) and 212.4 MT (2003-2004) and then to 208.3 MT (2006-2007).

2.12. Eleventh plan (2007-2012)

- Annual growth rate fixed as 4% per annum to achieve 10% GDP.
- Emphasize was given to achieve inclusive growth with aim of:
  - Improving accessibility of technology to farmers to increase production
  - Ensure optimum use of natural resources
  - Attractive public investment
  - Promoting diversified agriculture
  - Addressing issues pertaining to food security
  - Decentralize the decision making
  - Programmes implemented are
- **NFSM (2007-08)**-National Food Security Mission for over all production of rice, wheat and pulses.
- **RKVY (August, 2007)**-Rashtriya Krishi Vikas Yojana to encourage public invest in public agriculture and allied sectors.
- **MMA (2000-01)**-Macro Management of Agriculture-to improve agriculture productivity.

- **ISOPOM (2004-05)**-Integrated Scheme of Oilseeds, Pulses, Oil Palms and Maize-To target small and marginal farmers for oil seed cultivation in rainfed conditions.
- **NMSA**-National Mission for Sustainable Agriculture-to ensure food security and also protect natural resources like land, water and genetic resources.
- Problems faced by agricultural sector in eleventh plan
- Stagnation in production of major crops especially in wheat.
- Soil exhaustion due to of negative effect of green revolution was also noticed like depletion of nutrients due to mono cropping.
- Decrease in ground water table because of intensive usage in dry and rainfed farming
- Costly farm inputs
- Lack of organized market in rural areas, middle men and dealer dominated the market.
- Lack of storage facilities in rural areas.
- Global warming effect: increase of 2-3°C threatened the agriculture
- Total food production-241 million tonnes in 2010-11.

2.13. Twelfth year plan (2012-2017)

- Agricultural growth rate fixed was 4.0%
- **Formation of NITI-AAYOG ie., renaming of five year plan (1950) in January 1, 2015.**
- NITI – National Institution for Transforming India

3

Growth Pattern of Crops in Terms of Area, Production and Productivity

3.1. Tamil Nadu status (2017-18)

3.1.1 Agricultural crops

- Area: 46.03 lakh hectares (millets, cereals, pulses, oil seeds, cotton and sugarcane)
- Total food production – 100 lakhs metric tonnes (or) 10 million tonnes (millets, cereals and pulses)

Crop	Productivity (kg/ha)
Rice	3370
Millets	3656
Pulses	638
Oil seeds	2,400
Cotton	545 lakh bales
Sugarcane	103 metric tonnes

- Regarding productivity at National level, Tamil Nadu stands first position in maize, cumbu, groundnut, total oil seeds and cotton.
- Second position in coconut and rice
- Third position in sugarcane, sunflower and sorghum
- Fourth position in coarse cereals
- Eighth place in total pulses

3.1.2 Horticultural crops

Year	Total area (lakh hectares)	Production (lakh metric tonnes)	Productivity (metric tonnes/ha)
2016 – 17 (Estimated)	14.76	174.94	11.85
2017-18 (Programmed)	15.67	188.22	12.01

- **First place in area**
    - Banana (1.18 lakh ha)
    - Tapioca (2.21 lakh ha)
    - Cocoa (24,000 ha)
    - Flowers (55,000 ha)
- **First in production**
    - Banana (56.50 lakh metric tonnes)
    - Tapioca (49.7 lakh metric tonnes)
    - Plantation crops (48.42 lakh metric tonnes)
    - Loose flowers (3.44 lakh metric tonnes)
- **Productivity first place**
    - Papaya (198.70 metric tonnes/ha)
    - Pomegranate (32.70 metric tonnes/ha)
    - Sapota (32.80 metric tonnes/ha)
    - Vegetables (30 metric tonnes/ha)
    - Tapioca (41.30 metric/ha)
- **Second position in production**
    - Aromatic crops (1.62 lakh metric tonnes)
- **Third position in area**
    - Grapes (3000 ha)
    - Plantation crops (6.35 lakh hectares)
    - Pepper (4000 ha)

3.2. India status during 2016-17 and 2015-16.

Area (million ha)

Position	State	2016-17	2015-16
First	Uttar Pradesh	19.92	19.36
Second	Madhya Pradesh	17.03	15.56
Third	Rajasthan	14.11	12.19
	Tamil Nadu	2.99	3.75

Production (million tonnes)

Position	State	2016-17	2015-16
First	Uttar Pradesh	49.14	42.56
Second	Madhya Pradesh	32.98	30.39
Third	Rajasthan	27.99	28.4
	Tamil Nadu	6.22	11.48

Productivity (kg/ha)

Position	State	2016-17	2015-16
First	Punjab	4360	4269
Second	Haryana	3735	3648
Third	**West Bengal**	**2853**	**2823**
	Tamil Nadu	2084	3063

Area under irrigation (2014-15)

- Punjab-99.0%
- Haryana-92.7%
- Uttar Pradesh-80.4%
- Tamil Nadu-56.8%

4

Government Agricultural Policies

4.1. NAP (National Agricultural Policy) 2000

Objectives

- To achieve growth rate in excess of 4% per annum in agriculture
- Efficient use of natural resources and conservation of soil water and biodiversity
- To achieve equity growth across regions and farmers
- Demand driven growth
- To cater the domestic requirements and export of farm produces due to economic liberalization and globalization.
- Technological, environmental and economical friendly growth for sustainable agriculture.

4.2. National Policy for Farmers 2007 (NPF)

- M.S.Swaminathan, Chairman of the National Commission on Farmers submitted report in October 2006.
- **Major goals**-to improve economic viability of farming by increasing the net income of farmers.
- To protect land, water, biodiversity and genetic resources.
- To develop support services including provision for seeds, irrigation, power, machinery and implements fertilizers and credit.
- To strengthen bio-security of crops and other allied activities.
- To provide appropriate price and trade policy.
- To provide suitable risk management.
- **Fertilizer subsidy** was provided by the central government during green revolution period (1960).
- In 1973-74, Rs.33 crores subsidies was provided and it was Rs.72,079 crores in 2014-15.

- **Electricity subsidy:** Introduced in green revolution period for promoting irrigation. (Karnataka-free electricity for agriculture).
- **Credit policy for agriculture sector:** Banks provide short term and long term credit by Regional Rural Bank (RRB's), Scheduled commercial banks, Non banking financial institution (NBFI) and Self Help groups (SHG).
- RRB launched in 1975
- NABARD established in 1982 (National Banking for Agriculture and Rural Development)
- Public sector banks formulated special agricultural credit plan in 1994-1995 and extended to private sector banks in 2005 and 2006
- **Kisan credit cards (1998-1999):** For providing credit support from 65 banking system under single window to the farmers
    - March 2013: 12.03 crores KCC issued
    - March 31, 2018: 23.5 crores KCC issued
- **NAIS (National Agricultural Insurance Scheme)** introduced in rabi season of 1999-2001, to issue insurance for cereals, millets, pulses, oilseeds, commercial and horticultural crops
- So for 24 states and 2 union territories except Punjab and Andhra Pradesh implemented NAIS.
- **MNAIS: Pilot Modified National Agricultural Insurance scheme** launched in rabi season of 2010-11 in 50 districts.
- Interest Subvention Scheme launched in 2006-2007 to provide short term crop loan up to 3 lakhs at 7% (Against prevailing rate of 9%).

4.3. Agriculture Price Policy (Agricultural Price Commission 1965)

- To provide minimum support price to farmers for procurement of food grains and supply to consumers through PDS

Objectives

- To protect producers through MSP
- To increase aggregate agriculture output through adoption of hybrid seed and input usuage
- To induce market surplus production of food grains
- To protect consumers against an excessive rise in prices
- MSP is announced by **Commission for Agriculture Cost and Prices (CACP)** in March 1985.
- Presently 23 crops are covered under MSP scheme

4.4. Market Intervention Scheme (MIS)

- By Department of Agriculture and Co-operation
- For the procurement of horti-commodities which is perishable in nature not covered in MSP.
- Intervention price is fixed not exceeding the production cost.
- Loss incurred equally shared by central and state with restriction upto 25%
- Implemented when there is 10% increase in production and 10% decrease in ruling price.

4.5. Price Support Scheme (PSS)

- Implemented by Department of Agriculture and Co-operation
- Procurement of oilseed, pulses and cotton through NAFED (National Agricultural Co-operative Marketing Federation of India Limited).
- Implemented when price fall below MSP.

4.6. Pradhan Mantri Kisan Samman Nidhi (PM-KISAN) – 1.12.2018

- Implemented in 01.12.2018 to provide income support to all land holding eligible farmers family for procurement of various inputs to increase yield and farm income.
- Revised scheme approved in cabinet meeting held on 2019 to cover more beneficiaries of 14.5 crores.
- Total expenditure involved was Rs. 87,217.5 crores for 2019-2020.
- **Eligibility Condition:** Small and marginal landholder farmer families having 2 ha of cultivable land with a benefit of Rs. 6000 per annum/family in three equal installments (every four months).

4.7. Krishonnati yojana – 2016-17

- Green revolution – Krishonnati yojana is an umbrella scheme implemented in 2016-17 by clubbing schemes/missions.
- Budget – Rs.33, 269.976 crores for three financial year's ie., 2017-2020.

4.7.1. MIDH – Mission for Integrated development of Horticulture-2015

- Total central share: 7533.04 crores
- To promote horticulture sector growth

4.7.2. NFSM-National Food Security Mission-2007

- To increase production of rice, wheat, pulses, coarse cereals and commercial crops.
- Includes National Mission on Oilseeds and Oil palm (NMOOP) with central share of 6893.38 crores.

4.7.3. NMSA-National Mission on Sustainable Agriculture

- Total central share: 3980.82 crores.
- For integrated farming, soil health management and resource conservation technology.

4.7.4. SMAE-Sub-Mission on Agriculture Extension-2014

- Central share: Rs. 2,961.26 crores
- To strengthen ongoing extension mechanism of state government, local bodies for achieving food and nutritional security and socio economic empowerment of farmers.

4.7.5. SMSP-Sub-Mission on Seeds and Planting Material -2016

- Total central share: Rs. 920.6 crores.
- To increase certified seed production
- To increase SRR (Seed Replacement Rate)
- To upgrade the quality of farm saved seed.
- To promote new technologies and methods in seed production and processing seed testing etc.
- To promote, strengthen and modernizing infrastructure for seed production, storage and quality.

4.7.6. SMAM-Sub-Mission on Agricultural Mechanization-2014-15

- Total central share: Rs.3, 250 crores
- To increase the farm mechanization of small and marginal farmers and to promote custom hiring centres.

4.7.7. SMPPQ-Sub-Mission on Plant Protection and Plant Quarantine - 2014

- Total central share: Rs.1022.67 crores
- For minimizing yield and quality loss from insect and pests and diseases, weeds, nematodes and rodents etc.

4.7.8. ISACES-Integrated Scheme on Agricultural Census, Economic and Statistics

- Total central share: Rs.730.58 crores.
- To undertake agricultural census, cost of cultivation and research on agro-economic problems.

4.7.9. ISAC-Integrated Scheme on Agricultural Co-operation

- Total central share: Rs.1, 902.63 crores.
- To provide financial assistance to co-operatives for development in agricultural marketing, processing, storage, computerization, etc.

4.7.10. ISAM-Integrated Scheme on Agricultural Marketing

- Total central share: Rs.3863.93 crores.
- To develop agriculture marketing.
- To promote innovative and latest technologies and to improve marketing infrastructure.
- Market integration through common online market plat form.
- To promote **pan-India** trade.

4.7.11. NeGP-A -National e-Governance Plan-Agriculture

- Total central share: Rs.211.06 crores.
- To enhance reach and impact of extension services.
- To improve access of farmers to information and services and to integrate existing ICT initiatives of centre and state.

4.8. RKVY-Rashtriya Krishi Vikas Yojana

- Special additional central assistance scheme approved by NDC (National Development Council) and launched on 29.07.2007.
- To achieve 4% annual growth rate during 11th plan.
- Also known as NADP (National Agriculture Development Programme).

4.9. PMKSY-Pradhan Mantri Krishi Sinchayee Yojana-protective irrigation "Per drop more crop"

- To enhance physical access of water on the farm and expand cultivable area under irrigation.
- To integrate water source, distribution and its efficient use.
- To improve on-farm water use efficiency.

- Adoption of precision irrigation
- Recharging of aquifiers by water conservation measures.
- Development of water and soil conservation, water harvesting, arresting run off, recharging of ground water and providing livelihood.
- Reusing of treated municipal waste water for urban agriculture
- Attract private investment in irrigation

4.10. Pradhan Mantri Kisan SAMPADA Yojana

- SAMPADA-Scheme for Agro Marine Processing and Development of Agro processing clusters
- Approved May 2, 2017 for period of 2016-2020.
- Renamed **as PMKSY-Pradhan Mantri Kisan Sampada Yojana** in August 23, 2017.
- **Umbrella scheme** incorporating Mega food parks, Integrated cold chain and value addition infrastructure, Food safety and quality assurance infrastructure, Infrastructure for Agro processing clusters, Creation of backward and forward linkages and expansion of food processing and preservation capacities.

4.11. Livestock and Poultry related

- National Livestock Mission: 2014-15
- National Dairy Plan Phase I: 2011-12 to 2018-19
- Dairy Entrepreneurship Development Scheme: 1.09.2010
- National Programme for Bovine Breeding and Dairy Development: 2013-14 to 2016-17
- Fodder Development Scheme: 2005-06
- NDDB's Quality Mark for Milk and Milk Products: 20th July 2017
- Financial assistance schemes of AWBI (Animal Welfare Board of India-1962)
- DIDF - Dairy Processing and Infrastructure Development Fund: 21st Dec, 2017.

4.12. Fisheries related

- Blue revolution: 1985-90
- Intensive Aquaculture in Ponds and Tanks: 16th June, 2009
- Schemes for Ornamental Fisheries: 15th May, 2017
- Fisheries Development in Reservoirs

- Schemes for Coastal Aquaculture
- Schemes for Mariculture
- Seaweed Culture
- Development of Domestic Fish Marketing
- Infrastructure Development for Harbours and Landing Centres
- Deep Sea Fishing and Tuna Processing

4.13. Rural Employment related

- MGNREGA-**Mahatma Gandhi National Rural Employment Guarantee Act-2005 and launched in 2006.**
- **PMEGP (Prime Minister's Employment Generation Programme-** 25.9.2008: National level programme to generate employment opportunities.
- Agri Clinics and Agri Business Centres Scheme.
- **RGUMY (Rajiv Gandhi Udyami Mitra Yojana-April 1, 2015**
    - To provide handholding support and assistance to the potential first generation entrepreneurs.
- **SFURTI (Scheme of Fund for Regeneration of Traditional Industries-2005**
    - Fund for regeneration of traditional industries.
- **RISC (Rural Industry Service Centre** for Khadi and Village industry.
- Coir Udyami Yojana

4.14. Farmer awards

- Pandit Deen Dayal Upadhyay Antyodaya Krishi Puruskar.
- Haldhar Organic Farmer Award.
- N.G Ranga Farmer Award for Diversified Agriculture.
- Jagjivan Ram Innovative Farmer Award.

4.15. Agricultural development through NITI-AYOG

- NITI-AYOG-National Institution for Transforming India formed in January 1, 2015. (Chairman-President), replaced Planning Commission 1950.
- Vice chairman: Dr.Rajiv Kumar since September 1, 2017.
- **First Vice chairman-Mr.Arvind Panagariya.**
- **"Thing Tank"** of government of India.
- Main aim – doubling farming income in five years.

4.15.1. Land Leasing Law

- NITI Aayog Formulated Model Agricultural Land Leasing Act 2016.
- To recognize the rights of tenants and safegourd interest of land owners.
- Drafted by Dr.T.Haque (agricultural economist) for availing bank credit or loans by the tenants.
- To pay compensation for extreme weather shocks (hail storms, flooding etc) to owners of the land who are staying in big cities.
- Big Win–Win reform for states.
- Introduced land leasing laws that allow tenant to engage in written contracts with land owners.
- Unlocking the land in the middle of the contract with legal backups.

4.15.2. APMC Act reforms-Making of One India Agri Market

- Provisions for single point levy taxes.
- Uniformity in mandi taxes.
- Delisting fruits and vegetable out of APMC.
- Electronic trading and allowing private to have own market.
- Committee on doubling farmer's income headed by Ashok Dalwai.

4.15.3. Creating competition among states-Agricultural Marketing and Farmers Friendly Reforms Index

- Three key areas: 1. Agriculture market reforms, 2. Land lease reforms, 3. Forestry on private lands (felling and transit of trees)
- **Maharashtra** ranks first in implementation of various agricultural reforms.
- Gujarat holds second rank followed by Rajasthan and Madhya Pradesh

4.15.4. Task force on agriculture development

- Constituted on 16th March, 2015-Chairman: Dr. Arvind Panagariya.
- Raising agriculture productivity and making farming remunerative for farmers.
- Focus on five critical areas 1. Raising productivity 2. Remunerative prices to farmers, 3. Land leasing, Records, land titles, 4. Second green revolution focus on eastern states, 5. Responding to farmers distress.
- Task force submitted final report on 31st May, 2016
- Road map of NITI AYOG-to achieve quantitative frame work for **doubling farmers income by 2022.**

5

Export and Import

5.1. Export

- Amounts to US dollars 33.87 billion as on 2017 amounting to 10.5% of total export.

Basmati rice	Non basmati rice
India is leading exporter of basmati rice 2016-2017-India exported 400480 metric tonnes of Price Rs. 21,605 crores to Saudi Arabia, UAE, Iran, Iraq, kuwait, accounts to 45%	2016-17-4305592 MT. **Benin** is the largest importer. Nepal, Senegal, Gunea, Iraq-Top importers. Amount Rs. 10,27510 lakhs.

5.1.1. Fresh vegetables

- 2016-17 India exported Rs. 5922 crores
- Major importers-UAE, Malaysia, Bangladesh, Netherlands, Saudi Arabia, UK.

5.1.2. Fresh fruits

- 2016-17, Rs. 4448.88 crores-export value
- Pakistan, Qatar, Sri lanka, UAE are top importers

5.1.3. Groundnuts

- Total export was 725703.35 metric tonnes-2016-17
- Worth of Rs. 54, 4433 lakhs.
- **Indonesia** was the largest importer followed by Vietnam, Pakistan and Philippines.

5.1.4. Processed fruits and juices

- 2016-17, total export value was US$ 460 million dollars

5.1.5. Cereal preparation

- Export of biscuits. cornflakes, breads were 340873.188 MT with a profit of Rs. 3572.61 crores.
- Major importers - UAE, USA, Nepal.

5.2. Import: Tree Nuts

- India import value in 2005 was US 680 million dollars.
- In 2015, 2.4 billion dollars.
- India is the **world's fourth largest** consumer of **almond** and **fifth largest** consumer of **walnut.**
- USA is the fifth largest supplier of nuts. Value of almond and walnut import was 524 and 58 million US dollars respectively.
- In 2015, US enjoyed 73 percent market share for almonds by value.
- Australia is the main market competitor with 21% market share by value.

5.2.1. Pulses

- Leading producers, consumer and importer.
- 2015-Pulse import was 5.4 million metric tonnes.
- Peas accounts for 40%, lentils (mysore dhal-*Lens esculenta*) 21%, chickpea 13%, mung bean 11%.
- US is the leading supplier of peas and lentils followed by Canada and Australia.

5.2.2. Cotton

- India is the world's largest producer, third largest consumer, 6th largest importer of cotton by value.
- Also the world's largest producer and consumer of cotton seed meal and cotton seed oil.
- In 2015, US exported cotton to the value of 115 million dollars.

5.2.3. Apples and other fresh fruits

- India fresh fruit import value was US $ 589 million in 2015 (Apple and date palm representing 70%).
- Apple exports representing 90% of US fresh fruits exports by value.
- Total export value of apple was $ 215 million.
- **Grapes:** export value $ 12.86 million, fresh pears 3.6 million dollars.

6

Role of National Organizations

6.1. NSC (National Seed Corporation)

- Launched in March 7, 1963 at New Delhi.
- Chairman-Vinod Kumar Gaur

6.1.1. Main role

- Production and supply of foundation seeds and certified seed.
- Marketing of all classes of seed.
- Conduct biennial survey on seed demand, and market research also
- To maintain improved seed stocks of improved variety.
- Export and import of seeds
- Providing training facilities.

6.2. FCI (Food Corporation of India)

- Launched in1965 as per Food Corporation Act 1964, New Delhi.
- Chairman-Shri D.V. Prasad

6.2.1. Objectives

- Maintain buffer stock of food grains.
- To ensure national food security.
- Effective food price support operations by ensuring farmers remunerative prices.
- Distribution of food grains through public distribution system.

6.3. PDS (Public Distribution System)

- Comes under FCI
- Chain of fair price shops entrusted with distributing of basic food (wheat, rice, sugar, palm oil) and non food commodities (fuel and kerosene) to the disadvantaged group of the society at very low prices.

- Role of central government-procurement, storage, transportation, bulk allocation of food grains.
- State government role-distribution to the consumers through fair price shops.
    - Issue of ration cards to the families below poverty line.
    - Supervision and monitoring of PDS functions.
- **Ministry of Consumer Affairs, Food and Public distribution** is the apex authority.
- At state level-Civil Supply Department / corporations

6.3.1. Objectives

- To protect low income group.
- By guaranteeing supply of food grains at affordable price.
- Ensuring equitable distribution.
- Controlling the price rise of essential commodities in the open market.

6.3.2. Reforming PDS

- Automation of all ration shops across the country.
- Government of Odisha and Gujarat took initiative to perform this activity.
- The National Food Security Bill introduced in lok sabha on December 22, 2011, to make a provision for support of local distribution modes.
- **Aadhar card** is one of the major element to modernize the PDS by enrolling it in PDS.
- Adhaar-Unique Identification Authority of India (UIDAI) established in 2009 at New Delhi.
- PDS is operating in 36 states and union territories.
- As on now 5.27 lakhs fair shops available across the country [2018-data].
- **AAY-Antyodaya Anna Yojana** was initiated in December 2000, to benefit 1 crore poorest of poor family at subsidized rate for wheat Rs. 2/kg, rice Rs. 3/kg and coarse grain Rs.1/kg.

6.3.3. Points to be remembered

- Word green revolution- William gaud
- Ever green revolution- M.S. Swaminathan
- E-NAM-585 markets- 124 crops-21 mandies-8 states. First West Bengal.
- RKVY-2007-2008, NADP-1986, NAIS-1999, NAIP-2006

- NATP-1998
- Pradhan Mantri Krishi Sinchayee Yojna-1st July, 2015.
- Pradhan Manthiri Jan Dhan Yojana 20th August 2014. CACP recommends MSP's of 23 crops, 25 commodities.
- Sivaraman committee- NABARD
- NAFED- 1958, New Delhi., TRYSEM- 1978, ICAR 1929, HYVP 1966
- RRB- Narashimham committee
- Bankers bank – RBI

UNIT-II
FUNDAMENTALS OF CROP PRODUCTION
Part-1 Crop Production

1

Factors Affecting Crop Production

1.1. Internal factors (genetic factors)

- Eg. Crop yield, desirable characters, resistance mechanisms related to genetic makeup of plants
- Resistance to lodging
- Tolerance to drought, salinity, flood, insects, pest and diseases
- Early maturity
- High yielding capacity
- Chemical composition of grains (Oil and protein contents)
- Quality of grain (Fineness, Coarseness) and straw (Sweetness, Juiciness)
- Less influenced by environment

1.2. External factors

1.2.1. Climatic factors

- 50% of yield is attributed by climatic factors

1. Precipitation

- Includes rainfall, snow, hail, fog and dew.
- Amount and distribution of total precipitation affects the choice of cultivated crops.
- In heavy and evenly distributed rainfall areas, rice in plains and tea, coffee and rubber at higher elevation are grown.
- In low and uneven distributed areas, dry farming is common, drought resistant crops like pearl millet, sorghum, minor millets and pulses are grown.
- In desert areas, shrubs and grasses are common.

2. Temperature

- Measure of intensity of heat is known as temperature.
- Crops need range of temperature of 15-40°C

- Influences crop plants and vegetation distribution
- Also germination, growth and development
- Affects physical and chemical properties of soil
- Regulates solubility of different substances
- **Cardinal temperature:** The minimum, maximum and optimum temperature of individual's plant
- **Cardinal temperature for different crops.**

Crops	Minimum temperature (°C)	Opt. temperature (°C)	Maximum temperature (°C)
Rice	10	32	36-37
Wheat	4-5	20	30-32
Maize	8-10	20	40-43
Sorghum	12-13	25	40
Tobacco	12-14	29	29-35

3. Atmospheric humidity (RH)

- Invisible water present in the air
- RH-ratio between amount of moisture present in the air to saturation capacity of air at particular temperature.
- Measured by thermo hygrograph.
- 100% RH means entire space is filled with water-no soil evaporation and plant transpiration.
- Optimum RH for most of crops is 40-60%
- High RH leads to outbreak of pests and diseases.

4. Solar radiation

- Without which life will not exist
- Influences-photosynthesis, soil physical process including plant and environment.
- PAR-Photosynthetic active radiation
- 0.4-0.7 μ : Blue-Violet-Active
- 0.5-0.6 μ : Orange-Red-Active
- 0.5-0.6 μ : Green-Yellow-Low Active
- Photoperiodism-Plants response to day length.
- Short day-day length is less than 12 hours. Eg. Rice, sunflower, cotton

- Long day-day length is more than 12 hours. Eg. Barley, oats, carrot and cabbage.
- Day neutral-No influence of day length Eg. Tomato, maize
- Phototropism-Plant response to light direction Eg. Sunflower
- Photo intensive-Season bound

5. Wind velocity

- Basic function of wind to carry moisture, (precipitation) and heat
- Moving wind supplies moisture, heat and CO_2 to plant
- Suitable wind movement is 4-6 km/ hour.
- **Beneficial**
    - Transfer of pollen grains and seed dispersal, cleaning of farmers produces
- **Harmful**
    - Helps in pathogenic infection
    - Mechanical damage to crops
    - Causes soil erosion
    - Increases evaporation

6. Atmospheric gases

- CO_2-0.03%
- O_2-21% (20.95)
- N_2-78.09%
- Argon-0.93%
- Others-0.02%
- CO_2-Essential for photosynthesis-Taken by leaves through diffusion
- O_2-Essential for respiration-Organic matter decomposition release O_2 to atmosphere
- Nitrogen-as plant nutrient
- Nitrogen fixation by symbiotic (*Rhizobium*) and non symbiotic free living bacteria (Cyanobacteria, Azotobacter).
- Nitrogen also supplied by lightning and rainfall.
- Gases like SO_2, CO, CH_4, Chlorofluoro carbons are toxic to plants.

1.2.2. Edaphic Factors (soil)

1. Soil moisture (water)

- Essential for photosynthesis and plant growth
- Principal constituent of the growing plant
- Helps in chemical and biological activities of soil including mineralization.
- Moderates the soil temperature, regulates availability and mobility of nutrients.
- Moisture is more in clay soil than sandy soil.

2. Soil air

- Essential for water absorption by plants
- Inhibits germination in the absence of oxygen when water is stagnated
- Oxygen is needed for respiration and for microorganisms to help in proper decomposition
- Rice needs low level of oxygen due to water logging condition.
- Potato, tobacco, cotton, linseed, tea and legumes needs high O_2.

3. Soil temperature

- Affects the physical and chemical process of soil
- Regulates absorption of water and nutrients.
- Affects underground parts of tapioca and sweet potato.

4. Soil mineral

- Sources of plant nutrients like Ca, Mg, S, Mn, Fe and K

5. Soil organic matter

- Supplies major and micro nutrients.
- Improves soil texture.
- Increases water holding capacity.
- Source of food for microorganism

6. Soil organisms

- Helps in decomposition of raw organic matter
- N_2 fixation by microbes

7. Soil pH

- pH indicates soil reaction.
- Neutral soil having pH 7 is best for growth.
- Acidic soil has less than 7 pH.
- Saline and alkaline soil has more than 7 pH.
- Low pH soil is injurious to plant, due to high iron and aluminium toxicity.

1.3. Biotic Factors

1.3.1. Plants

- Competition for moisture, nutrients, sunlight.
- Complementary effect or synergistic effect. E.g. Cereals and legumes.
- Competitive crops-Striga in sorghum and sugarcane, Orobranche in tobacco and Cuscutta in lucerne.

1.3.2. Animals

- Soil fauna like protozoa, nematodes, snails, insects help in organic matter decomposition.
- Insects and nematodes causes crop damage.
- Honey bees and wasp helps in pollination.
- Borrowing of earth worms helps in aeration and drainage of soil as well as organic matter decomposition which results in additional nutrition of soil.
- Large animal cause damage to crops. E.g. animal grazing.

1.4. Physiographic factors

- Topography-nature of the surface earth (leveled and slope)
- Increase in latitude decreases the temperature and precipitation.
- Steep slope-favours run off and loss of top soil (nutrients)
- Slope-exposed to low intensity of light and strong wind results in poor yield.

1.5. Socio-economic factors

- Inclination towards farming
- Land holders choose only food and fodder crops.
- Non availability of drought, pest and disease resistance varieties
- Economic condition of the farmers like marginal, medium and small farmers find inability in resource mobilization.

2

Agricultural Seasons in India and Tamil Nadu

2.1. Three seasons in India

1. Kharif (May - July)
2. Rabi (September - October)
3. Zaid (December - January)

2.1.1. Kharif season

- Crop is sown at the beginning of the south west monsoon and harvest at the end of south west monsoon.
- Crops - Sorghum, pearl millet, rice, maize, cotton, groundnut, jute, sunhemp, sugarcane, tobacco.

2.1.2. Rabi season

- Crops requires cool climate during growth and warm climate during germination and harvesting
- Sowing season: October – December
- Harvesting: February – April
- Crops: Wheat, barley, gram (Bengal gram/Chick pea), linseed, mustard, peas, potatoes.

2.1.3. Zaid season

- Crops grown throughout the year through artificial irrigation

1. **Zaid kharif crops**: Sowing in August - September
   - Harvested in December - January
   - Crops - Rice, sorghum, cotton, oilseeds, rapeseed
2. **Zaid rabi crops**: Sown in February - March
   - Harvested in April and May
   - Crops - Watermelon, cucumber, leafy and other vegetables.

2.2. Tamil Nadu season

- **Paddy**
  - Navarai (December–January)
  - Sornavari (April-May)
  - Kar (May-June)
  - Kuruvai (June-July)
  - Early samba (July–August)
  - Samba (August)
  - Late samba/thaladi (September–October)
  - Late thaladi (October– November)
- **Maize**
  - Aadi pattam (July–August)
  - Puratassipattam (September–October)
  - Thaipattam (January–February)
- **Pulses**
  - Aadipattam (June–August)
  - Puratassipattam (September–November)
  - Markazhipattam–Thaipattam (winter irrigated season)
  - Rice fallows (January), Chithiraipattam (summer irrigated)
- **Sugarcane**
  - Main season (December–May)
  - Entire Tamil Nadu; early season (December–January)
  - Mid season (February–March)
  - Late season (April–May).
  - **Special seasons-June, July.** In parts of Trichy, Perambalur, Karur, Salem, Namakkal and Coimbatore districts also raised during June-September.
- **Cotton**
  - Winter irrigated (August–September)
  - Summer irrigated (February–March)
  - Rainfed irrigated (September–October)
  - Rice fallow cotton.

3

Cropping Pattern in India and Tamil Nadu

3. 1. Cropping pattern

- Defined as the proportion of area under various crops at a point of time.
- Yearly sequence and spatial arrangement of sowing and fallow on a given area.

3.2. Most important cropping patterns in India

Crop pattern	Region/State	Issues related to crop pattern
Rice-Wheat	UP, Punjab, Haryana, Bihar, West Bengal, Madhya Pradesh.	Stagnation in the production & productivity. Due to over mining of nutrients from the soil. Declining ground water table. Increases pest and diseases. Shortages of labour. Inappropriate use of Fertilizers.
Rice-Rice	Irrigated and Humid coastal system of Orissa, Tamil Nadu, Andhra Pradesh, Karnataka and Kerala.	Deterioration in soil physical conditions. Micronutrient deficiency. Poor efficiency of nitrogen use.
Rice-Groundnut	Tamil Nadu, Andhra Pradesh, Karnataka, Orissa & Maharashtra.	Excessive rainfall and water logging. Non-availability of quality seeds. Limited expansion of rabi groundnut in rice grown areas.
Rice-Pulses	Chhattisgarh, Orissa and Bihar.	Droughts and erratic rainfall Lack of Irrigation. Low coverage under HYV seeds. Weed Attacks.
Maize-Wheat	UP, Rajasthan, MP and Bihar	Sowing timing. Poor weed management. Poor plant varieties. Poor use of organic and inorganic fertilizers. Large area under rain fed agriculture.
Sugarcane-Wheat	UP, Punjab & Haryana accounts for 68% of the area under sugarcane. The other states which cover the crops are; Karnataka & MP.	Late planting. Imbalance & inadequate use of nutrients. Poor nitrogen use efficiency Build-up of *Trianthema portulacastrum* and *Cyperus rotundus* in sugarcane.

Cotton-Wheat	Punjab, Haryana, West UP, Andhra Pradesh, Karnataka, Tamil Nadu.	Delay planting. Stubbles of cotton create the problem of tillage operations. Cotton pest like boll worm and white fly. Poor nitrogen use efficiency.
Soya bean-Wheat	Maharashtra, MP and Rajasthan	Limited genetic diversity. Short growing period. Hinders availability of inputs. Rainfed nature of crop and water scarcity. Insect pests and diseases
Legume Based Cropping Systems (Pulses-Oilseeds)	MP, Gujarat, Maharashtra, Andhra Pradesh and Karnataka.	Lack of technological advancement. Loses due to erratic weather. Diseases and Pests. Low harvest index, flower drop, indeterminate growth habit and very poor response to fertilizers and water.

3.3. Food grains and their required agro-climatic conditions

Food Grains	Agro-climatic Condition
Rice	**Temperature:** 22-32 degree Celsius **Rainfall:** 150-300 cm **Soil Type:** Deep clayey and loamy soil
Wheat	**Temperature:** 10-15 degree Celsius (Sowing time) **Temperature:** 21-26 degree Celsius (Ripening & Harvesting) **Rainfall:** 75-100 cm **Soil Type:** Well-drained fertile loamy and clayey loamy
Millets	**Temperature:** 27-32 degree Celsius **Rainfall:** 50-100 cm **Soil Type:** Less sensitive to soil deficiencies, grown in inferior alluvial or loamy soil
Grams	**Temperature:** 20-25 degree Celsius (Mild cool & Dry Climate) **Rainfall:** 40-45 cm **Soil Type:** Loamy soil
Sugarcane	**Temperature:** 21-27 degree Celsius **Rainfall:** 75-150 cm **Soil Type:** Deep rich loamy soil
Cotton	**Temperature:** 21-30 degree Celsius **Rainfall:** 50-100 cm **Soil Type:** Black soil of Deccan and Malwa Plateau, also grows well in alluvial soils of the Sutluj-Ganga plain and red and laterite soils of the peninsular region.
Oilseeds	**Temperature:** 20-30 degree Celsius **Rainfall:** 50-75 cm **Soil Type:** Light sandy loams, red, yellow and black soils
Tea	**Temperature:** 20-30 degree Celsius **Rainfall:** 150-300 cm **Soil Type:** Well drained, deep friable loamy soil.
Coffee	**Temperature:** 15-28 degree Celsius **Rainfall:** 150-250 cm **Soil Type:** Well drained, deep friable loamy soil.

3.4. Regional distribution of crops in India

Cereals	Wheat	Uttar Pradesh, Punjab and Haryana
	Rice	West Bengal, Andhra Pradesh, Chhattisgarh & Tamil Nadu
	Gram	Madhya Pradesh and Tamil Nadu
	Barley	Maharashtra, Uttar Pradesh and Rajasthan
	Bajra	Maharashtra, Gujarat and Rajasthan
Cash Crops	Sugarcane	Uttar Pradesh and Maharashtra
	Poppy	Uttar Pradesh and Himachal Pradesh
Oil Seeds	Coconut	Kerala and Tamil Nadu
	Linseed	Madhya Pradesh and Uttar Pradesh
	Groundnut	Andhra Pradesh, Gujarat and Tamil Nadu
	Rapeseed and Mustard	Rajasthan and Uttar Pradesh
	Sesame	Uttar Pradesh and Rajasthan
	Sunflower	Maharashtra and Karnataka
Fibre Crops	Cotton	Maharashtra and Gujarat
	Jute	West Bengal and Bihar
	Silk	Karnataka and Kerala
	Sunhemp	Madhya Pradesh and Uttar Pradesh
Plantations	Coffee	Karnataka and Kerala
	Rubber	Kerala and Karnataka
	Tea	Assam and Kerala
	Tobacco	Gujarat, Maharashtra and Madhya Pradesh
Spices	Pepper	Kerala, Karnataka and Tamil Nadu
	Cashew Nuts	Kerala, Tamil Nadu and Andhra Pradesh
	Ginger	Kerala and Uttar Pradesh
	Turmeric	Andhra Pradesh & Odisha

3.5. Cropping pattern in Tamil Nadu

3.5.1. Rice based cropping system

Command areas

- Rice (August.-January) - pulses/sesame/maize (January-April)
- Rice (June-September) - rice (October-January) - pulses/gingelly/green manure (February-May)
- Rice (June-September) - rice (December-March)
- Rice (June-September) - green manure (October-November) - rice (December-March)
- Rice (August-January) - cotton (February-June)
- Rice (October-January) - pearl millet/vegetables/sesame/pulses/groundnut (February-May)
- Rice (April-August) - rice (September-March) – fallow

Well irrigated areas (Filter point well)

- Rice (August-January) - rice (January-April.) - groundnut (April-June)
- Rice (August-November)- rice (December-March) - rice(March-July)
- Rice/vegetables/marigold (June-October) - maize (October-January) - pulses (February-May)
- Rice (April-August) - groundnut (September-December) - vegetables / gingelly (January-March)
- Rice (August-January) - groundnut (February-April) - sesame/ pulses/maize (April - June)
- Rice (August-January) - cotton (February-June)/gingelly (February- May)
- Rice (June-October) - ragi / groundnut / gingelly (November- February)
- Rice (April-August) - rice (September-January) - fodder sorghum (January-March)
- Rice (October-February) - pulses (February-May)

Tankfed areas

- Rice (August-January) - rice (January-April)
- Rice (August-January) - maize (January-April)
- Rice/Vegetables/watermelon (August-January)- groundnut/ gingelly / pulses(February-May)
- Rice (June-September)-ragi (Sep-Dec.)-pulses (January-April)
- Rice (July-November) - groundnut / ragi (December-March) - fallow
- Rice (September-January) - cotton (February-August)
- Rice (September-January) - green manure (February-April)
- Rice (September-December) - senna* (January-March)
- Rice (June-September) - chillies (October-February)

Rainfed areas

- Upland rice/millets /pulses/groundnut/cotton/chillies (September-February)
- Dry rice (June-August) - pulses/vegetables/gingelly (September-January)
- Rice (July-November)

4

Production Technology of Cereals and Millets

4.1. Seasons and varieties

Crop	Season	Varieties
Rice	Navarai (December – January)	ADT 36, ADT 37, ASE 16, IR 64, ASD 18, ADT 42, ADT 43, MDU 5, ASD 20, IR 20, ADT 39, CO 43, CO 47, ASD 20, TRY (R) 2*.
	Sornavari (April – May)	ADT 36, IR 36, IR 50, ADT 37, ASD 16, ASD 17, IR 64, ASD 18, ADT 42, MDU 5, ASD 20, ADT 43, CO 47, TRY (R) 2*, ADT (R) 45, ADTRH 1, ADT (R) 47, ADT (R) 47.
	Early kar (April–May)	IR 50, ADT 36, IR 64, ADT 42, ADT 43, ADT 45, CO 47, ADT (R) 47.
	Kar (May–June)	IR 50. ADT 36, ASD 16, ASD 17, IR 64, ASD 18, ADT 42, ADU 5, ASD 20, ADT 43, CO 47, ADT (R) 45, TRY (R) 2*, ADTRH 1, ADT (R) 47, IR 36, ADT 37, Bhavani, IR 20, White ponni, CO 43, MDU 4, ASD 19, Paiyur 1.
	Kuruvai (June–July)	ADT 36, IR 50, IR 64, ASD 16, ADT 37, ASD 18, ADT 42, MDU 5, ADT 43, CO 47, ADT (R) 45, TRY (R)2*, ADTRH 1, ADT (R) 47, ADT (R) 48.
	Samba (August)	Ponmani, IR 20, White ponni, CO 43, ADT 40, Paiyur 1, PY 4, ADT 39, TRY 1, ASD 19, ADT(R) 44, CORH 2, CO 45, ASD 19, ADT 38, ADT (R) 46, CO 42, ADT 40, ADT 38, MDU 3, MDU 4, Bhavani.
	Late samba (September – October)	IR 20, White ponni, ADT 38, ADT 39, CO 43, CO 46, TRY 1, ADT (R) 46, CORH 2, Ponmani, ASD 19, ADT (R) 46, TPS 2, TPS 3, ASD 18, ASD 19, MDU 5
	Pishanam/Late pishanam (September – October)	ASD 18, ASD 16, ASD 19, CO 43, TRY 1, ADT (R) 46.
	Dry (July – August)	ADT 36, PMK 1, TKM 11, PMK (R) 3, TKM 10, TKM (R) 12.

Barley	Huskless barley - Karan 18 & 19, **Suited for hills:** Himani, Dolma, Kailash. **Suited for rainfed areas:** Ratna, Vijay, Azad, Ameru (best for malt). **Suited for irrigated areas:** Jyoti, Ranjit, Clipper (best for malt & brewing), Karan 18 & 19. **Dual purpose (fodder and grain):** Ratna, Karan2, Karan 5, Karan 10
Oats	Kent, Algerian, Bunker 10, Coachmen, HFO 114, UPO 50.
Rye	**Winter season:** Forage type - Athens, Common, Abruzzes; Grain type - Rosan, Dakold, Balba **Spring season** - Prolific, Merced.
Maize	**Tamil Nadu varieties:** TNAU maize hybrid Co 6, Baby corn Co(Bc) 1 CO 1, COH (M) 4,COH (M) 5, Amber, Sona, Vikram, Shakthi, Jawahar, Vijay, Kisan, Raja, Krishana, African tall, Panchganga, Ganga 5, Ganga 4, Ganga safed 2, Rattan, Agathi 76, Vivek Hybrid 27 (Central Maize VL Baby Corn 2), PusaVivek QPM-9, Naveen, Diara,
Bajra	ICMS-7703, CO 7, CO (Cu) 9, X 7, ICMV 221, CO 27, CO 10,
Sorghum	CO (S) 28, TNAU Sorghum variety CO 30, TNAU Sorghum Hybrid CO 5, CO 26, CO (S) 28, COH 4, PAIYUR 1, K Tall, K 11, BSR 1, Paiyur 2, APK 1
Finger Millet	CO RA 14, CO 13, CO 9 , TRY 1, Paiyur 1, INDAF 5, GPU 2
Thenai	Varieties in TN:CO 6, CO5
Samai	Varieties in TN: CO3, CO2, CO1, PAIYUR 2
Varagu	Varieties: CO3, APK -1
Panivaragu	Varieties: CO 3, CO 4 & K2

4.2. Production technologies

Crop	Season	Seed rate (kg/ha)	Spacing (cm)	Fert. Schedule (kg/ha)	Irrigation (no. & requirement)	Foliar spray	Herbicide	Yield (kg/ha)
Rice (*Oryza sativa*) Poaceae 2n = 24 Origin: South East Asia	Dec – Jan (Navarai), Apr – May (Sornavari), May – June (Kar), June – July (Kuruvai), July Aug (Samba), Sept – Oct (Late samba / pishanam), Oct – Nov (late thaladi).	Long duration: 30 Medium: 40 Short: 60 Hybrid: 20 SRI: 7- 8 Optimum age of seedlings Short: 18-22 DAS Medium: 25-30 DAS Long: 35-40 DAS SRI: 14 DAS	Short duration: 15 x10, Medium- 20 x 10, Long - 20 x 15, SRI- 25 x 25 cm	Short: 150:50:50 (Kavery Delta), Others: 120:40:40, Medium and long: 150:50:50, Hybrid: 175:60:60, FYM: 12.5 t/ha, Green leaf manure 6.25t/ha	1200-1400 mm	1% urea+ 2% DAP+ 1% KCl	Nursery – Pretilachlor + safener 0.3% kg/ha on 3 DAS Main field – Butachlor @ 1.25 kg/ha and Pendimethalin 3.0 lit/ha Anilophos @ 0.4 kg/ha mixed with 50 kg sand as per-emergence 2, 4-D sodium salt (Fernoxone 80% WP) @ 1.25 kg/ha on 3 weeks after transplanting (post-emergence)	4000 – 6000 Straw: 8000 – 10000

Wet nursery: Nursery area – 20 cents (800 m^2); seed treatment : treat the seeds with carbendazim @ 2 g/lit or *Pseudomonas fluorescens* @ 10 g/kg of seeds or *Azospirillum* 600 g/ha; seed bed: breadth 1.5-2 m, width 30 cm, length 8 to 10 cm.*

Nutrient management: 10 packets (2000 g/ha) *Azospirullum* and 10 packets of phosphobateria or 20 packets of Azophos with 25 kg FYM and 25 kg of soil; *Pseudomonas fluorescens* at 2.5 kg/ha mixed with 50 kg FYM and 25 kg of soil; broadcast 10 kg of BGA (Blue green algae) at 10 DAT, *Azolla* 250 kg/ha at 3-5 DAT; Gypsum 500 kg/ha.

Water management: 2.5 cm during puddling and maintain for 7 days for green manure decomposition (sun hemp –less fibrous) and 15 days for high fibrous green manures; 2 cm at transplanting and up to 7 DAT; 5 cm submergence throughout the growth period; last irrigation 15 days ahead of harvesting.

Crop	Season	Seed rate (kg/ha)	Spacing (cm)	Fert. Schedule (kg/ha)	Irrigation (no. & requirement)	Foliar spray	Herbicide	Yield (kg/ha)
Wheat Poaceae **Origin:** Asia minor, Abyssinia, Afghanistan 2n = 42	Second week of November to second week of December	100 125 (late sown crop)	Row - Row 22.5 to 23, 15-18 (late sown crop)	120:60:40 or 150:60:40	4-6 450-650 mm		Isoproturon @ 750-800g/ha; Pendimethalin @ 1000 g/ha at 3-4 DAS	North India :3000 Straw: 5000 Tamil Nadu: 2500

Mexican dwarf wheat : *Triticum aestivum* ; Emmer wheat : *T.dicoccum*; Macaroni wheat: *T.durum* (28); Common bread wheat: *T.vulgare* , Indian dwarf wheat: *T.spherococcum*

Barley *(Hordeum vulgare)* Poaceae 2n – 14 Abyssinia or South east Asia (China, Tibet, Nepal) or Near East region	Oct – Rainfed, Nov – irrigated, Hilly Apr- May	Irrigated : 100 Rainfed : 80-100	Irrigated 22.5 Rainfed 22.5 25	Irrigated -60:30:20 Malt barely -30:20:20 Rainfed - 40:20:20 FYM – 12.5 t/ha	2-3 200 – 300 mm	Isoproturon at 0.75 kg/ha + 0.5 kg 2, 4-D at 3-5 leaf stage (post emergence or Pendimethalin @1 kg/ha (pre emergence)	Rainfed : 1500-3000 Irrigated : 3000-3500 Straw: 4000-5000
Oats (*Avena sativa*) Poaceae 2n – 42 Asia minor / Asiatic	Oct – mid Nov	100	20 – 23 for fodder, 23-25 for grain	80:40:0 FYM: 12.5 t/ha	4-5 450-650 mm		Fodder : 50-60 t grain Grain: 3000-3500
Rye (*Secale cereale*) Poaceae 2n = 14 Western Asia to Southern Russia	Winter spring seasons, Forage – October, Grain November, Green manure – August.	Forage : 80 Grain : 60	Broad casting	50:40:65 FYM 12.5 t/ha	6		Fodder : 50-55 t/ha Dual crop: 25 t/ha Fodder grain : 2500 Straw : 2500
Triticale (Man made cereal) Wheat x Rye in Scotland during 1875	Spring or winter season; Name Tritical appeared in Germany in 1935; First fertile cross (wheat x rye) made in Germany in 1888; Cultivation practices similar to wheat; seed rate: 80 kg for dryland, 110 kg for irrigated.						
Millets							
Maize (Lea-Mays) Poaceae 2n = 20 Central America and Mexico	Adipattam (July August), Puratasipattam (September-October) Thaipattam (January - February)	Grain: 20 Baby corn: 25 Rainfed crop: Hybrid: 20 Variety: 25	Irrigated: 60 x 25 Rainfed: 45x20	Irrigated 135:62.5:50 Rainfed 60:30:30 FYM 12.5 t/ha	600-700 mm	Atrazine @500g/ha at 3 DAS Do not use Altrazine if pulse is raised as intercrop.	Grain: 5000 Straw: 10000 Baby corn: 6000 Green fodder: 25000 Rainfed: 3000 - Grain

Flint corn: *Zea mays indurata* ; **Dent corn:** *Zea mays indentata*; **Sweet corn:** *Zea mays saccharata*; **Flour corn:** *Zea mays amylaceae*;

Pop corn: *Zea mays everta*; **Waxy corn:** *Zea mays ceratina*

Seed treatment: With chlorpyriphos 20 EC @ 4ml + 0.5 gm gum in 20ml of water – to control stem borer; With metalxyl @ 2g/kg of seed – To control downy mildew & crazy top; With azospirillum @ 600 g/ha of seeds.

Water Management: 45-65 days after sowing Most critical phase; Critical stage: 6th leaf, late knee high, tasseling, 50% silking, dough stage

Sorghum (*Sorghum bicolor*) Poaceae 2n–20 Africa and India	*	Transplanted 7.5, Direct sown 10, Rainfed 15.	45 x 15	90:45:45 FYM: 12.5 t/ha Micronutrient: 12.5 kg/ha or 25 kg $ZnSO_4$ with sand.	450-650 mm Stress mitigation: spray 3% kaolin	Atrazine @500g/ha on 3 days after sowing	4000-6000

*Thaipattam(January – February); Chithiraipattam (April – May); Adipattam (June – July); Puratasipattam(September – October)

Seed treatment: Carbendazim @ 2g/kg of seed; Azospirillum @600g/ha; Azophos @1200g/ha. In case of intercrop with pulse-Sorghum 10 kg/ha and Pulse 10 kg/ha.

Pearl millet (*Pennisetum glaucum*) Poaceae 2n–24 Africa	Chithiraipattam (March–April) Masipattam (January-February)	Nursery 0.5 kg/cent, Direct sown: 5kg	45 x 15 Intercrop with pulses: 30 x 15	Varieties: 70:35:35 Hybrids: 80:40:40 FYM: 12.5 t/ha	300-400 mm	Atrazine @500g/ha on 3 DAS/DAT	Rainfed 1200-1600 Irrigated 2500-3500

1. Nursery: Area – 7.5 Cents for ha; **Seed treatment:** Carbendazim @ 2g/kg of seed; Azospirillum @600g/ha; Azophos @1200g/ha, Incorporate the nursery with 180 g phorate 10G or 600g Carbofuran 3G; Optimum age of seedlings-18 DAS, Treatment with 2% KCl, 3% NaCl.

Finger millet (*Eleusine coracana*) Poaceae 2n–36 Ethiopia and Ugandan highlands (East Africa)	June July	6-8	20-25	60:30:30	2-5 irrigation	Butachlor @1.25kg/ha 2,4D @0.5kg	Rainfed 1200-1500 Irrigated 4000-4500
Small millets	Rainy season	8-15, Line 10, Seed drill 12.5.	22.5 x 10	20-40:10-20:10-20		Isoproturan @500g/ha	**Barnyard millet** 800 **Kodo millet** 1000 **Panivaragu** 700 **Foxtail millet** 700 **Little millet** 1100

Samai little millet(*Panicum sumatrense*); Varagu kodo millet (*Paspalum scrobiculatum*); Kuthiraivali (Barnyard millet)-(*Echinochloa frumentacea*); Panivaragu-common millet *(Panicum miliaceum)*

5

Production Technology of Pulses

5.1. Varieties

Black gram	ADT 1, CO1, CO2, CO 3,KM 2, CO4, KM1, ADT 2, TMV 1, CO5, ADT 3, ADT 4, Vamban 1 to Vamban 8, APK 1, K1, CO 6, MDU 1, KKM 1, ADT 6
	Varietiy suitable for rice fallow pulse: CO 4, ADT 2, ADT 3, ADT 4, ADT 5, TMV 1
Green gram	Paiyur 1, ADT2, ADT 3, CO4, CO5, KM2, VBN 1, K1, VBN 2, VBN 3, VRM 2, CO 7.
Garden lab lab	CO 3 to CO 13
Horse gram	CO1, Paiyur 1 & 2
Urd bean	BR 68, HPU 6, Kulu 4, LBG 17, Naveen
Red gram	Grouped as short (100-150), medium (150-180) and long (180-300)days
	– N Hilly zone (hills of HP, J&K, UP)
	• T21-SD, UPAS 120-SD, ICPL 151-SD (Short Duration)
	– NW Plains (Delhi, Punjab, W.UP, N.Rajasthan & Haryana)
	• T21-SD, UPAS 120-SD, ICPL 151-SD
	– NE plains (C&E UP, Bihar, WB, Assam)
	• T21-SD, UPAS 120-SD, Bahar-LD (Long duration), SMR (Sterility mosaic resistant), AR (Alternaria resistant)
	– Central Zone (MP, Rajasthan, Maharastra & Gujarat)
	• T21-SD, ICPL 151-SD
	– Southern (Orissa, AP, Karnataka & TN)
	• SA1-LD, ICPL 87-SD, KM7-MD
	• For TN alone: COH 1-SD, CO 6-MD, COH 2-SD, Vamban 1-LD, CO7, VBN 3, APK1, BSR 1
Cow pea	CO2, 3, 4, 6,7, KM1, VBN 1 & 2, Paiyur 2

5.2. Production technologies

Crop	Season	Seed rate(kg/ha)	Spacing(cm)	Fert. Schedule (kg/ha)	Irrigation (no. & requirement)	Foliar spray	Herbicide	Yield (kg/ha)
Blackgram (*Vigna mungo*) India Leguminosae 2n=22,24	June – August Sep- oct Rice fallow- January	20-25, Treat with Carbendazim @ 2g/kg of seed, Trichoderma 4g/kg, Rhizobium 600g/ha.	Rainfed 25x10 Irrigated 30x10	Rainfed 12.5:25:12.5 + 10 S, Irrigated 25: 50:25 + 20 S, FYM: 12.5 t/ha	Irrigation: 280 mm	1% urea, 2% KCl- Stress tolerance 40 ppm NAA, 100 ppm Salicylic acid, 2% DAP flowering	Pendimethalin 3.3 lit/ha on 3 DAS + 100 ppm Boron	800-1100
Greengram (*Vigna radiata*) India Leguminosae 2n=24	June-July, Sep- Oct, Dec-Jan, Jan-Feb (Rice fallow), Feb- March	20, Rice fallow: 25, Treat with Carbendazim @ 2g/kg of seed, Trichoderma 4g/kg, Rhizobium 600g/ha.	30 x 10	Rainfed: 12.5:25:0, Irrigated: 25:50:0, FYM: 12.5 t/ha	Irrigation: 280 mm	40 ppm NAA, 100 ppm Solicylic acid, 2% DAP flowering	Fluchloralin (or) Pendimethalin 3.3 lit/ha	800-1000
Cowpea (*Vigna ungiculata*) Africa Leguminosae 2n=22,24	June -July, Sep –Oct, Dec-Jan	20, Rice fallow: 30, Treat with Carbendazim @ 2g/kg of seed, Trichoderma 4g/kg, Rhizobium 600g/ha.	30x15 45 x 15	Rainfed- 12.5:25:12.5 + 10 S, Irrigated: 25: 50:25 + 20 S			Pendimethalin 2 lit/ha	400-700
Horse gram (*Macrotyloma uniflorum*) India Leguminoseae 2n = 20, 22, 24	November	20	30 x 10	12.5 : 25:0				700 – 900 Green fodder 1000
Garden lablab (*Lablab purpureus var typicus*) Leguminosae India 2n = 22, 24	June – July, September- November	4-25	90 x 90 45 x 15 45 x 30	Rainfed:12.5:25:12.5, Irrigated: 25:50:0				Pandal 1200-1300 Bush 800-1000
Field lablab (*Lablab purpureus var lignosus*) Leguminosae	Throughout the year	20-25	90 x 30 45 x 15	Rainfed:12.5:25:12.5 +10S Irrigated: 25:50:25+20 S				Rainfed– 900 Irrigated 1400

Bengal gram (*Cicer arietinum*) Leguminoseae 2n = 16. South west Asia or Mediterranean region	November-Winter	75 – 90	30 x 10	Rainfed: 12.5:25:12.5 +10S Irrigated: 25:50:25+20 S			2000 – 3000
Redgram (*Cajanus cajan*) India Leguminoseae 2n=22, 44, 66,	Throughout the year	Long: 8-10 Medium: 10-12 Short: 12-15	Short: 40 x 20 Medium: 45 x 30 Long: 90x30	Rainfed 12.5:25: 0 + 20 S Irrigated 25: 50:0 + 20 S		Pendimethalin or fluchloralin at 0.75 kg/ha on 3 DAS	625
Soyabean (*Glycine max*) Leguminoseae 2n 40 China	June – July, Sep Oct, Feb – March.	70-80	30 x 5	20:80:40 – 40 S as Gypsum 220 kg/ha, and 25 kg Zinc sulphate	NAA 40 ppm, Salicylic acid 100 ppm, DAP 2%, Urea 2%	Pendimethalin @ 1 lit/ha Imazythypur @ 50g/ha 20 DAS	1800-3500

6

Production Technology of Oilseeds

6.1. Seasons and Varieties

Crop	Seasons	Varieties
Ground nut		TMV 2 – cosmopolitan – bunch (360g) TMV 7 – Bunch (360g) TMV 10 –semi spreading (540g) JL 24 – Bunch - rainfed / irrigated (460g) VRI 2 – Spanish bunch (500g) VRI 3 – Bunch (350g) VRI 4 – Bunch (408g) CO 1, 2, BSR 1 – Bunch TMV 13, VRI 7, ALR 3, CO 6.
Sesame		CO 1, TMV 3, TMV 4, TMV 5, TMV7, VRI 2, SVPR 1.
Sunflower	Adipattam (June-July)	TNAUSUF 7 (CO 4), COSFV 5, KBSH 1, KBSH 44, DRSH 1, TNAU Sunflower Hybrid CO 2
	Karthigaipattam (Oct-Nov)	TNAUSUF 7 (CO 4), COSFV 5, KBSH 1, KBSH 44, DRSH 1, TNAU Sunflower Hybrid CO 2
	Margazhipattam (Dec-Jan)	TNAUSUF 7 (CO 4), COSFV 5, KBSH 1, KBSH 44, DRSH1, TNAU Sunflower Hybrid CO 2
	Chithiraipattam (April-May)	TNAUSUF 7 (CO 4), COSFV 5, KBSH 1, KBSH 44, DRSH1, TNAU Sunflower Hybrid CO 2
Niger		Paiyur 1, No 71, IGP 76
Safflower		K1, CO1
Castor		SA 2, V 19 & RC 6. **Hybrids**: GCH 3, GAUCH 1, GCH 2 **In TN**: SA1, 2, TMV 4, 5, 6, TMVCH 1, CO 1, YRCH 1

6.2. Production technologies

Crop	Season	Seed rate (kg/ha)	Spacing (cm)	Fert. Schedule (kg/ha)	Irrigation (no. & requirement)	Foliar spray	Herbicide	Yield (kg/ha)
Groundnut (*Arachis hypogea*) 2n= 40 South America Brazil Leguminosae	Kharif: June-July, Rabi: Oct-Nov, Summer: Dec-Jan.	Rainfed: 140 Irrigated: 125 Treat with *T.viride* at 4g/kg	30-60 x 10-15: Spreading x Semi spreading, 45 x 10: Bunch type 15 x 15 (Ring mosaic)	Rainfed – 10:10:45 Irrigated 25:50:75 400 kg Gypsum (40 – 45 DAS) Sulphur: 60 kg/ha	10 irrigation, 400-600mm		Fluchloralin @2 lit/ha Metalachlor @1kg/ha (Pre emergence)	Irrigated – 800 Rainfed 600 Oil content : 48-50% Protein: 20%
Types: Bunch, semispreading, spreading, virgina, Spanish, Valencia.								
Seed management: Soak in 0.5% Ca_2Cl_2 for 6 hrs (50% by VOLUME), Spread over moist gunny bags and cover with moist gunny bag for 20-24 hrs, select viable seed with protrusion of radicle, Non viable seeds used for oil extraction.								
Sesame (*Sesamum indicum*) Africa India Pedaliaceae 2n – 26	June – July Nov Dec Feb – March	5 Treat with *T.viride* at 4g/kg, Azospirillum at 600 g/ha	30 x 30 30 x 15 45 x 10 (Kharif) 45 x 15 (Rabi)	Rainfed – 23:13:13 Irrigated 35:23:23 5 kg $MnSO_4$ 5 kg $ZnSO_4$	Irrigation 150 mm		Alachlor @ 1.25 kg/ha Fluchloralin @1kg/ha (Pre emergence)	Rainfed: 375-500 Irrigated : 500-575 Oil content : 46-52%
Sunflower (*Helianthus annuus*) North America Asteraceae 2n–34	Rainfed :June – July Kharif: Oct-Nov Irrigated: Dec – Jan Apr – May	5-30 depending on seed weight, 4-7 Treat with T.viride at 4g/kg, 2% Zn SO4,Azospirillu m 600g/ha	30 x 10 30 x 15 30 x 30 60 x 30 Depending on variety	TN:40:20:20 UP: 80:60:40 AP: 60:30:0 Hybrid: 60:90:30 Variety 30:60:30	Irrigation 450 mm		Fluchloralin or Pendimethalin @2 kg/ha (Pre emergence)	2000 Oil content:36-45%
Safflower (*Carthamus tinctorius*) Central Asia Africa Ethiopia Asteraceae 2n= 24	Sep – Oct Nov- Dec	7 – 20	45 x 15 45 x 20 60 x 30	Irrigated - 60:30:20 Rainfed – 40:20:0 FYM: 12.5 t/ha	250-300 mm		Fluchloralin - 0.75 – 1 kg Pendimethalin - 0.75 kg Oxadiazone - 0.75 – 1 kg	1500-2000 Oil content: 32-36%

Castor (*Ricinus communis*) Ethiopia Euphorbiaceae 2n= 20	June- August	10-Variety 5-Variety	120 x 90 90 x 90 90 x 60 60 x 45 60 x 35	Rainfed: Variety: 45:15:15 Hybrid: 60:30:30 Irigated: Variety:60:30:30 Hybrid: 90:45:45		Pendimethalin @ 1.5 kg/ha on 1-2 DAS	Irrigated :1500 -2000 Rainfed: 800 – 1000
Linseed (Flax) (*Linum usitatissimum*) Ethiopia Linaceae 2n= 30	Oct – Nov	25-50	Row: 20 -30 cm Plant: 7-10 cm	Irrigated-90:30:30 Rainfed -40:20:20	3-4 irrigation	Fluchloralin @1 kg/ha Methabenzthiozuran @1kg (Pre emergence) Dichlorfopmethyl@0.7kg (Post emergence)	Rainfed: 800-1000 kg/ha Irrigated: 1200-1500
Niger (*Guizotia abyssinica)* Ethiopia Asteraceae 2n – 30	Aug – Sept	5 – 8	Row: 20-30 cm Plant: 10/15	20:20:0 40:40:0			300-500
Rapeseed & Mustard (*Brassica sp*) Brassicaceae 2n=18	Aug-Nov (Early rabi)	4-6	30×10 30×15	Irrigated -60:40:40 Rainfed-30:20:20		Pendimethalin@0.5-1.5kg/ha Fluchloralin@1.25kg (Pre emergence) Isoproturon@0.75kg/ha (Post emergence)	Irrigated rapeseed 1500-2000 Rainfed rapeseed 1000-1500 Irrigated mustard 2000-2500 Rainfed mustard 1500-2000 Oil content: 35-45%

7

Production Technology of Cash and Forage Crops

7.1. Seasons and Varieties

Crop	Seasons	Varieties
Cotton		MCU 5, MCU 7,MCU 12, MCU 13, LRA 5166, MCU 5 VT, Supriya, Anjali, Surabhi, Sumangala, Sruthi, suvin, K11.
		TCHB 213, SVPR 2, SVPR 3, KC 2, KC 3, SVPR 4
Jute		**Capsularis :**JRC 212, JRC 321, JRC 7447
		Olitorius: JRO 524, JRO 878, JRO 7835
Sugarcane		COC 671, COC 771, COC 772, COC 773, COC 778, COC 779CO 419 COC 85 COC86062 COSi 86071 CO6304 C66191(COC 8001)COC 8201, COC 90063, CO 8021, COC 91061, COC 92061, CO 8362, COG93076, CO 8208, COG 94077, COG 95076, CO 85019, COSi 95071, CO Si 96071, CO 86010, COC 98061, CO Si 98071, CO 86249CoC 99061CoC, CO 86032, COC (SC) 22, COC 23, COC 24, TNAU SC Si7, TNAU SC Si8
Sweet sorghum	Kharif (June- July)	SSV 84, CSV 19 SS (RSSV 9)
	Rabi (Sept-Oct.)	SSV 84, RSSV 9
	Summer(Mar.- April)	SSV 84, RSSV 9

7.2. Production technologies

Crop	Season	Seed rate (kg/ha)	Spacing (cm)	Fert. Schedule (kg/ha)	Irrigation (no. & requirement)	Foliar spray	Herbicide	Yield (kg/ha)
Cotton (*Gossypium sp*) Malvaceae 2n=52 *G.barbadense*-Egypt,G.*hirsutum* – America, G. *arboretum*-India, G. *herbaceum* - South Africa.	North zone: May-June, Central zone: June-July, Irrigated-March, Southern zone: Karnataka: June, AP-June-July, TN: August.	Fuzzyseed: 7.5-15, Delinted: 2.0-15.0kg, Naked:6kg. **Seed treatment:** Carbendazim 2g/kg, Trichoderma 4g/kg, Azospirullum/phosphobacteria-600g Azophos: 1200 g	Variety: 75 x 30 45 x 15 90 x 45 60 x 30 Hybrid: 120 x 60	80:40:40, 60:30:30, 120:60:60	Irrigation 10-11, 600mm		Pendimethalin @3.3 lit/ha (Pre emergence)	1800-2000 (Kapas)
Tropical sugarbeet 2n – 16. Chenopodiaceae (*Beta vulgaris*)	Sep-Nov	3.6 (3.5 to 4)	50×20 45x15	150:75:75	800-850mm		Pretilachlor 0.5 Kg/ha Pendimethalin 3.75 lit/ha on 2 DAS	Root yield: 8000-10000 Root brix content15-18%
Sugarcane (*Saccharum officinarum*) Poaceae 2n=40–128 South east asia – *S. barberi*, India - *S. spontaneous*, *S. officianrum* - New guinea.	Early season: Dec-Jan, Mid season: Feb-Mar, Late season: Apr-May, Special: June-July	50,000 - 3 budded setts 75,000 - 2 budded setts 187500 - single budded setts	0.9m to 1.5m and 2.4m	Irrigated-270:112.5:60, Water Scarcity areas- 225:112.5:60 Jaggery-175:112.5:60	Irrigation: 1800-2200mm Adsali crop: 3200-3500mm		Atrazine @ 1.75kg/ha (Pre emergence) Gramoxone @ 2.5lit/ha (Post emergence)	100-150t/ha, optimum maturity for harvest 18-20% brix
Jute (*Corchorus capsulari* & Indo burma) *C.olitorius* – Africa, India) *2n= 14*	Mar-May	Capsularis- Line sowing: 5, Broadcast: 7, Olitorius: Line:7, Broadcast: 10	Olitorius-25×5, Capsularis - 30×5	Olitorius:40-80: 20-40:20-40, Capsularis: 60-80: 30-40: 30-40	Irrigation 2-3		Fluchloralin @1.5kg/ha	Fiber yield: 2 to 2.5t/ha Green Plant: 45-50t/ha

Tobacco (*Nicotiana tabacum*) Solanaceae 2n–48	1st Fortnight of oct, 15 Dec, Mid Oct-Nov	1g/ha	Chewing 75×75, Cigar 75×10, Cheroot 60×45.	100:100:50-100	1060kg/ha
Daincha (*Sesbania aculeata*) Fabaceae	All season March-apr	Green manure -50kg, Seed-20kg	45×20 (or) Broadcasting		Green biomass-25t/ha, Seed 500-600kg/ha.
Sesbania (*Sesbania speciosa*) Fabaceae	All season March-apr	Green manure- 30-40kg/ha, Seed-15kg/ha	45×20cm (or) Broadcasting		Green biomass-15-18t/ha, Seed 400-600Kg/ha.
Kolunchi (Tephrosia purpurea) Fabaceae	All season March-apr	Green manure-15-20kg/ha, Seed 10kg/ha	30×10		Green biomass-6-7t/ha, Seed 400-500kg/ha.
Sunnhemp (*Crotalaria juncea*) Fabaceae	All season March-apr	Green manure-25-35Kg/ha, Seed 20Kg/ha	45×20 30×10		Green biomass-13-15t/ha, Seed400kg/ha.
Pillipesara (*Phaseolus trilobus*) or (*Vigna trilobata*) Fabaceae	All season March-apr	Green manure-10-15kg/ha, Seed- 10kg/ha	30×10		Green biomass-6-7t/ha, Seed 400-500kg/ha
Guinea grass (*Panicum maximum*) Poaceae	Throughout the year	2.5kg/ha 66,000slips	50×30	50:50:40	175t/8 cuts 1st harvest 75-80DAS
Buffel grass/ Anjan grass (*Cenchrus glaucus*) Poaceae	North east monsoon	6-8kg/ha	50×30	25:40:20	40t/4cuts 70DAS
Cumbu Napier hybrid (*Pennisetum purpureum*) Poaceae	All season	Slips 40,000	50×50	50:50:40	250-400t
Deenanath grass (*Pennisetum pedicellatum*) Poaceae	Throughout the year	2.5kg	30cm row	40:60:40	40-45t R:20-25t

Para grass/water grass/buffalo grass Poaceae (*Brachiaria mutica*)	Throughout the year	40,000 slips	50×50	20:40:0	200-240t/ha (6-9cuts)
Lucerne (*Medicago sativa)* Fabaceae	All season	20kg	Row spacing 25 cm	25:120:40	70-80t/ha
Hedge Lucerne (*Desmanthus virgatus*) Fabaceae	All season	20kg	Row spacing 50 cm	10:60:30	80-100t/ha
Stylo (*Stylosanthes scabra*)	Jun-Jul Sep-oct	6kg	30×15	20:60:15	30t/ha
Fodder Cowpea	Jun-Jul	40kg/ha	30×10	25:40:20	18-20t/ha
Green leaf desmodium *(Desmodium Tortuosum)*	All season				
Subabul (*Leucaena leucocephala)*	Jun-Jul Sep-oct	10kg fodder 12.5kg fuel	100×20	10:60:30	80-100t/ha
Berseem (*Trifolium alexandrinum*)	Oct-Nov	20-25kg	20×20	20:40:0	50-100t/ha

8

Agro Climatic Zones of India and Tamil Nadu

8.1. India

- Planning commission (1950) demarked into 15 Agro Climatic Zones
- Renamed as NITI AYOG in January 2015

8.1.1. Western Himalayan region

- Srinagar and Ladakh, Himachal Pradesh, Uttaranchal, Uttar Pradesh.
- Skeletal soils of cold region, podsolic soils, mountainous soils, hilly brown soil, steep slope.
- Rainfall: 75-150cm
- Crops: Wheat, maize, barley, Oats, Horticulture crops.

8.1.2. Eastern Himalayan region

- Assam, Sikkim, West Bengal and all North Eastern states (Arunachal Pradesh, Manipur, Meghalaya, Tripura, Nagaland and Mizoram).
- High rainfall and dense forest cover area.
- Rainfall: 200-400cm
- Crops: Rice, maize, potato, tea
- One by third of cultivated area is under shifting cultivation, results in degradation of soils and forest.
- Heavy runoff soil erosion and flood.

8.1.3. Lower Gangetic plains

- West Bengal
- Alluvial soil and more prone to floods, high water table & ground water.
- Crop: Jute, maize, potato, pulses

8.1.4. Middle Gangetic plains

- UP, Gujarat, Bihar
- Irrigation area 39%
- Rainfall: 100-200 cm
- Crops: Rice, maize, millets, wheat, borley, gram, peas, mustard.

8.1.5. Upper Gangetic plains

- UP (Western & Eastern parts)
- Irrigation through canals and tube wells
- More exploitation of ground water
- Rainfall: 75-150cm
- Crops: Wheat, rice, sugarcane, millets, maize, gram, barley, oilseeds, pulses and cotton.

8.1.6. Trans Gangetic plain

- Punjab, Haryana, Delhi and Rajasthan
- Highest sown area, high irrigation
- High cropping intensity, high ground water cultivation
- Contributed more for green revolution and farm mechanization.
- Rainfall: 65-125cm
- Crop: Wheat, sugarcane, cotton, rice, gram, maize, millets, pulses and oilseeds.

8.1.7. Eastern Plateau and hills region

- Maharashtra, UP, Uttar Pradesh, West Bengal
- Irrigation – canals and tanks
- Shallow and medium depth soil
- Rainfall: 80-150cm
- Crops: Rice, millets, maize, oilseeds, gram, potato

8.1.8. Central Plateau and hills region

- Maharashtra, Madhya Pradesh, Rajasthan
- Average rainfall - 904 mm (50-100cm)
- Crop: Millets, wheat, gram, oilseeds, cotton, sunflower.

8.1.9. Western Plateau and hills region

- Southern part of Malwa plateau and Deccan plateau (Maharashtra)
- Wheat, gram, millets, cotton, pulses, groundnut, oilseeds, sugarcane, rice, oranges, grapes and banana
- Rainfall: 25-75cm

8.1.10. Southern Plateau and hills region

- Andhra Pradesh, Karnataka, Tamil Nadu
- Dry farming with more cropping intensity
- Rainfall: 50-100cm
- Crops: Millets, oilseeds, pulses, cotton, tea, cardamom, spices

8.1.11. East coast plains and hills region

- Tamil Nadu, Odisha, AP and Puducherry
- Irrigation – canals and tanks
- Rainfall: 75-150cm
- Crops: Rice, jute, tobacco, sugarcane, maize, millets, groundnut, oilseeds.

8.1.12. West coast plains and Guards regions

- Tamil Nadu, Kerala, Goa, Karnataka and Maharashtra
- Variety of cropping pattern, rainfall and soil type.
- Rainfall: >200cm
- Crops: Rice, coconut, oilseeds, sugarcane, millets, pulses, cotton.

8.1.13. Gujarat plains and hills region

- Gujarat
- Arid zone with low rainfall
- Irrigation by tube wells and open wells
- Rainfall: 50-100cm
- Crops: Groundnut, wheat, oilseeds, rice, millets, tobacco.

8.1.14. Western dry region

- Rajasthan
- Hot sandy dessert, low rainfall, high evaporation, sandy vegetation.

- Deep ground water and brackish water
- Famine and drought are very common.
- Rainfall: 25cm
- Crops: Wheat, gram, millets.

8.1.15. Islands zones

- Andaman and Nicobar islands and Lakshadweep
- 3000 mm rainfall
- Spread over 8-9 months.
- High forest zone with undulated land.
- Crops: Rice, maize, millets, pulses, arecanut, turmeric, cassava.

8.2. Tamil Nadu

- Seven agro – eco (climatic) zones

8.2.1. North Eastern zone

- Thiruvallur, Vellore, Kanchipuram, Thiruvannamalai, Villupuram, Cuddalore excluding Chidambaram and Kattumannar kovil, some parts of Perambalur including Ariyalur taluks and also Chennai.
- Mean annual rainfall – 1054 mm in 53 rainy days
- Both south west monsoon and north east monsoon contribute rainfall
- Maximum temperature – 28.2 -38.9°C
- Minimum temperature – 19.5 – 24.0°C
- Crops – Paddy, pulses, cotton, gingelly, sugarcane (Thiruvallur, Villupuram) and groundnut.

8.2.2. North Western zone

- Dharmapuri and Krishnagiri (excluding hilly areas), Salem, Namakkal (excluding Thrichengode taluk) Perambalur taluk of Perembalur district.
- Mean annual rainfall – 825 mm in 47 rainy days
- Benefitted by both south west and north east monsoon.
- Maximum temperature - 32-39°C
- Minimum temperature - 19 - 25.5°C
- Crops – Paddy, sorghum, maize, gingelly, sugarcane.

8.2.3. Western zone

- Erode, Coimbatore, Dindugal, Theni, Thiruchengode of Namakkal district, Karur taluk of Karur district and some western parts of Madurai district.
- Mean annual rainfall – 718 mm in 45 rainy days
- Mean maximum temperature - 30 - 35°C
- Mean minimum temperature - 19 - 24°C.
- Crops – Sorghum, maize, ragi, paddy, sunflower and cotton.

8.2.4. Kaveri Delta zone

- Thanjavur, Thiruvarur, Nagapattinam, Musiri, Trichy, Lalgudi, Duraiyur, Kulithalai taluk of Trichirappalli district, Arathangi taluk of Pudukottai district, Chidambaram and Kattumannar kovil of Cuddalore district.
- Mean annual rainfall – 1078mm
- Crop – Paddy, pulses, cotton and gingelly.

8.2.5. Southern zone (biggest zone)

- Sivagangai, Ramanathapuram, Viruthunagar, Tuticorin, Tirunelveli, Nattham taluk of Dindugal district, Melur, Thirumangalam, Madurai south and Madurai north of Madurai district, Pudukottai district excluding Aranthangi taluk.
- Surrounded by coastal areas on east, mountains on west.
- Rainshadow area and drought very often.
- Climate - Semi arid tropics
- Rainfall - 776 mm in 43 rainy days
- Mean maximum temperature - 28.5°C – 38.5°C.
- Minimum temperature - 21-27.5°C
- Crops – Paddy and minor millets.

8.2.6. High rainfall zone

- Kanyakumari district
- Monsoon tropic climate
- Mean annual rainfall - 4569 mm in 64 rainy days
- Mean maximum temperature - 28°C-33.5°C
- Minimum temperature - 22°C-26.5°C
- Major crops - Paddy.

8.2.7. High altitude and hilly zones

- Nilgiris, Shevroys, Kollimalai, Patchaimalai, Anamalai, Palani and Podhigaimalai and Elagiri-Javvadhu.
- Rainfall – 850 mm (Kalvarayan – 850 mm to 4500 mm in Anamalai hills)
- Plantation crops, Jack, hill banana

9

Minimal Tillage Practices

- Reducing the tillage operations to the minimum to ensure good seed bed.
- Omitting few operations that do not give much benefit.
- Combining two or more agricultural operations.
- Eg. Seed cum fertilizer drill.

9.1. Advantage

- *In-situ* decomposition of plant residues to improve soil conditions
- To improve high infiltration
- Less resistance to root growth
- Less soil compaction
- Less soil erosion

9.2. Disadvantage

- Low germination of seeds
- Application of more nitrogen as the rate of decomposition of organic matter is low
- Reduction in nodulation of legume crops
- Continuous use of herbicides cause pollution
- Difficulty in sowing

9.3. Different methods of minimal tillage

- **Row zone tillage** – after primary tillage with mould board plough
- **Secondary tillage** like disking and harrowing are reduced
- Secondary tillage is done in row zone only
- **Plough – plant tillage**: After the ploughing, special planter is used to pulverize the row zone and sow the seeds.

- **Wheel tracking planting** – tractor is used for sowing and the wheels pulverize the row zone.

9.4. Zero tillage (no tillage)

- Primary tillage is completely avoided.
- Secondary tillage is restricted to prepare seed beds in the row zone only.
- Till planting, only one method is followed in zero tillage.
- Machine does four task in one operations.
    - Clean a narrow strip over the crop row
    - Open the soil for seed sowing
    - Place the seed in the crop row
    - Cover the seed properly
- In zero tillage, herbicides like glyphosphate and paraquat are used to destroy the vegetation.
- Major advantage – reduces the soil erosion.

10

Stress Mitigating Technologies Including Microorganisms

- Stress due to unfavourable factors like deficit or excess moisture, high radiation, low and high temperature, water and soil salinity, nutrient deficiency or toxicity and atmospheric pollution.

10.1. Drought stress or water stress

10.1.1. Two types of drought

- Soil drought
- Atmosphere drought
    - Low atmospheric humidity, high wind velocity and high temperature cause more transpiration in the plants.

10.1.2. Physiological changes due to drought

- Stomata - loose its functions and may die because of starch denaturation in the guard cells due to wilting.

10.1.3. Reduction in photosynthesis

- Due to reduced CO_2 diffusion because of decrease in stomatal opening
- An increase in osmotic pressure of plant cell
- Decrease in carbohydrate metabolism

10.1.4. Methods to overcome drought

- **Three important categories of plant growing in drought** (Growing of drought tolerant varieties and species)- Ephemerals (short lived plants), Succulents and non succulents
- Adjust the time of sowing to harvest the crops before drought onset
- Seed hardening with KCl, KH_2PO_4 and $CaCl_2$
- Poorly established plants are thinned out

- Mulching to minimize evaporation loss
- Foliar spray of anti-transpirants like Kaolin, PMA (Phenyl Mercuric Acetate), waxes, silicone oils.
- Foliar spray of growth retardants CCC (Cycocel) and MH (Malic Hydrazide)

10.2. High moisture stress – flooding/water logging

- Water logging – creates poor aeration (O_2 deficiency) and anaerobic situation in the rhizosphere.
- Flood tolerant crops – rice
- Flood sensitive – tomato, soybean and sun flower.

10.2.1. Physiological changes

- Flooding induces stomatal closer in C3 plants which causes low water flow results in leaf dehydration and reduced root permeability
- Leaf decay because of reduced root permeability.
- Finally, wilting of leaves due to restricted water flow from shoots and roots.
- So, this results in accumulation of toxic substances like acetaldehyde/alcohol
- Reduction in endogenous level of GA and cytokinins in the roots.
- This promotes enhanced level of ABA and ethylene causing stomatal closure.

10.2.2. Management of high moisture stress

- Provide adequate drainage.
- Spray 500 ppm cycocel to arrest apical dominance and promotes growth of laterals.
- Foliar spray of 2% DAP + 1% KCl in the form of (MOP)
- Spray of 40 ppm NAA or 0.5 ppm brassinolide or 100 ppm salicyclic acid.
- Foliar spray 0.3% boric acid + 0.5% zinc sulphate + 0.5% ferrous sulphate + 1% urea during critical stage of stress.
- Nipping of terminal buds in cotton to arrest apical dominance and promote growth of sympodial branches.

10.3. Salt stresses

- Due to excess salt accumulation
- Decreasing in osmotic potential of soil results in exosmosis and wilting of plants

10.3.1. Saline soil

- EC is greater than 4 dSm-1
- Exchangeable Sodium Percentage (ESP) less than 15%
- pH less than 8.5
- Dominated by natural soluble solts, namely chlorides and sulphates of calcium, magnesium and sodium.

10.3.2. Alkaline soil known as sodic soil.

- EC less than 4 dSm^{-1}
- ESP is more than 15%
- pH is more than 8.5
- Soil dominated by carbonate and bicarbonate (CO_3 and HCO_3) of sodium ions

10.3.3. Classification of plants

- Based on tolerance to salt stress, plants are classified as
  - **Halophytes:** plants grow under high salt concentration
  - It is divided into **oligo halophytes** – moderate salt stress; **Euhalophytes** – tolerate extreme salt stress.
  - **Glycophytes** – cannot grow under high salt concentration.

10.3.4. Effect of salt stress

- Delays seed germination, reduction in seedling growth and reduce the vegetative growth because of very high transpiration rate which results in higher uptake of salt.
- Affects panicle initiation, spikelet formation, fertilization and pollen grain germination
- Decline in photosynthesis process due to drastic reduction in chlorophyll b.

10.3.5. Mitigation of salt stress

- **Relative salt tolerant crops** - Cotton, sugarcane, barley
- **Semi tolerant** – Rice, maize, wheat, oats, sunflower and soybean
- **Sensitive crop** – Cowpea, beans, groundnut and grams.
- Seed hardening with NaCl – 10 mM concentration

- Gypsum application
- Incorporation of daincha @ 6.25 ton/ha
- Foliar spray – 2% DAP + 1% KCl or 0.5 ppm brassinolide or 100ppm salicylic acid or 40 ppm NAA.
- Foliar spray – PPFM (Pink Pigmented Facultative Methanaotrops) at 10^6 as a source of cytokinin
- Split application N and K fertilizers.

10.4. Temperature stresses

10.4.1. Low temperature stress

- **Chilling injury:** Tropical plants when exposed to 0°C
- **Freezing injury** : Occurs at temperature below 0°C
- **Physiological effects :** Cell membrane lipids get solidified (liquid state to solid state)
- Cell membrane is weakened and leaky
- Inactivation of mitochondria
- Stoppage of protoplasm streaming
- Accumulation of respiratory toxic substances
- Inter cellular ice formation
- Growth of plants arrested.

1. Mitigation of low temperature stress

- Seed hardening with 0.01% ammonium molybdate and foliar spray of 0.1 % ammonium molybdate
- Foliar spray of 2% calcium nitrate or 2% DAP + 1% KCl or 500 ppm cycocel or 100 ppm salicylic acid or 0.5 ppm brassinolide.

10.4.2. High temperature stress

- **Heat injury:** When plant temperature is higher than the environmental temperature (exceeds 35°C).

1. Physiological changes

- Reduction in photosynthesis, respiration, solute transport, water and nutrient uptake.
- Arrest of general metabolic process, seedling growth and vigour.

- Decrease in fertilization and maturation
- Heat shock protein (HSP) – characterised by
  - Alteration of gene expression (synthesis of heat shock protein)
  - High level stress tolerance

2. Mitigation of high temperature stress

- Seed hardening with 0.5% $CaCl_2$
- Foliar spray 2% DAP + 1% KCl or 40 ppm NAA or 0.5% brassinolide or 100 ppm salicylic acid
- Foliar spray of 3% kaolinite
- Foliar spray of 0.5% zinc sulphate + 0.3% boric acid + 0.5% ferrous sulphate + 1% urea
- Foliar spray of PPFM @ 10^6 as a source of cytokinin

10.5. Microorganism and stress

- Gives defense mechanism against any stress and this known as IST (Induced Systemic Tolerance)
- Microorganism - alleviate abiotic stresses and generally they are called as PGPB (Plant Growth Promoting Bacteria)
- Eg. *Pseudomonas* sp, *Bacillus sp, Acetobacter* and *Pantoea* sp.

10.5.1. Mechanism of action

- A bacterium produces phytohormones and phytoelixins along with plant growth promoting substances helps in solubulization of phosphates and potassium, and nitrogen fixation.
- Enhances the uptake of macro and micro nutrients.
- Fungal pathogen produces enzymes, antibiotics, siderophores and osmolytes for improving texture and structure of soil.
- Microbes involved in solubilization of inorganic, insoluble phosphate by secreting organic acids to decrease soil pH are *Agrobacterium, Arthrobacter, Aspergillus, Azospirillum, Acetobacter, Bacillus, Penicillium, Pseudomonas, Rhizobium, Micrococcus*, Halococcus, *Erwinia*
- Production of phytohormones like auxins, IAA, cytokinins, gibberellin.
- Eg. *Micrococcus, Pantoea, Pseudomonas, Serratia, Methylobacterium, Cornybacterium, Bacillus, Arthrobacter.*

11

Nano Particles and Their Applications

- Nano means one billionth
- Materials measured in a billionth of metre are nano materials.
- Nano technology is science and technology of tiny things
- The materials are less than 100 nm in size (1 nm = 10^{-9}m)
- Nano technology combines solid state physics, chemistry, chemical engineering, biochemistry, bio physics and material science.
- Properties of nano particles overlap colloids which range from 1 nm – 1mm in diameter.

11.1. Properties

- Higher charge density and higher reactivity
- Activity of atoms on the surface of the particle becomes more than inside the particle
- More strength increase, heat resistance, decreased melting point and different magnetic properties of nano clusters.
- Its higher catalytic ability when structure is tedrahedral, followed by cubic and spherical structure.

11.2. Applications

- Nano fertilizers
- Zinc nano fertilizer to enhance pearl millet crop improvement
- Nano pesticides, fungicides, weedicides
- Nano sensors and nano biotechnology.
- Biosensor in aquaculture, use of nanotech in post harvest technology
- Nano food system and smart packaging
- Monitoring the identity and quality of farm produces

- Precision agriculture – No damage to soil and water, reduced nutrient loss due to leaching
- Nano seed, growth harmones.

Part-2 Agricultural Meteorology

1

Meteorology and Agro Meteorology

1.1. Meteorology and Agro meteorology

- Greek word - meteoro-means above the earth surface (atmosphere), logy means science
- Branch of science dealing with the atmosphere.
- Lower atmosphere extends upto 20 km from earth surface where frequent physical process takes place.
- Meteorology is the combination of both physics and geography.
- Principles of utilizes the physics to study behavior of air
- Study of weather is called as meteorology. Meteorology is the physics of the atmosphere

1.1.1. Weather

- Physical state of the atmosphere at a given place and time. Eg. Cloudy day.

1.1.2. Climate

- Long term regime of atmospheric variables of a given place or area. Eg. Cold season.

1.2.3. Agricultural meteorology

- Branch of applied meteorology deals with physical conditions of environment of growing plants.
- Relationship between weather and climate condition and agricultural production.
- Agro meteorology is the abbreviated form of agricultural meteorology and to study the interaction between meteorological and hydrological factors and agriculture.

1.2.4. Difference between meteorology and agricultural meteorology

Meteorology	Agricultural meteorology
Branch of atmospheric physics	Branch of applied meteorology deals with agriculture
It is a weather science	Product of agriculture and meteorology
Physical science and aims at weather forecasting	Bio-physical science. Improved quantity and quality of crop production through meteorological scales.
Weather service linking science to society	Agro advisory service linking science to farming community

1.2. Importance to crop production

- Helps in planning cropping patterns / systems.
- Selection of sowing dates for optimum crop yields.
- Helps for preparatory or primary cultivation, harrowing, disc ploughing, leveling and secondary cultivation after cultivation practices, thinning, weeding etc.,
- Effective irrigation management practices
- Efficient harvesting methods
- Reducing the losses of chemicals and fertilizers.
- Eliminating the outbreaks of pests and diseases.
- Managing weather abnormalities like cyclones, heavy rainfall, floods and drought.
- This can be achieved by protection, avoiding irrigation when rain forecast and apply irrigation when frost forecast.
- **Avoidance -** avoid fertilizer and chemical sprays when rain forecasting
- **Mitigation -** use shelter belts like cold and heat waves
- Effective environmental protection – avoid forest fires

1.3. Future scope

- Study climatic resource of given area for effective crop planning
- To evolve weather based effective farm operations, then to study the relationship between weather factors and incidence of pests and diseases.
- To delineate agro climatic zones for effective and fast transfer of technology for improving yield.
- To develop crop growth simulation models in different agro climatic zones.

- To monitor agricultural drought on cropwise
- To develop weather based agro advisories.
- To investigate micro climate aspects of crop canopy for increase in crop growth.
- Study the influence of weather on soil environment
- To investigate weather in protected environment (glass house)

1.4. Co-ordinates of India

- $8^{o}04'$ to $37^{o}06'$ – north latitude, $68^{o}07'$ to $97^{o}25'$ - east longitude
- Earth is elliptical in shape has three spheres.
- **Hydrosphere** – water portion
- **Lithosphere** – solid portion
- **Atmosphere** – gaseous portion

1.5. Atmosphere

- Colourless, odourless and tasteless physical mixture of gases, which surrounds the earth on all sides.
- Physical condition - mobile, compressable and expandable and contains huge number of solid and liquid particles called **aerosols.**
- At lower atmosphere, chemical composition of gas is uniform that is called **homosphere.**
- At higher levels, the chemical composition of air changes considerably that is called as **heterosphere.**

1.5.1. Uses of atmosphere

- Supplies O_2 which is used for respiration of animals and plants
- Supply CO_2 for photosynthesis.
- Nitrogen is essential for plant growth
- Medium for pollen transfer, protect crops from harmful UV rays.
- Maintains micro climate of the plant.
- Source of water to the field crops by way of precipitation (rainfall, glaciers, snow, dew).

1.5.2. Composition of atmosphere

- Nitrogen (N_2) - 78.08%
- Oxygen (O_2) - 20.95% (21%)
- Carbon dioxide (CO_2) - 0.03%
- Argon (Ar) - 0.93%
- Neon (Ne) - 0.0018%
- Helium (He) - 0.0005%
- Ozone (O_3) - 0.00004%
- Hydrogen (H_2) - 0.00006%
- Methane (CH_4) - 0.00018%
- Krypton (Kr) - 0.000114%

1.6. Vertical layers of atmosphere

- Based on temperature, divided into different spheres of layers

1.6. 1. Troposphere

- Trop means mixing or turbulence;- sphere means region
- Height 14 km above the mean sea level
- From equator: 16 km
- From poles: 7-8 km
- A cloud, thunder storms, cyclones, anti-cyclones occurs in this sphere.
- Also called as **seat of weather phenomena**.
- Decrease of temperatures due to increasing elevation at the rate of 6.5°C per km called as **lapse rate**
- Most of sun radiation received from sun is absorbed by earth surface, results in heating from below.
- 75% of total gas and most of moisture and dust particles are present in this layer.
- At the top of the troposphere, a shallow layer separating it from the stratosphere, called as **tropopause**.

1.6. 2. Stratosphere

- It is about 20 km to 50 – 55 km.
- Also known as **Seat of photo-chemical reactions**.

- Temperature remains constant around 20 km.
- This layer is iso - thermal, because air is thin, clear and cold and dried near tropopause.
- Temperature of this layer increases with height.
- The upper part of the stratosphere shows high temperature than earth surface, which is due to ultra violet radiation of sun observed by ozone in this region.
- Ozone produces high temperature.
- Maximum concentration of ozone is present between 30 and 60 km in this layer – known as **ozonosphere**.
- Ozone absorbs UV rays.
- If no ozone, most of the solar radiation are absorbed by earth, this resulted in no life on the earth.
- High temperature is absorbed in ozonosphere called as **chemosphere**, because of chemical process.
- Air density is much less.
- Less convection (warm at the top cold at bottom).
- High wind speed.
- The upper boundary of the stratosphere is called as **stratopause.**

1.6.3. Mesosphere

- Height – 50-80 km above the surface.
- Temperature increase with height at rate of 6.5°C/km.
- Coldest layer of atmosphere.

1.6.4. Thermosphere (Ionosphere)

- Lies beyond ozonosphere at the height of above 80 km above earth surface and extending up to 400 km.
- This layer is ionized enriched ion zones. So it is known as **ionosphere**.
- Temperature increases again is in the order of 1000°C.
- So, ionosphere reflects the radio waves because of presence of multiple reflections of short wave radio beams from the ionized shells results in long distance radio communication is possible.

1.6. 5. Exosphere

- Outer most of the earth atmosphere lies between 400-1000 kms.
- Density of the atoms is extremely low.
- Hydrogen, helium gases predominate.
- Collisions between the natural particles are rare.
- Decrease in air temperature with height is known as **normal/environmental lapse rate.**
- Decrease in temperature at 6.5°C per km.

1.6. 6. Adiabatic lapse rate

- Rate of change of temperature in an ascending and descending air mass through adiabatic process - Adiabatic lapse rate
- Adiabatic means thermodynamic transformation without exchange of heat between the system and environment.

2

Weather and Climate

2.1. Weather

- Condition of the atmosphere at a given place and at a given instant time.
- State of atmosphere at a particular time and expressed in numerical values. (World Meteorological Organization)
- Weather parameter observed are solar radiation, temperature, pressure, wind, humidity, rainfall, evaporation etc.,

2.2. Climate

- Generalized weather or summarized weather conditions over a given region during comparatively longer period.
- Sum of all statistical information of weather in a particular area during a seasons or a year or even a decade.
- Long term statistics in the form of mean values, variances, probability of extreme values are worked out generally for period of 30 years of zone or state or country.

2.3. Difference between weather and climate

Weather	Climate
Study of atmosphere condition for a short duration for limited area	Study of average weather conditions observed over a long period time for a larger area.
Predominant elements are temperature, humidity	Collective effect of all elements
Changes very often	More or less permanent
Concerns with small area of the country	Larger area of the continent
A place can experience different weather condition in a year	Place can experience only one type of climate

2.4. Factors affecting climate

2.4.1. Latitude and distance from the equator

- Place near the equator is warmer because sun rays falls vertically.
- Places far away from the equator is cooler because sunrays fall slanting. Eg. temperate or polar regions (7-8 kms)
- Malaysia which is near to the equator is warmer than England which far away from the equator.

2.4.2. Altitude

- Height from mean sea level.
- For a vertical rise of 165 metres, there is an average decrease in temperature at the rate of 1°C.
- Temperature decreases with increase in height. Eg. Shimla located at high altitude is cooler than Jalandhar, although both are almost on same latitude.

2.4.3. Conductivity

- Water is poor conductor of heat.
- Water takes longer time to cool and heat, because of this place near to the sea coast has a low range of temperature and high humidity.
- In the interior places of the continents, extreme temperature are observed Eg., Mumbai experiences low temperature and high rainfall than Nagpur although both are located on same latitude.

2.4.4. Nature of prevailing wind

- On-shore winds bring moisture from the sea and cause rainfall.
- Off-share winds coming from the land are dry and help in evaporation.
- Eg: In India, on-shore summer monsoon winds brings rainfall, while off - shore winter monsoon winds are generally dry.

2.4.5. Cloud cover

- Cloudless desert results in high temperature even under shade, because of hot day sunshine.
- At night, this heat radiates back from ground very rapidly.
- Cloudy sky brings rainfall – Trivandrum where the range of temperature is very small.

2.4.6. Ocean currents

- Brings large movement of water from a place of warm temperature to cool temperature and vice versa.
- So, a warm ocean current rise the temperature of the coast and brings rainfall.
- Cooler the current, lower the temperature and forms fog.
- Eg : Port Bergen in Norway, is free from ice even in winter due to warm north Atlantic drift, while Port Quebec (Canada) remains frozen during winter months due to chilling effect of cold labrador current even though Port Quebec located in much lower latitude than Port Bergen.

2.4.7. Direction of mountain chain

- Mountain chain acts as natural barrier for wind (heavy rainfall on windward side and low rainfall in the leeward side).
- Greater Himalayas check the moisture laden monsoon winds from crossing over to Tibet.
- Also check biting polar pole winds entering to India, because of this northern plains of India get heavy rain and Tibet remains rain shadow.

2.4.8. Slope

- Concentration of heat is more in gentle slope which raises the temperature, steeper slope lowers the temperature.
- Mountain slope facing the sun are warmer than the slopes far away from the sun's rays.
- Eg. Southern slopes of Himalayas are warmer than the northern slope.

2.4.9. Nature of soil and vegetation cover

- Stony and sandy soil – good conductor of heat of the sun quickly.
- Bare surface radiates heat and desert is hot in the day and cold in the night.
- The forest area having low range of temperature throughout the year when compared to other areas.

3

Monsoon and Rainfall

3.1. Monsoon

- Two main monsoon systems
  - South west monsoon
  - North east monsoon.
- Two more systems ie., winter rainfall and summer rainfall.
- Occurs due to series of cyclones that arise in Indian Ocean (Bay of Bengal – East) and Arabian sea – West).
- This cyclones travel from north east direction and enter into peninsular India along west coast.
- Most of these cyclones occur June to September result in south-west monsoon.
- Second rainy season from October – December (north east monsoon).
- Third rainy season from January – February, fourth from March – May.
- 80-95% of the total rainfall of the country is by **south west monsoon (June - September)**.

3.2. South west monsoon (June - September)

- Beginning of the year, temperature of the Indian peninsular regions rapidly rises under the increasing heat of the sun.
- A minimum barometric pressure is established in interior parts of peninsular by month of March.
- Westerly winds prevails on the west Kerala and south winds on the west of northern circas (Sarkar – Division of British India) Odisha and Bengal.
- During April-May, the region of high temperature shifted to north, namely upper Sindu, lower Punjab and western Rajasthan.
- This regions becomes low barometric pressure area, monsoon winds are directed.

- Western branch of south west monsoon touches north Karnataka, southern Maharashtra and then Gujarat.
- When the south west monsoon fully operated on western India, another branch of the same is acting in Bay of Bengal, it carries rain to Burma, Northern parts of East coast India, West Bengal, Assam and whole of north India in general.

3.3. North east monsoon (October - December)

- During September end, the south west monsoon penetrates to north western India region.
- On the account of increasing in barmetric pressure in north India there is a shift of barometric to south east.
- North eastern monsoon winds begins to flow on the eastern coast, this brings heavy and continuous rainfall to south east India.
- North east monsoon accounts for about 48% of annual rainfall of Tamil Nadu.

3.4. Winter rainfall (January – February)

- Restricted to northern India only and is received in the form of fog in Punjab, Rajasthan and central India.

3.5. Summer rainfall (March - May)

- May month has local storms, mostly received in south east of peninsular and in Bengal, western India does not receives this rainfall.
- Highest average rainfall receiving region
  - Mawsynram in Meghalaya
- Lowest average rainfall receiving region
  - Jaisalmer of Rajasthan
  - Leh in Ladakh

4

Influence of Weather on Crops

4.1. Temperature on plant growth.

- Important for growth and development of plant
- Optimum temperature is required for maximum dry matter accumulation.
- High night temperature helps in breaking down of food materials results in shoot growth.
- **Cardinal temperature points** - All plants have maximum, minimum and optimum temperature limits.
- Wheat – minimum temperature (3-4°C), optimum temperature (25°C), maximum (30-32°C)
- Rice – minimum temperature (10-12°C), Optimum temperature (30-32°C), maximum (36- 38°C).

4.2.1. Low temperature injury

- Affects survival, cell division, photosynthesis, water and nutrient transport, growth and final yield
- **Chilling injury** – If hot season crops are cultivated in low temperature, they are killed
- 15°C - show yellowing eg. Rice, sorghum, pearl millet
- **Freezing injury** – When plants are exposed to low temperature, water freezes into ice crystals in the intercellular spaces ie., cell dehydration of temperate crops. Eg. Potato, tea etc.,
- **Suffocation** – Formation of thick layer of ice/snow on the soil surface prevents the entry of oxygen which prevents respiration and lead to accumulation of harmful substance.
- **Frost heaving**–Lifting the plants along with soil from its actual position by ice crystals (Upward swelling of soil) and mechanical lifting of plants.
- **Frost damage** – Cell damage
- Two types of frost – **Advective frost** – attack of large masses of cold air from colder regions.

- **Radiation frost** – Release of clear calm night when heat is freely radiated from all exposable objects
  - **White frost and hoar frost** - Due to sublimation of ice crystals on tree branches
  - **Black frost** – Vegetation is frozen because of reduction of air temperature.
- **Management**
  - Frost free growing season
  - Adjusting sowing time
  - Growing frost resistant varieties
  - Adopt sprinkler irrigation.

4.2.2. High temperature injury

- Affects mineral nutrition and pollen development which result in low yield.
- Critical temperature above which plant gets killed that is called **thermal death point**.
- Temperature above 45°C kills annual crops.
- **Mineral nutrition** - Reduces the absorption and assimilation of nutrients.
- In maize, 28°C reduces calcium absorbtion.
- In rice, high temperature affects nutrients uptake and reduce nitrate reductase activity.
- **Shoot growth** - High temperature affects plant height, root elongation, pollen development.
- In wheat, high temperature affects boot leaf, pollen.
- More high temperature ie., above 27°C cause under development of anthers, loss of pollen viability.
- Above 37°C reduce grain set and yield.
- **Scorching** - Dehydration of leaves and then scorching - called as **sun scald.**
- Reduction in photosynthesis and respiration.
- **Burning off** – Young seedlings are killed
- At high soil temperature **stem girdling** – Stem scorching at ground level Eg: cotton

5

Weather Forecasting and Meteorological Instruments

- Prediction of weather for the next few days to follow.
- Useful for farmers

5.1. Weather forecasting services

i) General public

ii) Agriculture including forestry, horticulture, animal husbandry and fishing

iii) Mountaineering

iv) Cyclones, floods and drought

v) Government and post officials

vi) Shipping navy

vii) Off-shore drilling

viii) Defense service

ix) Aviation civil and military

5.2. Importance of forecast

- Weather has many socio-economic impacts. Weather plays an there important role in bringing 50% variations in crop production.
- Rainfall forecast – decides the crop production and also national economy.
- Planning for moisture conservation.
- Flood relief under strong monsoon conditions helps in effective sustainability and minimize the damage caused by unfavorable weather directly or indirectly.
- Forecast the incidence of pest and diseases to minimize the crop loss.
- Weather forecast decides the effective irrigation.

5.3. Types of weather forecast

Types of forecast	Validity period	Main uses	Predictions
1. Short range	Upto 72 hrs		
a) Now casting	0-2 hrs	Farmers, marine agencies, general public	Rainfall distribution (heavy rainfall)
b) Very short range	0-12 hrs		Heat and cold wave, thunderstorms
2. Medium range	Beyond 3 days and up to 10 days	Farmers	Rainfall and temperature
3. Long range	Beyond 10 days up to a month and a season	Planners	Whole Indian monsoon rainfall, expected deviation from normal condition
4. Extended range forecast	10-30 days	Farmers	Temperature, rainfall etc.

5.4. Meteorological instruments

- Thermometer - Thermograph: Minimum and maximum temperatures (degree celsivs or Fahrenheit).
- Barometer / barograph: Atmospheric pressure in millibars
- Hygrometer - temperature in degree Celsius or Fahrenheit and humidity (RH) - morning RH and evening RH in (%).
- Anemometer - direction of wind and speed of wind in miles/hour.
- Wind vane - direction of wind at any given point in time.
- Rain gauge - Rain in mm
- Hail pads - Size of hail that falls during a storm
- Cambell - Stokes recorder - Sunshine
- Psychrometer - RH
- Pyranometer - direct beam solar irradiation
- Altimeter - height/altitude
- Monometer - root pressure
- Potometer - transpiration
- Porometer - stomatal behavior
- Pycnometer - specific gravity of soil
- Lysimeter - evapotranspiration and leaching
- Peizometer - depth of water table
- Darosometer - dew
- Tensiometer - soil moisture
- Auxanometer/crescograph - growth of plant
- Lactometer - specific gravity of milk

6

Organizations Involved in Weather Forecast

6.1. IMD – Indian Meteorological Department

- Comes under Ministry of Earth Sciences
- Head quarters: New Delhi
- Regional offices: Mumbai, Kolkata, Nagpur and Pune
- 100 observation stations across India and Antartica.
- Main functions: Meteorological observations, weather forecasting, Seismology (study of earth quakes)

6.1.1. History

- First meteorological observatories established by British East India company.
- Calcutta observatory 1785.
- Madras observatory 1796.
- Colaba (Mumbai) observatory 1826.
- Cyclone hit in Calcutta during 1864 and famines during 1866-1871 (due to monsoon failure).
- Head quarters shifted to Shimla in 1905, then to Pune (1928) and finally New Delhi in 1944.
- IMD becomes member of WMO on 27th April 1949.
- **Director general - Dr. K. J. Ramesh on August 1, 2016.**
- 6 meteorological Regional Centres under Deputy Director General (Chennai, Guwahati, Kolkatta, Mumbai, Nagpur and New Delhi).
- Each IMD units has forecasting officers, agro meteorological officer, area cyclone warning centres.
- IMD collaborates with **IITM (Indian Institute of Tropical Meteorology)** established in November 17, 1962 in Pune, **National Centre for Medium Range Weather Forecasting and National Institute of Ocean Technology.**

6.2. NCMRWF-National Centre for Medium Range Weather Forecasting

- Head quarters: Noida (UP).
- Control agency: Ministry of Earth Sciences.
- Purpose: Provide medium range weather forecast through Agro Advisory Service(AAS) to the farmers.
- Offers research in numerical weather prediction, diagnostic studies, crop weather modeling and computer science.
- IMD issuing weather bulletins to the farmers, since 1945.
- Medium range forecast is for 3-4 days.

6.3. Agromet Advisory Services (AAS)

- Operated by NCMRWF
- Established 127 agro meteorological field units (AMFU) in all A to Z of India.
- Panel of experts – Agronomist, soil scientist, plant pathologist, entomologist, horticulturist, sericulturist and also from agriculture extension and animal husbandry.

6.4. IITM – Indian Institute of Tropical Meteorology

- Established in Pune on November 17, 1962 and also comes under Ministry of Earth Sciences, since July 12, 2006.
- Previously operated under Ministry of Tourism and Civil Aviation and then, Department of Science and Technology (DST) of Ministry of Science and Technology.

7

Climate Change and Its Impact

- Refers to variations in earths green house or regional climates over time.
- Ongoing changes in modern climate that is the average rise in surface temperature - known as global warming.

7.1. Global warming and climate change effects.

7.1.1. Agricultural effects

- Green house gases - CO_2 (carbon dioxide)
- CO - Carbon monoxide
- CH_4 – Methane
- Chloro fluoro carbon (CFC)
- N_2O - Nitrous oxide

1. Carbon dioxide

- CO_2 is confined to troposphere.
- Used for photosynthesis at optimum concentration
- CO_2 – at higher concentration, it is a pollutant
- Increase in CO_2 concentration forms a thick layer of gas
- This allows the sunlight to filter through, but prevents reradiation of heat
- Looks like a glass panel of green house because it prevents reradiation of UV rays into outer space. Because of this, inside the green house is always warmer than outside.
- Similar condition resulted in trophospere that turned into green house effect.
- CO_2 concentration increase due to industrial growth
- One hundred years CO_2 concentration was 275 ppm and today it is 413 ppm and in 2040, it is estimated to 450 ppm.

7.1.2. Global warming

- Now, it is predicted a global warming up to 42°C and over all increase in precipitation will be 10% by the year 2050.
- Create a more active effective hydrologic cycle, increasing cloudiness as well as precipitation.
- During 20th century, global sea level has risen by about 0.15m
- Some scientist estimated that, it was about 0.3m which will increase to 1.4m in 2030.
- A temperature rise by 1.5 to 4.5°C which will melt polar ice caps.
- A rise of 5°C would increase the sea level by 5m.
- In few decades, increase in evaporation of water noticed, thus reducing the grains yield.

1. Chlorofluorocarbon

- Destruction of ozone layer in recent past.
- Ozone layer is depleted by 2-3% at a global level.

2. Methane

- 15-20% contributes towards green house effect.
- **Sources of methane** – Generated by bacteria during the anaerobic decomposition, flooded rice field, guts of cattle, garbage dumps, leaking in the process of mining.
- Use of coal and natural gas
- Burning of bio-mass
- Swamps and rice field contribute 20-30% to the methane production
- About 90% of the world's paddy fields are located in Asia and about 60% in India and China.

3. Nitrous oxide

- Destruction of ozone layer.

4. Ozone

- Ozone is a secondary pollutant in troposphere.
- Ozone is formed as follows

$$NO_2 + \text{Hydro carbon} \xrightarrow{\text{UV rays (sun light)}} O_3 + H_2O_2 \text{ (Ozone + Hydrogen peroxide)}$$

- Ozone is also termed as **photo chemical oxidant (H_2O_2)**
- Photo chemical smog is ozone which has abundant oxidants.
- Ozone layer in the atmosphere protected all living things on earth from harmful UV radiation from sun.
- Depletion of O_3 layer by human activities (industrial activities, deforestation etc).

7.1.3. Harmful effects due to UV radiation

- If O_3 layer becomes thinner or has holes, it cause cancer (skin cancer) – **melanoma.**
- 10% decrease in stratosphere ozone will increase skin cancer by 20-30%
- Other diseases are catract, destruction of aquatic life, destruction of vegetation and loss of immunity.
- Photo chemical consists of O_3, NO, H_2O_2 organic peroxides, PAN (Peroxy Acetyl Nitrate).

7.1.4. Smog

- Combination of smoke and fog.
- Smog was a characterized air pollution episode in London, Glasgow, Manchester and other cities of UK where sulphur rich coal was used
- Smog was coined by **H. A. Des Voeux – 1905**.

7.1.5. Strong surges

- Increased occurrence in cyclone in Bay of Bengal especially in post monsoon period.
- Cyclone always associated with maximum wind speed

7.2. Effects on food production

- Creates food security problems and threaten the livelihood activities.
- Affects respiration, photosynthesis, plant growth and reproduction.
- Methane released from rice cultivation.
- About 54% methane emitted from agriculture process.
- 80% NO emission from fertilizer application.
- Affects the types of crops and crop yield.

7.3. Effects on human health

- Vector borne disease – malaria is the main concern, once in 5-7 years malaria epidemic takes place.
- World bank estimated about 5,77,000 lakhs daily were lost due to malaria in India in 1998-(disability - adjusted life years).
- NPL - National Physical Laboratory predicting the disease.

7.3.1. Heat and cold related diseases

- Cardio-vascular illness
- Vector borne diseases like filaria, malaria, Japanese encephalitis, dengue (bacteria, virus, mosquito, ticks), diarrhea, cholera
- Social-mental illness
- Malnutrition and hunger

7.4. NIO - National Institute of Oceanography

- Established in January 1, 1966.
- Head quarters: Goa
- Regional centres at Kochi, Mumbai, Vishakapattinam.
- Course studies - Ocean process, coastal studies, resource survey conservation and ocean engineering.
- Comes under CSIR (Council for Industrial Scientific and Research)

7.5. United Nations Climate Change Conferences

Conference of parties	Date and year of conference	Place
COP 1	28th March to 7th April 1995	Berlin, Germany
COP 2	18th July 1996	Geneva, Switzerland
COP 3	Dec-97	Kyoto, Japan
COP 4	Nov-98	Buenos Aires, Argentina
COP 5	25th Oct to 5th Nov 1999	Bonn, Germany
COP 6	13–25 Nov 2000	The Hague, Netherlands
COP 6	17–27 July 2001	Bonn, Germany
COP 7	29 Oct to 10 Nov 2001	Marrakech, Morocco
COP 8	**23 Oct to 1 Nov 2002**	**New Delhi**
COP 9	1–12 Dec 2003	Milan, Italy
COP 10	6–17 Dec 2004.	Buenos Aires, Argentina
COP 11/CMP 1	28 Nov - 9 Dec 2005	Montreal, Quebec, Canada.
COP 12/CMP 2	6–17 Nov 2006	Nairobi, Kenya
COP 13/CMP 3	3–17 Dec 2007	Nusa Dua, in Bali, Indonesia.
COP 14 /CMP 4	1–12 Dec 2008	PoznaD , Poland.

COP 15	7–18 Dec 2009	Copenhagen, Denmark
COP 16	28 Nov to 10 Dec 2010.	Cancún, Mexico
COP 17	28 Nov to 9 Dec 2011.	Durban, South Africa
COP 18	26 Nov to 7 Dec 2012	Doha, Qatar
COP 19/CMP 9	11 to 23 Nov 2013	Warsaw, Poland
COP 20/CMP10	1–12 Dec 2014	Lima, Peru
COP 21/CMP 11	30 Nov to 12 Dec 2015.	Paris
COP 22/CMP 12	7–18 Nov 2016	Marrakech, Morocco
COP 23/CMP13	6–17 Nov 2017	Bonn, Germany
COP 24/CMP 14	3–14 Dec 2018	Katowice, Poland
COP 25/CMP 15	11–22 Nov 2019	Santiago, Chile

7.6. Intergovernmental Panel on Climate Change

Session of IPCC	Date and year of session	Place
1 St Session	9 - 11 November 1988	Geneva, Switzerland
2nd Session	28 - 30 June 1989	Nairobi, Kenya
3rd Session	5 - 7 February 1990	Washington D.C., U.S.A
4th Session	27 - 30 August 1990	Sundsvall, Sweden
5th Session	13 - 15 March 1991	Geneva, Switzerland
6th Session	29-31 October 1991	Geneva, Switzerland
7th Session	10 - 12 February 1992	Geneva, Switzerland
8th Session	11 - 13 November 1992	Harare, Zimbabwe
9th Session	29 - 30 June 1993	Geneva, Switzerland
10th Session	10 - 12 November 1994	Nairobi, Kenya
11th Session	11 - 15 December 1995	Rome, Italy
12th Session	11-13 September 1996	Mexico City, Mexico
13th Session	22 & 25-28 September 1997	Maldives
14th Session	1 - 3 October 1998	Vienna, Austria
15th Session	15 - 18 April 1999	San Jose, Costa Rica
16th Session	1 - 8 May 2000	Montreal, Canada
17th Session	4 - 6 April 2001	Nairobi, Kenya
18th Session	24-29 September 2001	Wembley, UK
19th Session	17-20(Morning Only) April 2002	Geneva, Switzerland
20th Session	19-21 February 2003	Paris, France
21st Session	3-7 November 2003	Vienna, Austria
22nd Session	**9-11 November 2004**	**New Delhi, India**
23rd Session & Second Joint Session Working Group III	6-8 April 2005	Addis Ababa, Ethiopia
24th Session and Eighth Session of Working Group III	22-24 & 26-28 September 2005	Montreal, Canada
25th Session	26-28 April 2006	Port Louis, Mauritius

26th Session and 9th Session of Working Group III	30 April - 4 May 2007	Bangkok, Thailand
27th Session	12-17 November 2007	Valencia, Spain
28th Session	9-10 April 2008	Budapest, Hungary
29th Session	31 August - 4 September 2008	Geneva, Switzerland
30th Session	21-23 April 2009	Antalya, Turkey
31st Session	26-29 October 2009	Bali, Indonesia
32nd Session	11 - 14 October 2010	Busan, Republic of Korea
33rd Session	5 - 8 And 10 - 13 May 2011	Abu Dhabi, United Arab Emirates
34th Session	5 - 8 and18 - 19 November 2011	Kampala, Uganda
35th Session	6 - 9 June 2012	Geneva, Switzerland
36th Session	23 - 26 September 2013	Stockholm, Sweden
37th Session	14 - 18 October 2013	Batumi, Georgia
38th Session	25 - 29 March 2014	Yokohama, Japan
39th Session	7 - 12 April 2014	Berlin, Germany
40th Session	24 - 31 October 2014	Copenhagen, Denmark
41st Session	24 - 27 February 2015	Nairobi, Kenya
42nd Session	5 - 8 October 2015	Dubrovnik, Croatia
43rd Session	11 - 13 April 2016	Nairobi, Kenya
44th Session	17 - 20 October 2016	Bangkok, Thailand
45th Session	28 - 31 March 2017	Guadalajara, Mexico
46th Session	6 - 10 September 2017	Montréal, Canada
47th Session	13 - 16 March 2018	Paris, France
48th Session	1 - 5 October 2018	Incheon, Republic of Korea
49th Session	8 - 12 May 2019	Kyoto, Japan
50th Session	2 - 6 August 2019	Geneva, Switzerland

8

Clouds

- An aggregation of minute drops of water or ice or both suspended in the air at higher altitudes not touching the ground is called as cloud.
- Exist in a variety of shapes and sizes.
- Clouds are accompanied by precipitation, rain, snow, hail, sleet, and even freezing rain.
- Almost all cloud exists in the troposphere.

8.1. Development of clouds

- Water exists in three different states: as a solid, liquid or gas.
- Water existing as a gas is called water vapor.
- If the air is moist, that contains large amounts of water vapor.
- All clouds are a form of water.
- Smaller the drops in a cloud, the brighter the tops appear.

8.2. Cloud classification

- Clouds are grouped into 4 categories (viz., family A, B, C and D).
- Four clouds families are in different heights of the troposphere.
    - High level clouds (altitudes of 5-13 km)
    - Medium level clouds (2-7 km)
    - Low level clouds (0-2 km)
    - Clouds with large vertical extending (0-13 km)

8.2.1 Family A

- Clouds are high (7-12 km).

1. Cirrus (Ci)

- Ice crystals are present.
- Sun rays pass through these clouds.
- Does not produce precipitation.

2. Cirrocumulus (Cc)

- Ice crystals are present.

3. Cirrostratus (Cs)

- Ice crystals are present.
- Produces "Halo" (group of optical phenomenon in the form of rings, arcs, pillars or bright spots).

8.2.2. Family B

- Middle clouds: 2.5 – 7 km.
- 2 sub-categories.

1. Altocumulus (Ac)

- Ice water is present.

2. Alto-stratus (As)

- Water and ice are present separately.
- Rain occurs in middle and high latitudes.

8.2.3. Family C

- Lower clouds: 0 – 2.5 km.
- 3 sub-categorises.

1. Strato cumulus (Sc)

- Composed of water.

2. Stratus (St)

- Composed of water.
- Mainly seen in winter season and occasional drizzle occurs.

3. Nimbostratus (Ns)

- Composed of water or ice crystals.
- Gives steady precipitation.

8.2.4. Family D

- 0.5 – 16 km
- Two sub-categories are present.

1. Cumulus (Cu)

- Composed of water.
- Looks like cauliflower with wool pack.
- Develop into cumulo-nimbus clouds.

2. Cumulonimbus (Cb)

- Upper level clouds posses ice, lower level has water.
- Produce violent winds, thunder storms, hails and lightening during summer.

UNIT-III
NATURAL RESOURCE MANAGEMENT
Part-1 Soil Science

1

Soils, Components and Classification

1.1. Definition

- **Soil:** Thin layer of earth's crust made up of disintegrated and decomposed rocks, complex mineral compound, organic matter, water/air and living organism like bacteria, fungi, insects and worms and serves as the natural medium of growth of plants.
- Provides nutrients, moisture, anchorage (support) and air to plants.
- Parent materials are igneous rocks, sedimentary rocks and metamorphic rocks.
- **Soil Science:** Science dealing with soil (Natural resource) on the surface of the earth, including pedology (Soil genesis, classification and mapping), physical, chemical, biological and fertility properties and their relation to management for crop production.

1.2. Components of soil

- Soil is made up of the following components:
  - Mineral particles,
  - Dead organic matter or humus,
  - Soil atmosphere,
  - Soil water,
  - Biological system or soil microorganisms.

1.2.1. Mineral components

- Derived from the parental rocks or regolith.
- Found in the form of particles of different sizes; from clay (0.0002 mm or less in diameter) to large pebbles and gravels.
- Represent about 90% of the total weight of the soil.
- Consists of O_2, Si, Fe, Al, N, P, K, Ca, Mg, C, H, etc.

1.2.2. Organic matter or humus

- Presence of dead remains of plants and animals and decomposed through metabolic activities of living organisms.
- Final product of microbial decomposition is humus, which is dark coloured, jelly-like amorphous substance composed of residual organic matters not further decomposed by soil microorganisms.
- Formation process is called humification.
- Chief elements found in humus are carbon, hydrogen, oxygen, sulphur and nitrogen.
- Compounds found in it are carbohydrates, phosphoric acid, some organic acids, fats, resins, urea, etc.
- Tree litter (Very little decomposed dead matter) also contains lime, potash, Mn, Mg, silica, Cu, Al, Ga, Na, K, etc.
- Humus is not soluble in water.
- Present in soil in the form of organic colloids.
- Less in arid soils and very high in humid soils.
- In top layer of the soil, humus quantity is greater than in the deep layers.

1.3. Classification based on soil taxonomy

- Order: Entisols: Suborder: Fluvents
- Great group: Torrifluvents; Subgroup: Typic Torrifluvents
- Family: Fine-loamy, mixed, super active, alluvium.

1.3.1. Alluvial soils

- **Area: 15 lakh sq km (43%); light grey to ash grey in colour**
- Formed by transportation in streams and rivers and are deposited in flood plains or along the coastal belts.
- Occur in the basins of Indus, Ganges, Brahmaputra, Godavari, Krishna, Cauvery and Tambiraparani deltas and spread in U.P., Bihar, West Bengal, Gujarat, Punjab, Rajasthan, Andhra Pradesh, and Tamil Nadu.
- Newer alluvium is called as Khadar - sandy, light colour and less Kankar nodules.
- Older alluvium is called as Bhangar - full of clay, dark colour and more Kankar nodules.
- Alluvial soils of high altitude are acidic and plains are neutral to alkaline.

- Medium in phosphorous content and high in potassium content.
- Rich in nutrients and are fertile
- **Crops**: Rice, wheat, cotton, maize, sugarcane, vegetables, jute, oil seeds, millets, pulses and fruits.

1.3.2. Black soil

- **Area : 5.46 lakh sq km (16.6 %); deep black to light black in colour; Rich in Fe, lime, Ca, K, Al, Mg; Deficient in N, P.**
- Colour due to clay-humus complex.
- Also called black cotton soil.
- Montmorillonite clay is present.
- Found in Maharashtra, Madhya Pradesh, South Odisha, South and Coastal Andhra Pradesh, North Karnataka and parts of Tamil Nadu.
- Contains high proportion of clay (30-40%).
- Water holding capacity is high.
- Typical characteristics are swelling (During wet period) and shrinkage (Dry period). While dry, it forms a very deep cracks of more than 30-45 cm.
- In Kovilpatti (Tamil Nadu) areas, the cracks may extend to 2 to 3 m long with a width of 1 to 6 cm.
- Contain high proportion of calcium and magnesium carbonates.
- Poor in N, medium in P and medium to high in K.
- In Tamil Nadu, black soils have high pH (8.5 to 9) and rich in lime (5-7%), have low permeability, more cation exchange capacity (40-60 m.e./100 g).
- **Crops:** Cotton, bengal gram, mustard, millets, pulses, oilseeds (sunflower, safflower).
- Most of the soils come under rainfed areas.

1.3.3. Red soil

- **Area: 3.5 lakh sq km (10.6%); Deficient in lime, P, Mn, K, humus; and $CaCO_3$ (kankar) absent**
- Red colour is due to presence of ferric oxides.
- Formed from granites and other metamorphic rocks.
- Mostly found in semi-arid and low rainfall areas.
- Colour varies from red to yellow.

- Light textured, with kaolinite type of clay.
- Well drained with moderate permeability.
- Low cation exchange capacity and low water holding capacity.
- Present in Gujarat, Tamil Nadu, Karnataka, Andhra Pradesh, North and East of Arunachal Pradesh, Madhya Pradesh, Parts of Bihar and Uttar Pradesh.
- Shallow in depth because of degraded or drained soil.
- Lesser clay and more sandy than vertisol.
- Acidic nature.
- **Crops:** Millets, pulses, oil seeds (Ground nut, gingelly, castor) and tuber crops (Cassava).

1.3.4. Laterites and lateritic soil

- **Area: 2.48 lakh sq km; acidic in nature; Rich in Fe and Al; Deficient in N, P, K and lime**
- Formed due to the process of laterisation. i.e., leaching of all cations leaving Fe and Al oxides.
- Mostly found in hills and foothill areas.
- Formed under high intensive down pour of rainfall.
- Modified form of red soil, clay content is minimum.
- Rich in organic matter content and fertility, and medium water holding capacity.
- **Crops:** Acid loving crops (Plantation crops) and fruits (Pineapple, avacado) are more cultivated. Tea, rubber, coconut, arecanut, cinchona, pepper, spices are cultivated. At lower elevation places, rice is grown.

1.3.5. Desert soil

- **Area: 1.42 lakh sq km (4.32% of total area); Deficient in N and P is normal**
- Found in desert regions of Rajasthan (Thar desert), parts of Haryana and Punjab.
- More sand is found and sand dunes are common.
- Clay content is < 8% only.
- Poor fertility, poor water holding capacity and susceptible to soil erosion.
- Presence of sodic salts (High Na content) leads to alkalinity.
- **Crops:** Date palm, cucumber, millets.

1.3.6. Peaty and organic soil

- Heavy soil, black in colour.
- Rich in organic matter.
- Found in Kerala, coastal regions of West Bengal, Odisha, South and East coast of Tamil Nadu.
- Not suitable for majority of crops.
- **Crops:** Rice is mostly cultivated in coastal area.

1.3.7. Problem soil

1. Saline soils

- Contain excess amounts of neutral soluble salts dominated by chlorides and sulphates of Na, Ca and Mg.
- White encrustation of salts and hence, called white alkali.
- EC: >4dSm^{-1}, ESP: < 15; pH: < 8.5.
- Soil needs leaching and drainage before cropping.
- **High salt tolerant**: Sesbania, rice, sugarcane, oats, berseem, lucerne, indian clover and barley.
- **Medium salt tolerant**: Castor, cotton, sorghum, pearl millet, maize, mustard and wheat.
- **Low salt tolerant**: Pulses, peas, sunnhemp, gram, linseed and sesame.

2. Sodic/Alkali soils

- High content of carbonates and bicarbonates of Na.
- High exchangeable sodium percentage (ESP) with dark encrustation, hence called as black alkali.
- Rich in $NaHCO_3$.
- pH: > 8.5; EC : < 4dSm-1 ; ESP : > 15.
- Use gypsum ($CaSO_4$, $2H_2O$ @ 400 kg per ha) as amendment for reclamation.
- Iron pyrites (FeS_2), bulky organic manures (Especially green manure) and crop residues which produces weak organic acids.
- **Tolerant crops:** Karnal/rhodes/para/bermuda grass, rice and sugar beet.
- **Semi tolerant:** Wheat, barley, oats, berseem and sugarcane.
- **Sensitive:** Cowpea, gram, groundnut, lentil, peas and maize.

3. Acid soils

- Low pH with high amounts of exchangeable H^+ and Al_3.
- Occur in regions with high rainfall.
- Significant amount of partly decomposed organic matter exist.
- Have low CEC and high base saturation.
- Liming and judicious use of fertilizers are the management measures.
- **Suitable crops:** Acedophytes (Like potato).
- **Tolerant crops:** Pepper, bell pepper, potato, radish, sweet potato, parsley, beans, cabbage, carrot, onion, cucumber, tomato, turnip.

4. Comparison of three types of soils

Parameters	Saline soil	Saline alkali	Alkali soil
EC (dSm^{-1})	>4	>4	<4
ESP (%)	<15	>15	>15
pH	<8.5	>8.5	>8.5

1.3.8. Forest soils

- Occurs in high rainfall area and acidic.

1.3.9. Mountainous soil

- Immature soil, low humus, acidic.

1.4. Soils of Tamil Nadu

Type of soil	Areas in Tamil Nadu
Red loam (79.8 L. ha and 61.7 %) Further classification • Red loamy (30 %) • Red sterile (6 %) • Red sandy (6 %) • Thin red (2 %) • Deep red loamy soils (8 %)	Parts of Kancheepuram, Cuddalore, Salem, Dharmapuri, Coimbatore, Tiruchirappalli, Thanjavur, Ramanathapuram, Madurai, Tirunelveli, Sivagangai, Thoothukudi, Virudhunagar, Dindigul and Nilgiris.
Laterite soil (3.8 L.ha and 2.9 %)	Parts of Nilgiris.
Black soil (15.0 L. ha and 11.6 %)	Parts of Kancheepuram, Cuddalore, Vellore, Thiruvannamalai, Salem, Dharmapuri, Madurai, Ramanathapuram, Tirunelveli, Sivagangai, Thoothukudi, The Nilgiris, Virudhunagar and Dindigul.
Sandy coastal alluvium (9.8 L. ha and 7.6 %)	Coastal areas of Ramanathapuram, Thanjavur, Nagapattinam, Cuddalore, Kancheepuram and Kanyakumari.
River alluvium (21.0 L. ha & 16.2%)	All river deltaic areas (Cauvery, Vaigai, Tambiraparani).

1.4.1. USDA system of soil classification (Taxonomy)

- Entisols (6 %): Young river alluvium, sandy and eroded red and laterite soils.
- Inceptisols (50 %): Moderately deep red, laterite and black soils are included under inceptisol.
- Alfisols (30 %): Deep red and laterite soils.
- Mollisols (Negligible)
- Ultisols (1 %): Highly weathered laterite soils.
- Vertisols (7 %): Deep black cotton soils and old alluvial soils are classified under vertisol.

2

Physical Properties of Soil

2.1. Soil structure

- Important physical properties of soils
- Soil texture
- Soil structure
- Surface area
- Soil density
- Soil porosity
- Soil colour
- Soil consistence
- Soil temperature
- Soil air

2.1.1. Definition

- Arrangement and organization of primary and secondary particles in a soil mass.

2.1.2. Formation of soil structure

- Soil particles present either as single individual grains or as aggregate i.e. group of particles bound together into granules or compound particles known as secondary particles.
- **Single grain**
  - Individual particle
  - Sandy or silty soil
  - Solid
  - Structure less
- **Secondary particles**
- Aggregate/compound particle

- Clayey soil
- Porous/spongy
- Granulated or crumby structure

2.1.3. Types of structure

1. Plate-like (Platy)

- Aggregates are arranged in relatively thin horizontal plates or leaflets.
- Horizontal axis or dimensions are larger than the vertical axis.
- When the units/ layers are thick they are called **platy**.
- When they are thin, it is **laminar.**
- Noticeable in the surface layers of virgin soils, but may be present in the subsoil also.
- Inherited from the parent material, especially by the action of water or ice.

2. Prism-like

- Vertical axis is more developed than horizontal, giving a pillar like shape.
- Length 1- 10 cm.
- Occur in sub soil horizons of arid and semi arid regions.
- When the tops are rounded, the structure is termed as **columnar.**
- When the tops are flat / plane, level and clear cut, it is termed as **prismatic**.

3. Block like

- Irregularly six faced with their three dimensions more or less equal.
- When the faces are flat and distinct and the edges are sharp angular, the structure is named as **angular blocky**.
- When the faces and edges are mainly rounded, it is called sub **angular blocky**.
- Confined to the sub soil and favourable for drainage, aeration and root penetration.

4. Spheroidal (Sphere like)

- Rounded aggregates (Peds).
- Not exceeding an inch in diameter.
- Sphere-like structure and not affects infiltration, percolation and aeration.
- Aggregates of this group are termed as **granular** which are relatively less porous.

- When the granules are more porous, it is termed as **crumb.**
- Specific to surface soil and high in organic matter/ grass land soils.

2.1.4. Class of soil structure as differentiated size of soil peds

Class	Platy	Prismatic	Columnar	Blocky	Sub angular blocky	Granular	Crumb
Very fine or Very thin	< 1	< 10	< 10	< 5	< 5	< 1	< 1
Fine or Thin	1-2	10-20	10-20	5-10	5-10	1-2	1-2
Medium	2-5	20-50	20-50	10-20	10-20	2-5	2-5
Coarse or thick	5-10	50-100	50-100	20-50	20-50	5-10	-
Very coarse or Very thick	> 10	> 100	> 100	> 50	> 50	> 10	-

2.2. Soil texture

- **Definition:** Refers to the relative proportion of particles or it is the relative percentage by weight of the three soil separates viz., sand, silt and clay.

  Simply refers to the size of soil particles.
- Soil separates are defined in terms of diameter in millimeters.
- Soil particles less than 2 mm in diameter are excluded from soil textural determinations.
- Fine earth: Less than 2mm
- Components of fine earth: Sand, silt and clay (Soil separates).
- Gravels : 2 – 4 mm
- Pebbles : 4 – 64 mm
- Cobbles : 64 – 256 mm
- Boulders : > 256 mm

2.2.1. Textural classification

- **USDA: United States Department of Agriculture**

Soil separates	Diameter (mm)
Clay	< 0.002
Silt	0.002 - 0.05
Very fine sand	0.05 - 0.10
Fine sand	0.10 - 0.25
Medium sand	0.25 - 0.50
Coarse sand	0.50 - 1.00
Very Coarse sand	1.00-2.00

- BSI: British Standard Institute

Soil separates	Diameter (mm)
Clay	< 0.002
Fine silt	0.002 - 0.01
Medium silt	0.01 - 0.04
Coarse silt	0.04 - 0.06
Fine sand	0.06 - 0.20
Medium sand	0.20 - 1.00
Coarse sand	1.00 - 2.00

- ISSS: International Society of Soil Science

Soil separates	Diameter (mm)
Clay	< 0.002
Silt	0.002 - 0.02
Fine sand	0.02 - 0.2
Coarse sand	0.2 - 2.0

- **European system**

Soil separates	Diameter (mm)
Fine clay	< 0.0002
Medium clay	0.0002 - 0.0006
Coarse clay	0.0006 - 0.002
Fine silt	0.002 - 0.006
Coarse silt	0.02 - 0.06
Fine sand	0.06 - 0.20
Medium sand	0.20 - 0.60
Coarse sand	0.06 - 2.00

Sand

- Consists of quartz, but may also contain fragments of feldspar, mica and occasionally heavy minerals viz., zircon, tourmaline and hornblende.
- Has uniform dimensions and spherical with jagged surface.

Silt

- Particle size intermediate between sand and clay
- Smaller and more surface area.
- Coated with clay.
- Has the physico - chemical properties as that of clay to a limited extent.
- Sand and Silt forms the SKELETON.

Clay

- Particle size less than 0.002 mm.
- Plate like or needle like in shape.
- Belong to alumino silicate group of minerals.
- Sometimes considerable concentration of fine particles which does not belong to alumino silicates. Eg. Iron oxide and $CaCO_3$.
- Secondary minerals derived from primary minerals in the rock.
- Flesh of the soil.

2.2.2. Land use capability and methods of soil management depends on texture

2.2.3. Particle size distribution / determination

- Determination of relative distribution of the individual soil particles below 2 mm diameter is called as particle size analysis or mechanical analysis.
- Two steps are involved.

i) Separation of all the particles from each other i.e. complete dispersion into ultimate particles.

ii) Measuring the amount of each group.

- **Measurement:** Once the soil particles are dispersed into ultimate particles, measurement can be done.
- **Coarse fraction:** Sieves are used in mechanical analysis. For 2mm, 1mm, 0.5 mm, circular sieves and for smaller sizes, wire mesh screens are used.
- **Finer fractions:** By settling in a medium. Settling or the velocity of the fall of particles is influenced by viscosity of medium. Difference in density between medium and falling particles.

Separation

S.No.	Aggregation agents	Dispersion method
1.	Lime and oxides of Fe and Al	Dissolving in HCl
2.	Organic matter	Oxidizes with H_2O_2
3.	High concentration of electrolytes (Soluble salts)	Precipitate and decant or filter with suction
4.	Surface tension	Elimination of air by stirring with water or boiling

- **Stokes' law:** Particle size analysis is based on a simple principle i.e. when soil particles are suspended in water, they tend to sink. Because, there is little variation in the density of most soil particles, their velocity (V) of settling is proportional to the square of the radius ‘r’ of each particles.

- Thus, V = kr 2, where k is a constant. This equation is referred to as Stokes’ law. Stokes (1851) was the first to suggest the relationship between the radius of the particles and its rate of fall in a liquid. He stated that the velocity of a falling particle is proportional to the square of the radius and not to its surface. The relation between the diameter of a particle and its settling velocity is governed by Stokes law.

$$V = \frac{2gr^2}{9n}(ds - dw)$$

Where,

V - velocity of settling particle (cm/sec.)

g - acceleration due to gravity (cm/sec^2) (981)

ds - density of soil particle (2.65)

dw - density of water (1)

n - coefficient of viscosity of water (0.005 at 4°C)

r - radius of spherical particles (cm)

2.3. Soil colour

- Inherited from its parent material and referred as **lithochromic,** e.g. red soils developed from red sandstone.
- Also develops during soil formation through different soil forming processes and referred as **acquired or pedochromic colour**, e.g. red soils developed from granite or schist.

2.3.1. Factors affecting soil colour

- **Organic matter:** High organic matter show colour variation from black to dark brown.
- **Iron compounds:** Higher iron compounds impart red, brown and yellow tinge colours.
- **Silica, lime and other salts:** Higher silica and lime or both imparts white or light colour.
- **Mixture of organic matter and iron oxides:** Imparts brown.
- **Alternate wetting and drying condition:** Due to heavy rain, reduction of soil occurs and during dry period, oxidation of soil also takes place.
- **Variegated or mottled:** Due to residual products of these processes especially iron and manganese compounds.

- **Oxidation-reduction conditions:** Presence of ferrous compounds in waterlogged soils impart bluish and greenish colour.
- Dark coloured soils absorb more heat than light coloured soils.

2.3.2. Determination of soil colour

- Determined by the comparison with the Munsell colour chart.
- Colour of the soil is a result of the light reflected from the soil.
- Soil colour notation is divided into three parts:
- **Hue** - Denotes the dominant spectral colour (Red, yellow, blue and green).
- **Value** - Denotes the lightness or darkness of a colour (The amount of reflected light).
- **Chroma** - Represents the purity of the colour (Strength of the colour).
- Munsell colour notations are systematic numerical and letter designations of each of hue, value and chroma.
- For example, the numerical notation 2.5 YR5/6 suggests a hue of 2.5 YR, value of 5 and chroma of 6. Soil colour for this Munsell notation is 'red'.

2.4. Soil water

2.4.1. Importance of soil water

- Serves as a solvent and carrier of food nutrients for plant growth.
- Yield of crop is more often determined by the amount of water available.
- Acts as a nutrient itself.
- Regulates soil temperature.
- Influences soil forming processes and weathering.
- Essential for microorganisms for metabolic activities.
- Helps in physical, chemical and biological activities of soil.
- Principal constituent of the growing plant.
- Plant protoplasm has 85-95%.
- Essential for photosynthesis.
- Plants absorb some water through leaf stomata (Openings), but most of the water used by plants is absorbed by the roots from the soil.
- Cultivated loam soil contains approximately 50% solid particles (Sand, silt, clay and organic matter), 25% air and 25% water.

2.4.2. Structure of water

- Simple compound, its individual molecules containing one oxygen atom and two much smaller hydrogen atoms.
- Bonded together covalently, each hydrogen or proton sharing its single electron with the oxygen.
- Atoms being arranged linearly (H-OH); the hydrogen atoms are attached to the oxygen as a 'V' shaped.

2.4.3. Factors affecting soil water

- **Texture:** Finer the texture, more is the pore space and also surface area, greater is the retention of water.
- **Structure:** Well-aggregated porous structure favors better porosity, which in turn enhance water retention.
- **Organic matter:** Higher the organic matter, more is the water retention in the soil.
- **Density of soil:** Higher the density of soil, lower is the moisture content.
- **Temperature:** Cooler the temperature, higher is the moisture retention.
- **Salt content:** More the salt content in the soil, less is the water available to the plant.
- **Depth of soil:** More the depth of soil, more is the water available to the plant.
- **Type of clay:** 2:1 type of clay increases the water retention in the soil.

2.4.4. Classification of soil water

2.4.4.1. Physical classification

1. Gravitational water

- Occupies the larger soil pores (Macro pores) and moves down readily under the force of gravity.
- Water in excess of the field capacity is termed gravitational water.
- No use to plants, because it occupies the larger pores.
- Reduces aeration in the soil.
- Soil moisture tension at gravitational state is zero or less than 1/3 atmosphere.

2. Capillary water

- Held in the capillary pores (Micro pores).

- Retained on the soil particles by surface forces.
- Held so strongly and gravity cannot remove it from the soil particles.
- Free and mobile and present in a liquid state.
- Evaporates easily at ordinary temperature.
- Plant roots are able to absorb it.
- Known as **available water**.
- 1/3 and 31 atmosphere pressure.

3. Hygroscopic water

- Held tightly on the surface of soil colloidal particle.
- Non-liquid and moves primarily in vapour form.
- 31 to 10000 atmosphere.
- Plants cannot absorb it.
- Microorganism may utilize hygroscopic water.
- Hygroscopic water cannot be separated from the soil unless it is heated.

2.4.4.2. Biological classification

1. Available water

- Lies between wilting coefficient and field capacity.
- Obtained by subtracting wilting coefficient from moisture equivalent.

2. Unavailable water

- Includes the whole of the hygroscopic water plus a part of the capillary water below the wilting point.

3. Super available or superfluous water

- Water beyond the field capacity stage is super available.
- Includes gravitational water plus a part of the capillary water.
- Unavailable for use of plants.
- Harmful to plant growth, because of the lack of air.

2.5. Particle density and bulk density

2.5.1. Particle density

- Also termed as **true density**.
- Weight per unit volume of the solid portion of soil.

- Particle density of normal soils is 2.65 grams per cubic centimetre.
- Particle density is higher, if large amount of heavy minerals such as magnetite, limonite and hematite are present in the soil.
- Increase in the organic matter, the particle density decreases.

1. Particle density of different soil textural classes

Textural class	Particle density (g/cm^3)
Coarse sand	2.655
Fine sand	2.659
Silt	2.798
Clay	2.837

2.5.2. Bulk density

- Oven dry weight of a unit volume of soil inclusive of pore spaces.
- Bulk density of a soil is always smaller than its particle density.
- Bulk density of sandy soil is about 1.6 g / cm^3.
- Organic matter is about 0.5 g / cm^3.
- Low bulk densities have favorable physical conditions.

1. Bulk density of different textural classes

Textural class	Bulk density (g/cc)	Pore space (%)
Sandy soil	1.6	40
Loam	1.4	47
Silt loam	1.3	50
Clay	1.1	58

2. Factors affecting bulk density

- **Pore space:** More pore space-low bulk density; less pore space-compact.
- **Texture:** Silt loams, clays and clay loams generally have lower bulk densities than sandy soils.
- **Organic matter content:** More the organic matter content, high pore space there by shows lower bulk density of soil and vice-versa.

2.6. Porosity

- Part of a soil volume that is not occupied by soil particles or organic matter.
- Pore space of a soil is the space occupied by air and water.
- In sandy soils, the particles are arranged closely with low pore space.
- In clay soils, the particles are arranged in porous aggregates with high pore space.
- Presence of organic matter increases the pore space.

2.6.1. Factors influencing pore space

- **Soil texture:** Sandy surface soil: 35 to 50% pore space.
- Medium to fine textured soils: 50 to 60% pore space.
- Compact sub soils: 25 to 30 % pore space.
- **Crops/vegetation:** Blue grass increases the porosity to 57.2% from the original 50%.
- Virgin soils have more pore space.
- Continuous cropping reduces pore space
- More the number of crops per year, lesser will be the pore space.
- Conservation tillage and no tillage reduces porosity than conventional tillage.

2.6.2. Size of pores

1. **Macro pores (Non - capillary pores):** Diameter > 0.05 mm
2. **Micro pores (Capillary pores):** Diameter < 0.05 mm

- In macro pores, air and water moves freely due to gravitation and mass flow.
- In micro pores, the movement of air and water is very slow and restricted to capillary movement and diffusion.
- Sandy soils have more macro pores and clay soils have more micro pores.
- Loamy soils will have 50% porosity and have equal portion of macro and micro pores.

2.7. Soil temperature

- Affects plant growth directly and also indirectly by influencing moisture, aeration, structure, microbial and enzyme activities, rate of organic matter decomposition, nutrient availability and other chemical reactions.
- Apple grows 18°C, potato 16 to 21°C, maize 25°C.

2.7.1. Sources of Soil heat

- Solar radiation (External source), heat released during microbial decomposition of organic matter and respiration by soil organisms including plants (Internal source).
- Rate of solar radiation reaching the earth's atmosphere is called as **solar constant** and has a value of 2 cal cm^{-2} min^{-1}.

2.7.2. Factors affecting soil temperature

- Average annual soil temperature is about 1°C higher than mean annual air temperature.

1. **Environmental factors**

- **Solar radiation:** Amount of heat received from sun on earth's surface is 2 cal cm^{-2} min^{-1}.
- Heat transmission into soil depends on the angle on incident radiation, latitude, season, time of the day, steepness and direction of slope and altitude.
- Insulation by air, water vapour, clouds, dust, smog, snow, plant cover, mulch etc., reduces the amount of heat transferred into soil.

2. **Soil factors**

- **Thermal (Heat) capacity of soil:** Amount of energy required to raise the temperature by 1°C is called **heat capacity.**
- When it is expressed per unit mass (Calories per gram), then it is called as **specific heat.**
- Specific heat of water is 1.00 cal g^{-1}
- Specific heat of a dry soil is 0.2 cal g^{-1}.
- Increasing the water content in soil, increases the specific heat of the soil.
- Dry soil heats up quickly than a moist soil.
- **Heat of vaporization:** Evaporation of water from soil requires a large amount of energy, 540 kilocalories kg^{-1} soil.
- Specific heat of a wet soil is higher than dry soil.
- **Thermal conductivity and diffusivity**: Refers to the movement of heat in soils.
- In soil, heat is transmitted through conduction.
- Heat passes from soil to water about 150 times faster than soil to air.
- Movement of heat will be more in wet soil than in dry soil.

- Thermal conductivity of soil forming materials is 0.005 units.
  - Air is 0.00005 units.
  - Water is 0.001 units.
- Dry and loosely packed soil will conduct heat slower than a compact and wet soil.
- **Biological activity:** Respiration by soil animals, microbes and plant roots evolve heat.
- More the biological activity, more will be the soil temperature.
- **Radiation from soil**: Radiation from high temperature bodies (Sun) is in short waves and that from low temperature bodies (Soil) which is in long waves (0.3 to 2.2 μ).
- Longer wavelengths have little ability to penetrate water vapour, air and gas and hence, soil remains warm during night hours, cloudy days and in glass houses.
- **Soil colour:** Produced due to reflection of radiation of specific wavelength.
- Dark coloured soils radiate less heat than bright coloured soils.
- Ratio between the incoming (Incident energy) and outgoing (Reflected energy) radiation is called **albedo.**

$$\text{Albido} = \frac{\text{Reflected energy (Outgoing energy)}}{\text{Incident energy (Incoming energy)}}$$

- Larger the albido, the cooler is the soil.
- **Soil structure, texture and moisture:** Compact soils have higher thermal conductivity than loose soils.
- Natural structures have high conductivity than disturbed soil structures.
- Mineral soils have higher conductivity than organic soils.
- **Soluble salts:** Indirectly affects soil temperature by influencing the biological activities, evaporation etc.

2.7.3. Effect of soil temperature on plant growth

- **Soil temperature requirements of plants:** Temperature at which a plant thrives and produces best growth is called **optimum range**.
- Range of temperature under which plant grows including optimum range is called **growth range**.
- Maximum and minimum temperatures beyond which the plant will die are called **survival limits**.

Range	Maize(°C)	Wheat (°C)
Optimum range	25 - 35	15 - 27
Growth range	10 - 39	5 - 35
Survival limits	0 - 43	0 - 43

- **Availability of soil water and plant nutrients:** Low temperatures reduce the nutrient availability, microbial activities and root growth and branching.

2.7.4. Soil temperature management

- **Use of organic and synthetic mulches:** Mulches keep soil cooler in hot summer and warm in cool winter.
- **Soil water management:** High moisture content in humid temperate region, lowers soil temperature.
- **Tillage management:** Tilling soil to break the natural structure reduces the heat conductance and heat loss.
- A highly compact soil looses heat faster than loose friable soil.

2.7.5. Methods of measuring soil temperature

- Mercury soil thermometers of different lengths, shapes and sizes with protective cover are buried at different depths to measure the temperature.
- Infra-red thermometers measure the surface soil temperature.
- Automatic continuous soil thermographs record the soil temperatures on a time scale.
- The International Meteorological Organization recommends standard depths of 10, 20, 50 and 100 cm to measure soil temperatures.

2.8. Soil air

- Constant state of motion of air from the soil pores into the atmosphere and from the atmosphere into the pore space.
- Constant movement or circulation of air in the soil mass resulting in the renewal of its component gases is known as **soil aeration**.

2.8.1. Composition of soil air

- Consists of nitrogen, oxygen, carbon dioxide and water vapour.

1. Composition of soil and atmospheric air

	Nitrogen (%)	Oxygen (%)	Carbon dioxide (%)
Soil air	79.2	20.6	0.3
Atmospheric air	79.9	20.97	0.03

2.8.2. Factors affecting the composition of soil air

- **Nature and condition of soil:** The quantity of oxygen in soil air is less than that in atmospheric air.
- Oxygen content of the air in lower layer of soil is usually less than that of the surface soil.
- Light texture soil or sandy soil contains much higher percentage than heavy soil.
- Concentration of CO_2 is usually greater in subsoil.
- **Type of crop:** Soils with crops contains more CO_2 than fallow lands.
- Amount of CO_2 is usually much greater near the roots due to respiration by roots.
- **Microbial activity:** Microorganisms in soil require oxygen for respiration.
- Decomposition of organic matter produces CO_2 .
- Soils rich in organic matter contain higher percentage of CO_2.
- **Seasonal variation:** Quantity of oxygen is usually higher in dry season than monsoon.
- **Temperature:** High temperature during summer season encourages microorganism activity results in higher production of CO_2.

2.8.3. Causes of poor aeration

- **Excess moisture:** Flooded or waterlogged condition develops poor soil aeration in which the most of the plants cannot grow.
- Aeration is improved by drainage or by controlled run off.
- **Gaseous Interchange:** Inadequate interchanges of gases between the soil and the free atmosphere depend on two factors:
  - Rate of biochemical reactions influencing the soil gases.
  - Actual rate at which each gas is moving into or out of the soil.

2.8.4. Exchange of gases between soil and atmosphere

- **Mass flow:** With every rain or irrigation, a part of the soil air moves out into the atmosphere as it is displaced by the incoming water.
- As and when moisture is lost by evaporation and transpiration, the atmospheric air enters the soil pores.
- **Diffusion:** Oxygen present in the atmospheric air (Partial pressure of O_2 is greater), therefore, diffuses into the soil till equilibrium is established.
- Oxygen and carbon dioxide are the two important gases that take in diffusion.

2.8.5. Importance of soil aeration

- **Plant and root growth:** Supply of oxygen to roots in adequate quantities and the removal of CO_2 from the soil atmosphere are very essential for healthy plant growth.
- When the supply of oxygen is inadequate, the plant growth either retards or ceases completely.
- Most noticeable on the root crops.
- Abnormally shaped roots are common on the compact and poorly aerated soils. Penetration and development of root are poor and cannot absorb sufficient moisture and nutrients from the soil
- **Microorganism population and activity:** Deficiency of air (Oxygen) in soil slows down the rate of microbial activity.
- Decomposition of organic matter is retarded and nitrification arrested.
- Microorganism population is also drastically affected by poor aeration.
- **Formation of toxic material:** Development of toxin and other injurious substances such as ferrous oxide, H_2S gas, CO_2 etc.
- **Water and nutrient absorption:** Under poor aeration condition, plants exhibit water and nutrient deficiency.
- **Development of plant diseases:** Wilt of gram and dieback of citrus and peach occur.

3

Soil Chemical Properties

3.1. Soil colloids

- Clay fraction of the soil contains particles less than 0.002 mm in size.
- Particles less than 0.001 mm size possess colloidal properties and are known as **soil colloids**.

3.1.1. General properties of soil colloids

1. **Size:** Inorganic and organic colloids are extremely small size - smaller than 2 micrometers in diameter.

- Seen only with an electron microscope.

2. **Surface area:** Soil colloids have a larger external surface area per unit mass.

- External surface area of 1 g of colloidal clay is 1000 times that of 1g of coarse sand.
- Total surface area of soil colloids ranges from 10 m^2/g for clays with only external surfaces to more than 800 m^2/g for clays with extensive internal surfaces.
- Colloid surface area in the upper 15 cm of a hectare of a clay soil is as high as 700,000 km^2g^{-1}.

3. **Surface charges:** Both external and internal surfaces of soil colloids carry negative and/or positive charges.

- Most of the organic and inorganic soil colloids carry a negative charge.

4. **Sources of negative charge**

- **Ionizable hydrogen ions:** Al-OH or Si-OH portion of the clay ionizes the H and leaves an unneutralized negative charge on the oxygen ($-AlO^-$ or $-SiO^-$).
- **Isomorphous substitution**: Due to the substitution of a cation of higher valence with another cation of lower valence, but similar size in the clay crystal structure.

- Dominantly, clays have Si^{4+} in tetrahedral sites and Al^{3+} in octahedral sites.
- Other ions present in large amounts during clay crystallization can replace some of the Al^{3+} and Si^{4+} cations.
- Common substitutions are Si^{4+} replaced by Al^{3+}, and replacement of Al^{3+} by Fe^{3+}, Fe^{2+}, Mg^{2+} or Zn^{2+}.
- As the total negative charge from the anions (Oxygen) remains unchanged, the lower positive charge of the substituted cations result in excess negative charges on clay crystals.

5. **Adsorption of cations:** As soil colloids possess negative charge, they attract and attach the ions of positive charge.

- They attract cations like H^+, Al^{3+}, Ca^{2+} and Mg^{2+}.

6. **Adsorption of water:** Some water molecules are attracted to the adsorbed cations and the cation is said to be in hydrated state.

- Other water molecules are held in the internal surfaces of the colloidal clay particles and play a critical role in soil physical and chemical properties.

7. **Cohesion:** Indicates the tendency of clay particles to stick together due to attraction of clay particles for water molecules held between them.
8. **Adhesion:** Attraction of colloidal materials to surface of any other substance when it comes in contact.
9. **Swelling and shrinkage:** Some soil clay colloids belonging to smectite group like Montmorillonite swell when wet and shrink when dry.

- After a prolonged dry spell, soils high in smectite clay (e.g. Black soil - Vertisols) often show cris-cross wide and deep cracks.
- These cracks first allow rain to penetrate rapidly.
- Later, because of swelling, the cracks will close and become impervious.
- Soils dominated by kaolinite, chlorite, or fine grained micas do not swell or shrink.
- Vermiculite is intermediate in swelling and shrinking.

10. **Dispersion and flocculation:** If colloidal particles loose their negative charge, the particles coalesce, form flock or loose aggregates and settle down known as **flocculation**.

- Reverse process of breaking up of flocks into individual particles is known as de-flocculation or dispersion.

11. **Brownian movement:** Refers to oscillation of particle due to collision of colloidal particles when they are suspended.

12. **Non permeability:** Colloids cannot pass through a semi-permeable membrane.

3.1.2. Types of soil colloids

Four major types of colloids

1. **Layer silicate clays:** Also known as **phyllosilicates** (Phyllon – Leaf) because of leaf like or plate like structure.

- Made up of two kinds.
- One dominated by silicon and other by aluminium and/or magnesium.
- **Silica tetrahedron:** Composed of one silicon atom surrounded by four oxygen atoms
- Four sided configuration.
- **Aluminium octahedron:** Aluminium and/or magnesium ions are surrounded by six oxygen atoms or hydroxyl group giving eight sided building block.

2. **Iron and aluminum oxide clays (Sesquioxide clays):** Extensive leaching by rainfall and long time intensive weathering of minerals in humid warm climates, results in leaching silica and alumina.

- Remnant materials, which have lower solubility are called **sesquioxides.**
- Sesquioxides (Metal oxides) are mixtures of aluminum hydroxide, $Al(OH)_3$, and iron oxide, Fe_2O_3, or iron hydroxide, $Fe(OH)_3$.
- Latin word sesqui means one and one-half times more oxygen than Al and Fe.
- Examples: Gibbsite ($Al_2O_3.3H_2O$) and geothite ($Fe_2O_3.H_2O$).

3. **Allophane and other amorphous minerals:** Silicate clays are mixtures of silica and alumina.

- Amorphous in nature.
- Common in soils forming from volcanic ash (eg. Allophane).
- Have high anion exchange capacity or even high cation exchange capacity.

4. **Humus (Organic colloid):** Amorphous, dark brown to black, nearly insoluble in water, but mostly soluble in dilute alkali (NaOH or KOH) solution.

- Composed basically of carbon, hydrogen, and oxygen.
- Negative charges of humus are associated with partially dissociated enolic (-OH), carboxyl(-COOH) and phenolic groups.

5. Difference between organic and inorganic colloids

Humus	Clay
Made up of C,H,O	Made up of Si, Ai, O
Complex amorphous organic colloid	Inorganic and crystalline
More dynamic, formed & destroyed more rapidly	Clays are stable relatively
Complex structure not well known	Clays have definite & well known structure

3.1.3. Classification of layer silicate clays

1. 1:1 type minerals

- Made up of one tetrahedral (Silica) sheet combined with one octahedral (Alumina) sheet.
- **Kaolinite group** is the most prominent 1:1 clay mineral, which includes kaolinite, hallosite, nacrite and dickite.
- Hexagonal in shape.
- Kaolinite exhibits very little plasticity (Capability of being molded), cohesion, shrinkage and swelling.

2. 2:1-type minerals

- Characterized by an octahedral sheet sandwiched between two tetrahedral sheets.
- Three general groups
- **Expanding minerals:** Smectite group and vermiculite.
- Smectite group of minerals is noted for their interlayer expansion and swelling when wetted.
- **Montmorillonite** is the most prominent member of 2:1 type mineral
- Beidellite, nontronite, and saponite are also found in soils.
- Each layer is made up of an octahedral sheet sandwiched between two tetrahedral (Silica sheets)
- Show high cation exchange capacity, swelling and shrinkage properties
- Wide cracks commonly form in smectite dominated soils (Eg. Vertisols) when dried.
- **Vermiculites** are also 2:1 type minerals in that an octahedral sheet occurs between two tetrahedral sheets.
- Octahedral sheet is aluminium dominated (Di-octahedral).
- Magnesium dominated (Tri-octahedral) vermiculite are also present.

- Higher cation exchange capacity.
- Vermiculite crystals are larger than smectite, but much smaller than kaolinite.
- **2:1 Non-expanding minerals:** Micas type minerals, Eg. Muscovite and biotite.
- Hydration, cation adsorption, swelling, shrinkage and plascticity are less in fine grained micas than in smectite, but are more than kaolinite.
- Specific surface area varies from 70 to 100 m^2g^{-1}about one eighth that for the smectite.

3. **2:1:1 type minerals**

- This silicate group is represented by chlorites.
- Chlorites are basically iron magnesium silicates with presence of some aluminum.
- Negative charge of chlorites is about the same as that of fine grained mica and less than smectite or vermiculite.

4. **Mixed and interstratified layers**

- Chlorite-vermiculite and mica - smectite are mixed layer minerals.
- In some soils, they are more common than single structured minerals such as **montmorillonite**.

5. **Carbonate and sulfate minerals**

- Carbonate minerals: Calcite ($CaCO_3$) and dolomite ($CaMg\ (CO_3))_2$.
- Major sulfate mineral: Gypsum.

6. Comparative properties of silicate clay minerals

Property	**Montmorillonite**	**Illite**	**Kaolinite**
Structure	2:1 lattice	2:1 lattice	1:1 lattice
	Substitution in octahedral sheet by Mg or Fe	Substitution in tetrahedral sheet by Al	No substitution
Shape	Irregular flakes	Irregular flakes	Hexagonal crystals
Total surface area (m^2g^{-1})	700-800	100 - 120	5 - 20
Cohesion plasticity and swelling capacity	High	Medium	Low
External surface Internal surface	Very high	Medium	Not at all
Cation exchange capacity	0 - 100	15 - 40	3 - 15
Anion exchange capacity	Low	Medium	High

3.2. Ion-exchange reaction

- Ion exchange phenomenon was first identified by Harry Stephen Thompson in England during 1850.
- When soil was leached with ammonium sulphate, calcium sulphate was detected in the leachate. Ammonium ion in the solution replaced calcium in the soil.

  $(NH_4)_2SO_4$ + Soil Ca → Soil $(NH_4)_2$ + $CaSO_4$
- Ion exchange is a process by which ions are exchanged between solid and liquid phases and /or between solid phases, if in close contact with each other.
- Common exchangeable cations are $Ca^{2+}Mg^{2+}$ H^+,K^+, NH_4^+ and Na.
- Common anions are SO_4^{2-}, Cl^- PO_4^{-3} and NO_3^-.
- Ion exchange is due to the presence of residual positive and negative charges on the soil colloids.
- Negative charges attract positively charged ions and the positive charges attract negatively charged ions
- Ions thus, attracted are reversible
- Exchange of cation is called **cation exchange**.
- Exchange of anion is called **anion exchange.**
- Cation exchange phenomenon was first discovered by **Thomasway (1850).**
- Ion exchange is the second most important reaction. First one is photosynthesis.
- Capacity of the soil to hold cation is called **cation exchange capacity** (CEC) (Unit is **C mol (P^+) / kg**).
- Capacity to hold anion is called anion exchange capacity (AEC) (Unit of is **C mol (e^-) / kg).**

3.2.1. Mechanism of cation exchange

- Clay colloids have negative charges.
- Cations are attracted to the clay particles and held electrostatistically.
- They are held by small particles of clay and organic matter.
- These small particles are called **Micelle (Micro cell).**
- When calcium is added to an acid soil, the following reaction takes place.

Micelle	H^{+} + Ca -------	Micelle	Ca^{+} $2H^{+}$
H^{+}			

- When H^{+} is added to the soil solution through organic matter or acidic materials, Ca^{2+} is replaced from the exchange complex by H^{+}.

Micelle	Ca^{2+} + 2H ----------	Micelle	H + Ca^{2+}
		H	

$2XNa^{+} + Ca^{2+} \rightarrow XCa^{2+} + 2Na^{+}$

X= exchangeable

3.2.2. Cation exchange capacity (CEC)

- Sum total of the exchangeable cations that a soil can adsorb.
- Also defined as the amount of cationic species bound at pH 7.0.
- Some authors consider pH 4.0 as the appropriate point.

1. CEC of different textural classes

Textural class	CEC (C mol (P^{+}) / Kg)
Sand	0-5
Sandy loam	10-May
Loam	15-Oct
Clay loam	15-30
Clay	30

2. CEC of important clay minerals

Clay minerals	CEC (C mol (P^{+}) / Kg)
Kaolinite	7-10
Montmorillonite	80-100
Vermiculite	100-150
Illite	25-30
Chlorite	25-30
Fe and Al oxides	5
Humus	200-400

3. Factors influencing CEC

- **Soil texture:** Increasing clay content will increase the CEC.
- **Organic matter:** CEC increases with increase in organic matter content.

- **Nature of clay:** Montmorillonite and vermiculite have higher CEC than kaolinite, chlorite or illite.
- **Soil reaction:** CEC increases with increase in soil pH.

4. Replacing power of ions

- Divalent cations have more replacing power than monovalent ions.
- H ions are adsorbed more strongly than other monovalent or divalent ions.
- Replacing power of monovalent cations increases in the order of Li < Na < K < Rb < Cs < H
- Divalent cations: Mg < Ca < Sr < Ba.
- Mixture of monovalent and divalent cations replacing power increases in the following order:

  Na < K < NH4 < Mg < Ca < H
- Na is more easily replaced than K and K more easily than NH_4.
- In general, the power of replacement is H > Ca > Mg > NH4 > K > Na

5. Base saturation

- Percentage of CEC that is satisfied by the base forming cations

$$\% \text{ Base saturation} = \frac{\text{Exchangeable base forming cations (C mol / kg)}}{\text{CEC (C mol / kg)}} \times 100$$

- Aluminium and hydrogen are acid forming ions called **acidoids.**
- Calcium, magnesium, potassium and sodium are base forming ions called as **besoids**.
- **Exchangeable sodium percentage (ESP):** Percentage of sodium in the total CEC.

  [(Na/CEC) x 100]
- **Usefulness:** Helps in determining the quantity of lime required to raise the pH of acid soils.
- Indicates the proportion of plant nutrients in CEC. It is an index of soil fertility.
- For a fertile soil, base saturation percentage should be more than 80.

3.2.3. Anion exchange

- Refers to replacement of one anion by another anion on the positively charged colloids.

- Positive charges are due to OH of iron and aluminium, 1:1 clays and allophone (Amorphous clays).
- Lower the pH, greater is the anion exchange.
- Kaolinite dominant clay have higher anion exchange capacity than montmorillonite or illite.
- Relative order of anion exchange is OH > H_2PO_4 > SO_4 > NO_3 > Cl.

1. Importance of anion exchange

- **Important for the release of fixed P in the soil:** In acid soils the phosphorus is fixed as insoluble Al-Phosphate. Liming the acid soils release fixed P.
- Here the OH ion replaces H_2PO_4 from $Al(OH)_2H_2PO_4$

  $Al(OH)_2H_2PO_4 + OH \rightarrow Al(OH)_3 + H_2PO_4$
- Availability of other nutrients like NO_3, SO_4 and Cl are influenced by anion exchange.

  Soil colloid NO_3 + Cl → Soil colloid Cl + NO_3
  Soil solution Soil solution

3.3.4. Significance of iron exchange

- Next to photosynthesis ion exchange is the most important reaction in the world.
- Plants take up their food material from the soil through ion exchange only.
- Plant roots which are in contact with the soil solution exchange the nutrients from soil solution due H^+ ions present on the surfaces of root hairs.
- **Effect on soil fertility:** Soil is considered to be fertile when the base saturation percentage is more than 80.
- Each per cent of humus contributes about 2 C mol /kg of CEC.
- Montmorillonite contributes about 1 C mole/kg.
- Kaolinite contributes about 0.08 C mol/kg for every one per cent.
- **Availability of applied nutrients**: Soils with high CEC can adsorb higher amounts of nutrients.
- Clay soils have very high CEC and larger quantities of fertilizers is applied in a single dose.
- Since sandy soils have very low CEC, fertilizers can be applied in splits.
- **Effect of adsorbed cations**: When the soil exchange complex has calcium, the soil will have desireable physical properties.

- Activity of soil microorganisms, ammonification and nitrification processes are also determined by the cations of exchange complex.
- **Toxic ions:** When the exchange complex had adsorbed metals like cadmium, nickel and lead, they are toxic to the crop plants.
- **Effect on soil pH:** Clays with H are acidic and with Na are alkaline.

3.3.5. Milliequivalent concept of cation exchange capacity (CEC)

- Defined as one milligram of hydrogen or the amount of any other ion that will combine with or displace it. Milliequivalent weight of a substance is one thousandth of its atomic weight.

3.3.6. Buffering of soils

- Power to resist a change in pH is called **buffer action.**
- **Importance:** Stabilization of soil pH.
- Amounts of amendments necessary to affect a certain change in soil reaction or soil pH.

4

Soil Microorganisms

4.1. Role of soil microorganisms

- Soil N utilization by plants.
- Dead organic matters decomposed.
- Nutrient transformation.
- Physical and chemical properties of soil.
- CO_2 content.
- Soil formation.

4.2. Classification

- Soil organisms are classified broadly into soil flora and soil fauna.
- Subdivided into micro and macro flora.
- Micro flora again classified into bacteria, actinomycetes, fungi and algae.

4.2.1. Bacteria

- Single celled.
- Rod shaped and spherical.
- About 1 μm wide and up to 3 μm long and about 2 μm in diameter.
- Most abundant microorganism in the soil.

1. Classification of bacteria

a) Based on O_2 requirement

- **Aerobic facteria:** *Bacillus subtilis, Azotobacter, Arthrobacter, Mycobacterium.*
- **Anaerobic bacteria:** *Actinomycetes, Clostridium, Propionibacterium, Bacteroides, Fusobacterium, Bifidobacterium.*

b) Based on temperature:

- **Facultative:** Optimal. An organism that can use only O_2, but has anaerobic method of energy production.

- Facultative aerobes – *Chlorobium limicola.*
- Facultative anaerobes – *Escherichia coli, Bacillus anthracis, Pseudomonas, Klebsiella, Staphylococcus spp, Streptococcus spp, Salmonella, Listeria.*
- **Psychrophiles:** Grows best at moderate temperature close to freezing.
- *Arthrobacter, Psychrobacter, Chryseobacterium greenlandensis.*
- **Mesophiles:** Grows best at moderate temperature.
- *Listeria, Thiobacillus novellus, E.coli, Clostridium*
- **Theromophile:** Grows best at high temperature.
- *Chloroflexus aurantiacus, Thermus thermophilus, Pyrolobus fumarii*

c) **Based on their food preparation:**

- **Autotroph:** Able to form nutritional organic substances from inorganic substances such as carbon dioxide.
- Cyanobacteria, Green and purple sulfur bacteria, Nitrosomonas.
- **Heterotroph:** Derives nutritional requirements from complex organic substances.
- Clostridium, Mycobacterium, *Rhizobium.*
- **Chemoautotroph:** Obtains energy by the oxidation of reduced compounds.
- Nitrosomonas, *Thiobacillus thiooxidans.*
- **Obligate chemoautotrophs:** Prefer specific substrates,
- Nitrobacter – Nitrite as substrate,
- Nitrosomonas – Ammonia as substrate,
- Thiobacillus - Converts sulphur compounds to SO_4.
- Ferrobacillus – Converts ferrous to ferric,

d) **Based on symbiotic relationship**

Symbiotic N fixers:

- Associated with a host plant.
- Both the host and the bacteria get the benefits.
- Fix atmospheric N.
- *Rhizobium* (Legume association)
- *Bradyrhizobium* (Soybean association)
- *Azorhizobium* (*Sesbania rostrata* association)

- *Anabaena azollae* (Azolla-Anabaena)
- Frankia (Casuarina association)

Non symbiotic N fixers:

- Bacteria present without the association of a plant.
- But fix atmospheric N.
- Symbiotic, non symbiotic and cellulose decomposers come under Heterotrophs.
- Nitrifiers, denitrifiers nitrate formers and sulphur oxidizers are autotrophs.
- *Azotobactor, Beijerinckia* (Named after scientist Beijerinck), *Clostridium, Chlorobium, Rhodopseudomonas, Desulphovibrio, Thiobacillus, Nostoc.*

2. Role of bacteria

- Bacteria carryout the decomposition of organic matter and synthesis of humus.
- Enzymatic transformations are carried out by bacteria
- Bacteria oxidizes or reduces many chemical reactions such as N fixation, sulphur oxidation, nitrification etc.,

4.2.2. Actinomycetes

- Unicellular like bacteria.
- Have same size as bacteria.
- Filamentous and profusely branched.
- Mycelial threads are smaller than those of fungi.
- No nuclear membrane as in bacteria.
- Also called as filamentous.
- Sensitive to acid soils.
- Potato scab, a disease due to actinomycetes, can be controlled by lowering the soil pH by applying sulphur.
- Heterotrophic - optimum temperature 25-30°C, pH 6.5 – 8.0.
- Actionomycetes are important for organic matter decomposition.
- Chitin and phospholipids are reduced to simple compounds.
- Aroma of freshly ploughed land at certain times of the year is probably due to actinomycetes as well as certain molds.

- Actinomycete population in soil exceeds all other organisms except bacteria.
- Their proportion increases with soil depth. Their population and biomass are almost equal to that of bacteria.
- Eg. Frankia family (*Actinobacteria*).
- *Streptomyces flavovirens.*

4.2.3. Fungi

- Soil fungi may be parasitic or saprophytic.
- Possess filamentous mycelium composed of individual hyphae which are 5-20 μm in diameter and several centimeters in length.
- Most fungi are heterotrophic and hence, they depend on the organic matter content of the soil.
- Dominant in acid soils and some can tolerate a pH upto 9.0.
- Strictly aerobic.
- Classified into phycomycetes, ascomycets, basidiomycetes and fungi imperfecti.
- Fungi may also may be classified as molds, yeast and mushrooms.
- **Molds:** Filamentous microscopic molds develop vigorously in all types of soils.
- In acid forest soils, it involves in decomposing of organic matter.
- Common mold genera - *Mucor, Fusarium* and *Aspergillus*.
- Average population is 10-200 billion / m^2.
- In humus formation and aggregate stabilization, molds are more important than bacteria.
- Continue to decompose complex organic substances, after bacteria and actinomycetes have stopped their function.
- **Yeast:** Group of fungi which exist as an unicellular organism.
- Reproduce by fission or budding.
- Used as food supplement and also for the production of alcoholic beverages.
- Yeast is not common in soils, but produce several plant diseases.
- **Mushroom:** Present in forests and grasslands where ample moisture and organic residues are present.
- Some mushrooms are edible.

- Mushrooms are also not common in cultivated soils.
- Their fruiting body is above the ground.

4.2.4. Algae

- Filamentous, 10μm in diameter. Population in soil is around 1-10 billion / m^2.
- Their mass in soil may be 50-600 kg/ha.
- Algae are photoautotrophs.
- Divided into 4 general groups - blue green, green, yellow green and diatoms.
- Blue green algae are numerous in rice soils.
- Blue green algae growing within the leaves of aquatic fern.
- Azolla can also fix atmospheric N.

5

Organic Matter and Its Decomposition

5.1. Organic matter

- In sandy soil, the sticky and slimy organic materials cement the sand particles together to form aggregates.
- In clayey soil, it modifies the properties of clay by reducing its cohesiveness and making clay more crumby.

5.2. Composition of organic residues

- Plant residues contain 75% moisture and 25% dry matter.
- This 25% is made up of carbon (10-12%), oxygen (9-10%), hydrogen (1.5-2.5%), N (1-2%) and mineral matter (1-3%).

5.3. Composition of plant tissues

- Celluloses 20-50%
- Hemicellulose 10-30%
- Starch, sugar 1-5%
- Proteins 1-15%
- Fats, waxes, tannins 1-10%
- Lignins 10-30%

5.4. Organic matter classification

1. Based on solubility

a) Water insolubles:

- Proteins, peptides.
- Nitrogenous peptones and S containing materials.

b) Water soluble:

- NO_3, NH_4 compounds.

- Non nitrogenous carbohydrates (celluloses, hemicellulose, starch, sugar etc.,).

c) **Ether soluble:**

- Fats, oils, waxes, resins lignins etc.

2. Based on rate of decomposition

a) **Rapidly decomposed:** Sugars, starches, proteins etc.

b) **Less rapidly decomposed:** Hemicelluloses, celluloses etc.

c) **Very slowly decomposed:** Fats, waxes, resins, lignins etc.

5.5. Decomposition of soil organic matter

- Enzymatic oxidation of the bulk with the release of CO_2, water, energy and heat.
- Essential elements are released (N, P, S etc.,) and immobilized by a series of reactions.
- Formation of compounds which are resistant to microbial action.
- **Aerobic condition:**

  When glucose is decomposed under aerobic conditions, the reaction is as under:

  Sugar + Oxygen $\rightarrow CO_2 + H_2O$

  CO_2, NH_4, NO_3 , H_2PO_4, SO_4, H_2O and essential plant nutrients like Ca, Mg, Fe, Cu, Zn etc.,

- **Anaerobic conditions:**

  Sugar + Oxygen → Aliphatic acids (Acetic, formic etc.,) or hydroxy acids (Citric, lactic etc.,) or alcohols (Ethyl alcohol etc.,)

  $C_6H_{12}O_6 + 2O_2 \rightarrow 2CH_3COOH + 2CO_2 + 2H_2O$

  $2C_6H_{12}O_6 + 3O_2 \rightarrow C_6H_8O_7 + 4H_2O$

  $C_6H_{12}O_6 + 2O_2 \rightarrow 2C_2H_5OH + 2CO_2$

  CH_4, organic acids like lactic, propionic, butyric, NH_4, various amine residues ($R\text{-}NH_2$) H_2S, ethylene ($CH_2{=}CH_2$) and humic substances.

a) Decomposition of soluble substances

i) Ammonification

Organic N → Polypeptides → Amino acids → NH_3 or NH_4

- Transformation of organic nitrogenous compounds (Amino acids, amides, ammonium compounds, nitrates etc.,) into ammonia.

- Hydrolytic and oxidative enzymatic reaction under aerobic conditions by heterotrophic microbes like *Bacillus, Clostridium, Proteus, Pseudomonas* and *Streptomyces.*

ii) Nitrification:

- Conversion of ammonia to nitrites (NO_2) and then to nitrate (NO_3)
- Aerobic process by autotrophic bacteria

$NH_4 + O_2 \rightarrow NO_2 + 2H^+ + H_2O + \text{energy}$

$NO_2 + O_2 \rightarrow NO_3 + \text{energy}$

$$\underset{\text{Ammonia}}{NH_4} \xrightarrow{\text{Nitrosomonas}} \underset{\text{Nitrite}}{NO_2} \xrightarrow{\text{Nitrobactor}} \underset{\text{Nitrate}}{NO_3}$$

iii) Denitrification:

- Conversion of nitrates into gaseous nitrogen or nitrous oxide.

$\text{Nitrate} \xrightarrow{\text{Pseudomonas}} \text{Nitrogengas}$

$$\text{Pr oteins} \xrightarrow[\text{Aminization}]{\text{Hydrolysi}} \text{Peptones}$$

$\text{Ammonia peptides} \rightarrow \text{Amides}$

b) Decomposition of insoluble substances

i) **Breakdown of protein**: Proteins are first hydrolyzed to a number of intermediate products.

- **Aminization**: Conversion of proteins to amino acids.
- **Ammonification**: Conversion of amino acids and amides to ammonia.

ii) **Breakdown of cellulose**: Decomposition of the most abundant carbohydrates.

$$\text{Cellulose} \xrightarrow[\text{(Cellulase)}]{\text{Hydrolysis}} \text{Cellobiose} \xrightarrow[\text{(Cellobiase)}]{\text{Hydrolysis}}$$

$$\text{Glucose} \xrightarrow{\text{Oxidation}} \text{Organic acids} \rightarrow CO_2 + H_2O$$

iii) **Breakdown of hemicelluloses:** Hydrolyzed in to sugars and uronic acids.

- Sugars are converted to organic acids, alcohols, carbon dioxide and water by microbes.
- Uronic acids are broken down to pentose and CO_2.

iv) **Breakdown of starch:** First hydrolyzed to maltose by the action of amylases.

- Maltose is next converted to glucose by maltase.

$(C_6H_{10}O_5)n + nH_2O \rightarrow n(C_6H_{12}O_6)$

c) **Decomposition of ether soluble substance**

Fats → Glycerol + Fatty acids

Glycerol → CO_2 + Water

d) **Decomposition of lignin**: Complete oxidation gives rise to CO_2 and H_2O.

e) **Sulphur containing organic compounds** converted to $SO_4^{-2} + H^+$ + energy by sulphur oxidizing bacteria

f) **P containing organic compounds**

- **Mineralisation:** Biological conversion of organic forms of C, N, P and S to inorganic or mineral forms.
- **Immobilization:** Conversion of inorganic forms of C, N, P and S by the soil organism into organic forms.

5.6. Carbon cycle

- Carbon is a constituent of all organic matter
- Fixed as organic form by photosynthesis.
- Fixed carbon becomes unavailable for use in the generation of new plant life.
- Carbon containing materials need to be decomposed and returned to the atmosphere for the survival of the higher organisms.
- **Carbon cycle:** revolves about CO_2 and its fixation and regeneration.
- Final decomposition and production of CO_2 from humus and the rotting tissues.
- Involves two processes namely immobilization and mineralization.
- **Immobilization:** conversion of inorganic forms of nutrients to organic forms.
- **Mineralization:** conversion of organic forms of C, N, P and S into inorganic or mineral forms.

5.6.1. C: N ratio

- Refers to ratio between the nitrogen content in the microbes and organic residues, and to the carbon content.
- When fresh plant residues are added to soil, they are rich in carbon and poor in N. Results in wider C: N ratio (40:1).
- Decomposition of organic matter in soil changes to humus resulting in narrow C: N ratio (10:1).

- When materials with high carbon are added to soil, microbial population increased due to plentiful supply of food materials.
- During this process the microorganisms utilize the soil N for their body build up and there is a temporary block of N.
- When decomposition of fresh organic residues takes place, C: N ratio is 20:1 due to increase in the availability of N.
- C: N ratio of cultivated soils ranges from 8:1 to 15:1, with an average of 10:1 to 12:1.
- Legumes and farm manures: 20:1-30:1.
- Straw: 100:1
- Saw Dust: 400:1
- In microorganisms: 4:1 to 9:1
- Humus: 10:1
- C: N ratio is lower in soils of arid regions than humid regions.
- C: N ratio is smaller in subsoils.

5.7. Role of organic matter

- Creates a granular condition of soil which favours aeration and permeability.
- Water holding capacity of soil is increased by reducing surface runoff, erosion etc., results in good infiltration.
- Surface mulching with coarse organic matter lowers wind erosion and soil temperatures in the summer and keeps the soil warmer in winter.
- Organic matter serves as a source of energy for the microbes
- Reservoir of nutrients, hormones and antibiotics that is essential for plant growth.
- Supplies food for earthworms, ants and rodents.
- Makes soil P readily available in acid soils.
- Organic acids released from decomposing organic matter help to reduce alkalinity in soils
- Humus (A highly decomposed organic matter) provides a storehouse for the exchangeable and available cations.
- Acts as a buffering agent which checks rapid chemical changes in pH and soil reaction.

5.8. Factors affecting soil organic matter

1. Climate:

a) **Temperature:** Decomposition of organic matter is accelerated in warm climates as compared to cooler climates.

b) **Rainfall:** Increase in organic matter with an increase in rainfall.

2. **Natural vegetation:** Total organic matter is higher in soils developed under grasslands than those under forests.

3. **Texture:** Fine textured soils are generally higher in organic matter than coarse textured soils.

4. **Drainage:** Poorly drained soils because of their high moisture content and relatively poor aeration are much higher in organic matter and N than well drained soils.

5. **Cropping and tillage:** Cropped lands have much low N and organic matter than comparable virgin soils.

6. **Rotations, residues and plant nutrients:** Crop rotations of cereals with legumes results in higher soil organic matter.

5.9. Factors affecting decomposition

1. **Temperature:** Warm summers may permit plant growth and humus accumulation.
2. **Soil moisture:** Extremes of both arid and anaerobic conditions reduce plant growth and microbial decomposition.
3. **Nutrients:** Lack of N slows down decomposition.
4. **Soil pH:** Microbes grow best at pH 6 to 8. Severely inhibited below pH 4.5 and above pH 8.5.

6

Humus

6.1. Humus

- Fraction of organic matter.
- Complex and resistant mixture of brown or dark brown amorphous and colloidal organic substance.

6.2. Humus formation

1. **Decomposition:** Chemicals in the plant residues are broken down by soil microbes.

- Simpler chemicals are metabolized into new compounds in the body tissue of soil microbes.
- New compounds are subjected to further modification and synthesis as the microbial tissue is subsequently attacked by other soil microbes.

2. **Synthesis:** Breakdown of lignin as the phenols and polyquinines by polymerization.

- They interact with N-containing amino compounds and forms a significant component of resistant humus such as humic group and non-humic group.

6.3. Soil organic matter fractions

6.3.1. Humic group

- Soluble portion of humus.
- 60-80% of the soil organic matter.
- Most complex and resistant to microbial attack.
- Formed by decomposition, synthesis and polymerization.
- Includes polyphenols and polyquinines.

Humic substances are classified based on resistance to degradation and solubility in acids and alkalis into

- Humic acid

- Fulvic acid
- Humin

6.3.2. Non humic group

- Insoluble part of humus.
- 20-30% of the organic matter in soil.
- Less complex and less resistant to microbial attack.
- Polysaccharides, polymers having sugar like structures and polyuronides.
- Include proteins, carbohydrates, lignins, fats, waxes, resins, tannins and some compounds of low molecular weight.

6.3.3. Theories of humus formation

- **Lignin theory:** Proposed by Waksman (1936).
- Humic substances are formed due to incomplete degradation of lignin.
- **Kononovas theory:** Humic substances are formed by cellulose decomposing mycobacteria earlier to lignin decomposition.
- **Polyphenol theory:** Flaig and Sochtig (1964).
- Humic substances are formed by the condensation of phenolic materials.
- Microbial products are amino acids, nucleic acid and phospholipids.

6.3.4. Properties of humus

- Tiny colloidal particles composed of C, H and O_2.
- Impart black colour to soils.
- Negatively charged (-OH, -COOH or phenolic groups) has very high surface area, higher CEC (150-300 Cmol/kg), 4-5 times higher WHC than silicate clays.
- Favorable effect on aggregate formation and stability.

7

Plant Nutrients

- Involved in plant metabolic functions and the plant cannot complete its life cycle without the element.

7.1. Plant nutrient concentration and plant growth

1. Deficient

- Lowest concentration of an essential element that limit yield severely and cause distinct deficiency symptoms.

2. Critical range

- Nutrient concentration in the plant at which a yield response to added nutrient occurs.

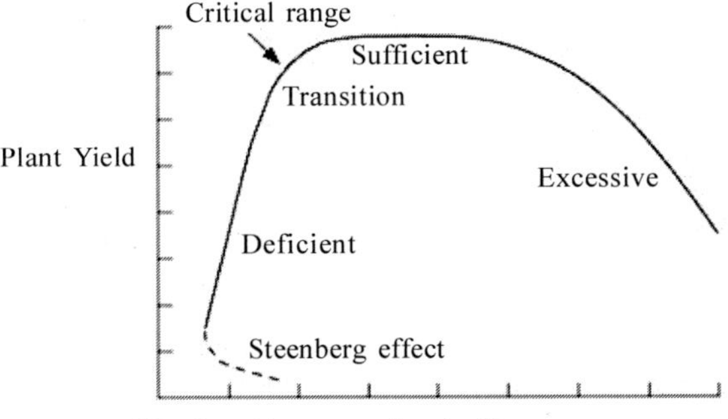

3. Sufficient

- Concentration range at which added nutrients will not increase yield, but can increase nutrient concentration.

4. Luxury consumption

- Indicates that the plant is absorbing nutrient above that needed for maximum yield.

5. Excessive or toxic

- When the concentration of essential elements is high enough to reduce plant growth and yield and cause an imbalance in other essential nutrients.
- **Steenberg effect**
- Yield is severely affected when a nutrient is deficient. When deficiency is corrected, growth increases more rapidly. Under severe deficiency, rapid increase in growth with added nutrient can cause a small decrease in nutrient concentration due to dilution effect. This is called the steenburg effect.

7.2. Mechanism of nutrient transport in plants

- Three major mechanisms of movement of ions from soil to roots.

1. Root interception

- Occurs when a nutrient comes into physical contact with the root surface.
- Enhanced by mycorrhizal fungi, which colonize roots and increases root exploration into the soil.
- Responsible for uptake of Ca, Mg, Zn and Mn.
- CEC of roots for monocots is 10 - 30 meq/100 g and takes up monovalent cations more readily
- Dicots is 40 - 100 meq/100 g and takes up divalent cations more readily.

2. Mass flow

- Nutrients are transported to the surface of roots by the movement of water in the soil (i.e. percolation, transpiration, or evaporation).
- Most of N, Ca, Mg, S, Cu, B, Mn and Mo move to the root by mass flow.
- Nutrients supplied by mass flow are mobile nutrients. e.g. N, S, B

3. Diffusion

- Movement of nutrient from an area of higher concentration to an area of lower concentration.
- Nutrients absorbed by diffusion are P, K, Zn and Fe.
- Nutrients supplied by diffusion are immobile nutrients e.g. P, K.

8

Essential Elements/Nutrients: Functions and Deficiencies

8.1. Classification of essential nutrients

8.1.1. Based on the amount required by the plant

- Major nutrients-Required in large quantities eg. N, P, K.
- Secondary nutrients-Required in lesser quantities compared to major nutrients eg. Ca, Mg, S.
- Micronutrients-Required in trace quantities eg. Fe, Mn, Zn, Cu, B, Mo.

8.1.2. Classification based on the role of element in plant system (According to TRUOG, 1954)

- Accessor structural elements: N, P, S
- Regulator and carriers: K, Ca, Mg
- Catalyst and activators: Fe, Mn, Zn, Cu, Mo, Cl, B.
- Seventeen elements are considered essential to plant growth.
- Carbon (C), hydrogen (H) and oxygen (O) are the most abundant elements in plants.
- Carbon, H and O are not considered as mineral nutrients.

8.2. Essential nutrients and their principle forms for uptake

Nutrient	Chemical symbol	Principle forms for uptake
Carbon	C	CO_2
Hydrogen	H	H_2O
Oxygen	O	$H_2O.O_2$
Nitrogen	N	$NH^{+4}.NO^{-3}$
Phosphorus	P	H_2PO^{-4}, HPO_2^{-4}
Potassium	K	K^+
Calcium	Ca	Ca^{2+}
Magnesium	Mg	Mg^{2+}

Sulphur	S	SO_2^{-4}, SO_2
Iron	Fe	Fe^{2+}, Fe^{3+}
Manganese	Mn	Mn^{2+}
Boron	B	H_3BO_3
Zinc	Zn	Zn^{2+}
Copper	Cu	Cu^{2+}
Molybdenum	Mo	MoO_2^{-4}
Chlorine	Cl	Cl^-
Nickel	Ni	Ni^{2+}

8.3. Functions of essential nutrients in plants

Nutrient	Function
Carbon (C)	Basic molecular component of carbohydrates, proteins, lipids and nucleic acids
Oxygen (O)	Occurs in all organic compounds of living organisms.
Hydrogen (H)	Hydrogen plays a central role in plant metabolism.
	Important in ionic balance and as main reducing agent.
	Plays a key role in energy relations of cells.
Nitrogen (N)	Component of proteins to nucleic acids and chlorophyll.
	Essential role in plant growth.
	Also food in soils.
Phosphorus (P)	Central role in plants is in energy transfer during respiration ADP + Pi ATP and protein metabolism.
	Root growth and photosynthesis.
Potassium (K)	Helps in osmotic and ionic regulation.
	Cofactor or activator for many enzymes of carbohydrate and protein metabolism.
	Resistance to cold, disease and insects.
Calcium (Ca)	Involved in cell division and maintenance of membrane integrity.
	Signal transduction pathway.
Magnesium (Mg)	Component of chlorophyll.
	Cofactor for many enzymatic reactions.
Sulphur (S)	Involved in plant cell energy.
Iron (Fe)	Essential component of many heme and nonheme Fe enzymes.
	Involved in key metabolic function such as N fixation, photosynthesis and electron transfer.
	Chlorophyll production.
Zinc (Zn)	Essential component of several dehydrogenases and peptidases, including carbonic anhydrase, alcohol dehydrogenase, glutamic dehydrogenase and malic dehydrogenase.
	Synthesis of protein, hormones, enzymes.
Manganese (Mn)	Involved in the O_2 evolving system of photosynthesis.

	Component of the enzymes like arginase and phosphortransferases.
	Chlorophyll production.
Copper (Cu)	Constituent of enzymes like cytochrome oxidase, ascorbic acid oxidase and laccase.
Boron (B)	Involved in carbohydrate metabolism and synthesis of cell wall components.
	Pollen production, fertility and fruit formation.
Molybdenum (Mo)	Required for the normal assimilation of N in plants.
	Essential component of nitrate reductase as well as nitrogenase (N_2 fixation enzymes).
Chlorine (Cl)	Essential for photosynthesis and as an activator of enzymes involved in splitting water.
	Involved in osmoregulation of plants growing on saline soils.
Nickel (Ni)	Involves in N metabolism and nitrogen fixation

8.4. Deficiency symptoms of nutrients

8.4.1. Nitrogen (N)

- Promotes rapid vegetative growth
- Gives plants healthy green colour
- **Deficiency:**
- Stimulated growth, pale yellowish colour.
- Burning tips and margins of leaves starting at the bottom of the plant.

8.4.2. Phosphorus (P)

- Stimulates early growth and root formation.
- Hastens maturity.
- Promotes seed production.
- Makes plants hardy.
- **Deficiency:**
- Small root growth, spindly stalk, delayed maturity.
- Purplish discoloration of leaves.
- Dying of tips of older leaves, poor fruit and seed development.

8.4.3. Potassium (K)

- Improves plant ability to resist disease and cold.
- Aids in the production of carbohydrates.

- **Deficiency:**
- Slow growth.
- Margin of leaves develop a scorched effect on the older leaves.
- Weak stalk, shriveled seed and fruits.

8.4.4. Calcium

- Absorbed by plants as Ca^{2+} and concentration in plant ranges from 0.2 to 1.0%.
- **Deficiency:**
- Young leaves of terminal buds dieback at the tip and margins.
- Root may become short, stubby and brown.
- Causes acidity of soil.
- Cell may become rigid and brittle.
- Young leaves of cereals remain folded.

8.4.5. Magnesium

- Absorbed as Mg^{2+} and concentration in plant varies between 0.1 and 0.6%.
- **Deficiency:**
- Interveinal chlorosis.
- Stiff, brittle, twisted, wrinkled and distortion of leaves.
- Cotton – lower leaves develop a reddish purple finally nicrotic **(Reddening of leaves)**.
- In brassica, chlorosis with interveinal mottling uniformly distributed in older leaves and other vascular tissues remain green called **Puckering.**

8.4.6. Sulphur

- Absorbed by plant roots as SO_4^{2-} ions.
- Concentration in plants ranges between 0.1 and 0.4%.
- **Deficiency:**
- Stunted growth, pale green to yellow colour.
- Immobile in plants and plants symptoms start first at younger leaves.
- Poor seed set in rapeseed.
- **Tea –tea yellows**.

8.4.7. Iron

- Absorbed by plants roots as Fe^{2+}, Fe^{3+}.
- Concentration in plant tissue is 50-250 ppm.
- **Deficiency:**
- Occurs in younger leaves since Fe is immobile element.
- Occurs in calcareous or alkaline soils and poorly drained water logged soils.
- Younger leaves develop interveinal chlorosis.
- Entire leaf turns yellow colour.

8.4.8. Manganese

- Mn concentration in plant ranges from 20 to 500 ppm.
- **Deficiency:**
- Immobile in plant and deficiency starts in the younger leaves.
- Interveinal chlorosis occurs.
- **Oats - Gray specks/streaks**
- **Peas - Marsh spot**
- **Sugarbeet - Speckled yellow**
- **Sugarcane - Pahala blight**

8.4.9. Zinc

- Normal concentration in plant is 25 to 150 ppm.
- **Deficiency:**
- Light yellow or white areas between the veins of older leaves.
- Death of tissue, discoloured.
- Malformation of fruits.
- **Cotton : White bud (or) little leaf**
- **Citrus : Mottle leaf**
- **Potato : Fern leaf**
- **Fruit trees : Rosette (Abnormal growth)**
- **Paddy : Khaira**

8.4.10. Copper

- Absorbed Cu^{2+}.
- Normal concentration in plants is 5-20 ppm.
- **Deficiency:**
- Chlorosis, withering and distortion of terminal buds.
- **Guava-cracking of fruits and terminal bud dieback.**

8.4.11. Boron

- Since it is immobile, deficiency occurs in terminal bud growth.
- Poor pollen production and fertility.
- Restricted flowering, fruit development and fruit set.
- Sterility and mal formation of reproductive organs.
- Discolouration, cracking or rotting of fruit, tubers or roots.
- In **apple, internal cracking**.
- Break down of internal tissue in root crops gives rise to darkened areas - **brown heart / black heart.**
- **Cotton - weeping disease.**

8.4.12. Molybdenum

- Absorbed as molybdate (MoO_4).
- Plant contains <1 ppm Mo.
- **Deficiency:**
- Inhibits flower formation.
- **Whiptail in cauliflower.**

8.4.13. Chlorine

- Absorbed by plants as Cl^- through roots and aerial parts.
- Normal concentration in plant is 0.2-2.0%.
- **Deficiency:**
- Partial wilting and loss of turgidity.
- Necrosis, leaf bronzing and reduction in growth.

8.4.14. Nickel

- Content in plant is 0.1 - 1.0 ppm.
- Legumes: whole leaf chlorosis along with necrotic leaf tips in cowpea.

9

Nutrient Deficiency Management

9.1. Nutrient deficiency of corrective measures

9.1.1. Major and secondary nutrients

S. No.	Crop	N	P	K	Ca	Mg	S
Cereals							
1	Rice	Soil application of urea @ 330kg/ha. Foliar application of Urea 1% at weekly interval till the symptoms disappear.	Application of phosphobacteria as a seed coating, or as a seedling dip. Application of P fertilizer 15-30 kg P ha^{-1} to the soil, Rock phosphate broadcast/ before flooding when soil pH islow.	Foliar spray of KCl @ 5g/lit at 15 days interval up to the disappearance of symptoms	Apply $CaCl_2$ or Ca containing foliar sprays for rapid treatment of severe Ca deficiency. Apply gypsum in Ca-deficient high pH soils, e.g., on sodic & high K soils.	Foliar application of liquid fertilizers containing Mg (e.g., $MgCl_2$ 2%)	For moderate S deficiency, apply 10 kg S ha-1 application of 20-40 kg S ha^{-1}. Use slow acting S forms (gypsum, elemental S) if leaching is likely to be a problem. Soil application of calcium silicate: 120-200 kg ha^{-1} or Potassium silicate: 40-60 kg ha^{-1}.
Millets							
2	Cumbu	Foliar spray of Urea 1% or DAP 2%.	Foliar spray of DAP 2% 2-3 sprays.	Foliar spray of KCl 1%.	Foliar application of $CaSO_4$ 2% twice.		Foliar spray of $CaSO_4$ 2%
3	Maize	Foliar spray of 0.5% urea (5 g /lit) for twice at 10 days interval.	Foliar spray of DAP 2%.	Foliar spray of KCl 1%.	Foliar application of $CaSO_4$ 2% twice.	Foliar spray of $MgSO_4$ 2%.	Foliar spray of $MgSO_4$ 1%
4	Sorghum	Foliar spray of Urea 1% or DAP 2%.	Foliar spray of DAP 2% 2-3 sprays.	Foliar spray of KCl 1%	Foliar application of $CaSO_4$ 2% twice.	Foliar spray of $MgSO_4$ 2%.	Foliar spray of $CaSO_4$ 2%.
Pulses							
5	Bengalgram		Foliar spray of 1% DAP	Foliar spray of 1% KCl			Foliar spray of $CaSO_4$ 0.5%.
6	Blackgram					Foliar spray of $MgSO_4$ 1% at fortnightly intervals.	Folia spray of $CaSO_4$ 0.5-1.0%.
7	Cowpea			Foliar spray of KCl 1% can control this deficiency.		Foliar spray of 1% $MgSO_4$ at fortnightly intervals.	

8	Greengram	10-20kg N/ha has been recommended or Foliar spray of DAP 2% at weekly intervals.		Soil application of Murate of potash spraying of 1% potash at flowering stage	Foliar spray of $CaSO_4$ 1% at fortnightly intervals.		
9	Redgram		Foliar spray of DAP 2% at fortnightly interval.	Foliar spray of KCl 1% at weekly intervals.		Foliar spray of $MgSO_4$ 2% and 1% Urea	Foliar spraying of Calcium Sulphate 0.5 % can control sulphur deficiency.
10	Soyabean	Foliar spray Urea 1% at fortnightly interval		Foliar spray of KCl 1% at fortnightly interval		Foliar spray of $MgSO_4$ 2% at fortnightly interval	Foliar spraying of Calsium Sulphate 0.5-1.0 % can control the deficiency.
Oilseeds							
11	Groundnut	Foliar spray of Urea 1 to 2% at fortnightly intervals.		Foliar spray of KCl 1 to 2% at fortnightly intervals	Soil application of gypsum 200 kg/ha.		Soil application of gypsum 200 kg/ha.
12	Sesame		Soil application of single super phosphate 2% DAP foliar spray at fortnightly intervals.		Soil application of gypsum 50 kg/ha	Foliar spray of 2% $MgSO_4$ at fortnightly intervals.	Foliar spray 0.5 - 1% of calcium sulphate
13	Sunflower	Foliar spray of 1% urea at weekly intervals.	Soil application of SPP or foliar application of DAP 2%.	Soil application of murate of potash 20 kg/ha at foliar spray of KCl 1%.		Foliar spray of $MgSO_4$ 1-2%	Application of 25 kg S/ha or 80 kg N+25 kg S.
Fibre crops							
14	Cotton	Urea 1% foliar spray or DAP 2% can control this deficiency	Foliar application of 2% DAP.	Foliar spray of 1 % KCl	Soil application of gypsum 50 kg/ha.	Foliar spray of $MgSO_4$ @ 1 %	Foliar spray of $MgSO_4$ @ 1 %
Sugar crops							
15	Sugarcane	Soil application of N fertilizer or foliar spray of urea 1-2% twice at weekly interval.	Foliar spray of DAP 2% twice at fortnight interval Applying large amounts upto 1 ton/ha of rock phosphate. Application of triple super phosphate @ 0.5 to 0.75 kg /ha	Foliar spray of KCl 1% twice at fortnightly interval	Soil application of 100kg/ha of gypsum.	Soil application of $MgSO_4$ 25kg/ha or foliar spray of $MgSO_4$ 2% twice at fortnight interval.	Foliar spray of K_2SO_4 1% twice at fortnight interval.
16	Sugarbeet	Foliar spray of urea 1% twice at weekly interval.	Foliar spray of DAP 2% twice at fortnight interval.	Foliar spray of KCl 1% twice at weekly interval	Foliar spray of $CaCl_2$ 0.5% twice at fortnight interval	Foliar spray of $MgSO_4$ 1% twice at fortnight interval.	

9.1.2. Micro nutrients

S.No.	Crop	Fe	Mn	Zn	Cu	B	Mo
Cereals							
1	Rice		Apply $MnSO_4$ (5-20 kg Mn ha^{-1}) in bands along rice row. Apply foliar $MnSO_4$ (1-5 kg Mn ha^{-1} in about 200 l water ha^{-1}.	Broadcast $ZnSO_4$ in nursery seedbed. Dip seedlings or presoak seeds in a 2-4% ZnO suspension. Apply 5-10 kg Zn ha^{-1} as Zn sulfate, apply 10-25 kg ha^{-1} $ZnSO_4$ 7 H_2O. Apply 0.5-1.5 % $ZnSO_4$ ha^{-1} as a foliar spray. At tillering (25-30 DAT), 2-3 repeated applications at intervals of 10-14 days. Zn chelates (e.g., Zn – EDTA) can be used for foliar application.	Soil application $CuSO_4$ at 1-5 kg or apply cupric sulfate solution as foliar spray.	Soil application of borax at 0.5-3 kg ha^{-1} or as foliar spray during vegetative growth.	
Millets							
2	Cumbu	Soil application of 20-25 kg $FeSO_4$ or foliar spray of 1% $FeSO_4$ at weekly interval.		Soil application of $ZnSO_4$ 20-25 kg/ha or foliar spray of $ZnSO_4$ 0.5%	Foliar spray of $CuSO_4$ 0.2%	Foliar spray of borax 0.5 % at fortnightly intervals	
3	Maize	Soil application of 20-25 kg $FeSO_4$ or foliar spray of 1% $FeSO_4$ at weekly interval	Foliar spray of $MnSO_4$@2%	Soil application of $ZnSO_4$ 20-25 kg/ha or foliar spray of $ZnSO_4$ 0.5%.	Foliar spray of $CuSO_4$ @ 0.2%.	Foliar spray of borax 0.5 % at fortnightly intervals.	
4	Sorghum	Soil application of 20-25 kg $FeSO_4$ or foliar spray of 1% $FeSO_4$ at weekly interval.	Foliar spray of $MnSO_4$ 0.2%	Soil application of $ZnSO_4$ 20-25 kg/ha or foliar spray o f $ZnSO_4$ 0.5%	Foliar spray of $CuSO_4$ 0.2%	Foliar spray of borax 0.5 % at fortnightly intervals	
Pulses							
5	Beans	Foliar spray of $FeSO_4$ 0.5%.		Soil appln. of 5-10 kg $ZnSO_4$ / ha or foliar appln. of 0.5% $ZnSO_4$.		Foliar spray of borax 0.2%	

6	Horsegram	One hundred (10 g / litre) ferrous sulphate through the spray solution by three times in the space of ten days or apply ferrous sulphate by 25 kg per hectare as basal.				
7	Bengalgram	Foliar spray of $FeSO_4$ @ 0.5%		Foliar spray of $ZnSO_4$ @ 05.%		
8	Blackgram		Spraying of 1% $MnSO_4$ during 20, 30, 40 DAS or application of 10 kg of $MnSO_4$ as a basal dose	Application of basal dose of $ZnSO_4$ at 25 kg per ha. Spraying of 0.5% $ZnSO_4$ during 20, 30, 40th day after sowing.		Foliar spray of borax 0.2% at fortnightly intervals.
9	Cowpea	Foliar spray of $FeSO_4$ @ 0.5%	Application of Manganese Sulphate 5 kg/ha in soil. Foliar spraying of manganese sulphate twice at 10 days interval.			Foliar spray of borax @ 0.2%
10	Greengram	Spraying of 1% $FeSO_4$ – 10 day intervel of 3 times. Application of 25 kg of $FeSO_4$ as basal dose.	Spraying of 1% $MnSO_4$ during 20, 30, 40 DAS or application 10 kg of $MnSO_4$ as a basal dose.	Application of basal dose $ZnSO_4$ at the rate of 25 kg per ha. Spraying of 0.5% $ZnSO_4$ during 20, 30, 40th day afte r sowing.		Foliar spray of borax 0.2% at fortnightly intervals.
11	Redgram	Foliar spray of $FeSO_4$ 0.5% at weekly intervals.	Spraying of manganese sulphate (5g / Litre) at 10 days interval.	Foliar spray of $ZnSO_4$ at 0.5% at fortnightly interval or soil application of $ZnSO_4$ 10-15 kg/ha.		Spraying of broron 0.2 % at two week interval by foliar spray.
12	Soyabean	Foliar spray of $FeSO_4$ 1% at fortnightly intervals or soil application of $FeSO_4$ 5 to 10 kg/ha	Foliar spray of $MnSO_4$ @ 0.5% at fortnightly intervals or soil application of $MnSO_4$ @ 20 to 25 kg/ha	Foliar spray of $ZnSO_4$ 1% at fortnightly intervals or soil application of $ZnSO_4$ 20 to 25 kg/ha		Spraying of foliar application of borax @ 3 g/litre twice at 10 days interval. Application of borax @ 5g/ha

Oilseeds							
13	Groundnut	Foliar spray of 1% $FeSO_4$ on 30,40 and 50 days after sowing.	Foliar spray of $MnSO_4$ 0.5% at fortnightly interval / soil appln. of $MnSO_4$ 5-10 kg/ha.	Apply 25 kg $ZnSO_4$/ha (basal) or foliar spray of $ZnSO_4$ 1% at fortnightly intervals.	Foliar spray of $CuSO_4$ 0.2% at fortnightly intervals.	Apply borax 10 kg + gypsum 200 kg/ha at 45 days after sowing.	
14	Sesame	Foliar application of ferrous sulfate 0.5% at weekly interval.	Foliar spray of 0.2 – 0.3% $MnSO_4$ solution 2-3 times at weekly intervals or soil application of $MnSO_4$ 10 kg/ha.	Foliar application of $ZnSO_4$ 0.5% or soil application of $ZnSO_4$ 10 kg/ha.		Foliar spray borax 0.2% at fortnightly intervals.	
15	Sunflower	Foliar spray of $FeSO_4$ @ 0.5%	Foliar spray of 0.2-0.3% $MnSO_4$ 2-3times at weekly interval or soil application of $MnSO_4$ 10 kg/ha.	Foliar spray of $ZnSO_4$ @ 0.5%	2 to 3 sprays of 0.2% copper sulphate solution at weekly intervals.	Spray of borax (0.2%) on capitulum at ray floret opening stage for increasing seed filling, yie ld and oil content.	
Fibre crops							
16	Cotton	Soil application of $FeSO_4$ @ 5 kg/ha or foliar spray of 0.5% $FeSO_4$.	Soil application of $FeSO_4$ @ 5 kg/ha or foliar spray of 0.5% $FeSO_4$.	Soil application of $ZnSO_4$ 5 kg/ha or foliar application of $ZnSO_4$ 1%.		Soil application of borax 0.5 kg/ha or foliar spray of borax 0.2	
Sugar crops							
17	Sugarcane	Foliar spraying of 250-500g of ferrous sulphate dissolved in 100 lit of water or soil application of 25kg/ha of ferrous sulphate or application of 100 kg of ferrous sulphate mixed with 12.5 tons of farmyard manure for one hectare or Alternatively foliar spraying of 5 kg of ferrous sulphate with	Foliar spray of $MnSO_4$@1-2%	Foliar spray of $ZnSO_4$ @ 0.5%	Foliar spray of $CuSO_4$ @ 2%	Foliar spray of borax @ 0.2-0.5%	Foliar spray of sodium molybdate @ 2mg/lit

		2.5 kg of urea in 500 litres of water for one hectare should be done. The foliar spraying may be repeated at an interval of 7-10 days depending upon on the severity of the disorder.				
18	Sugarbeet	Foliar spray of $FeSO_4$ 0.5% twice at fortnight interval	Foliar spray of $MnSO_4$ @ 1%		Foliar spray of $CuSO_4$ 0.25%.	Foliar spray of borax @ 0.5%.

10

Fertilizers

- Any material of natural or synthetic origin added to the soil to supply one or more plant nutrients.

10.1. Classification of fertilizers

10.1.1. Straight fertilizers

- Supply only one primary plant nutrient, namely nitrogen or phosphorus or potassium.
- Eg. Urea, ammonium sulphate, potassium chloride and potassium sulphate.

10.1.2. Complex fertilizers

- Contain two or three primary plant nutrients of which two primary nutrients are in chemical combination.
- Granular form.
- Eg. Diammonium phosphate, nitrophosphates and ammonium phosphate.

10.1.3. Mixed fertilizers

- Physical mixtures of straight fertilizers.
- Contain two or three primary plant nutrients.
- Made by thoroughly mixing the ingredients either mechanically or manually.

10.1.4. Solid fertilizers

- Powder (Single superphosphate)
- Crystals (Ammonium sulphate)
- Prills (Urea, diammonium phosphate, superphosphate)
- Granules (Holland granules)
- Supergranules (Urea supergranules)
- Briquettes (Urea briquettes).

10.1.5. Liquid fertilizers

- Urea ammonium nitrate
- Super phosphoric acid
- Ammonium polyphaste

10.2. Nitrogenous fertilizers

- More than 80 % of the fertilizers used in this country are nitrogenous fertilizers, particularly urea.

10.2.1. Ammoniacal fertilizers

- Nutrient nitrogen in the form of ammonium or ammonia (NH_4).
- Readily soluble in water
- Readily available to crops.
- Resistant to leaching loss,
- Ammonium ions get readily absorbed on the colloidal complex of the soil.

1. Ammonium sulphate [$(NH_4)_2$ $S0_4$]:

- 20.6 % nitrogen and 24.0 % sulphur.
- Used in rice and jute.

2. Ammonium chloride (NH_4Cl):

- 26.0 % nitrogen.
- Not recommended for tomato and tobacco as may be injured by chlorine.

3. Anhydrous ammonia (NH_4):

- 82.0 % nitrogen.
- Becomes liquid (Anhydrous ammonia) under suitable conditions of temperature and pressure.

10.2.2. Nitrate fertilizers

- Nitrogen in the form of NO_3.
- Lost by leaching because of the greater mobility of nitrate ions.

1. Sodium nitrate ($NaNO_3$):

- 15.6 % nitrogen.
- Soluble in water and readily available for the use of plants.
- Easily lost by leaching and denitrification.

- Useful for acidic soils.

2. **Calcium nitrate [Ca $(NO_3)_2$]:**

- 15.5 % nitrogen and 19.5 % calcium.
- Useful for maintaining a desirable soil pH.

3. **Potassium nitrate (KNO_3):**

- 13.0 % nitrogen and 36.4 % potassium.

10.2.3. Ammoniacal and nitrate fertilizers

- Nitrogen in both ammonia and nitrate forms.

1. **Ammonium nitrate (NH_4N0_3):**

- White, water soluble and hygroscopic crystalline salt.
- 35 % nitrogen, half as nitrate nitrogen and half in the ammonia.
- Cannot be easily leached from the soil.
- Highly hygroscopic and not fit for storage.
- Explosion hazard.

2. **Calcium ammonium nitrate (CAN) [NH_4NO_3 + $CaCO_3$*$MgCO_3$]:**

- Fine free-flowing, light brown or grey granular fertilizer, containing 26 % of nitrogen.
- Ammoniacal form and half is in nitrate form.

3. **Ammonium sulphate nitrate [$(NH_4)2S0_4$ NH_4NO_3]:**

- 26 % nitrogen, three fourths of it in the ammoniacal form and the rest (6.5%) as nitrate nitrogen.
- 12.1 % sulphur.
- White crystalline form or as dirty white granules.
- Readily soluble in water.

10.2.4. Amide fertilizers

- Amide form of nitrogen is easily changed to ammoniacal and then, to nitrate form in the soil.

1. **Urea [CO $(NH_2)_2$]:**

- 46 % nitrogen.
- White crystalline substance readily soluble in water.
- Urea sprays are readily absorbed by plants.

2. Calcium cyanamide ($CaCN_2$):

- 20.6 % of nitrogen.
- Greyish white powdery material.

10.3. Phosphatic fertilizers

- Absorbable form (Phosphate anions).

1. Super phosphate [$Ca(H_2PO_4)_2$]:

- 16 % P_2O_5 in available form.
- Grey ash like powder.

2. Triple super phosphate [$Ca(H_2PO_4)$]:

- 46 % P_2O_5.
- Suitable for all crops and all soils.

10.4. Potassic fertilizers

1. Potassium absorbed form is K^+.
2. Water soluble and readily available to plants.

1. Potassium chloride (KCl):

- White or red, crystal.
- 60.0 % K_2O.
- Chlorine content is about 47.0 %.
- Chlorine content is objectionable to tobacco and potato.

2. Potassium sulphate (K_2SO_4):

- White salt.
- 48 % K_2O.
- Preferred for fertilization of tobacco, potato etc.,

10.5. Secondary major nutrient fertilizers

1. Magnesium sulphate ($MgSO_4$):

- Contains 10 % Mg and 14 % S.

2. Calcium chloride ($CaCl_2$ $6H_2O$):

- 15 % calcium.
- Highly water soluble.

3. Ferrous sulphate ($FeSO_4$ $7H_2O$):

- Water soluble, 20 % Fe.
- 15 % S.

4. Manganous sulphate ($MnSO_4$ $7H_2O$):

- Water soluble.
- 24 % Mn.
- 17 % S.

5. Zinc sulphate ($ZnSO_4$ $7H_2O$):

- Water soluble.
- 23 % Zn.
- 15 % S.
- **Zinc oxide (ZnO)** - 70 % Zn.

6. Copper sulphate ($CuSO_4$ $5H_2O$):

- 25 % Cu.
- 13 % S.
- **Copper sulphate ($CuSO_4$ H_2O)** – 36 % Cu.

7. Borax (Na_2B_4O $10H_2O$):

- 11% B.
- Water soluble white salt.
- **Boric acid (H_3BO_3):** 18% B, white crystalline powder.

8. Sodium molybdate ($Na_2MoO_4.2H_2O$):

- 40 % Mo.

9. Ammonium molybdate ($(NH_4)_6MO_7O_{24}.4H_2O$):

- 54 % Mo.

10.6. Fertilizer grade

- Guaranteed minimum percentage of nitrogen (N), phosphorus (P) and potash (K) contained in fertilizer material.
- Numbers representing the grade are separated by hyphens and are always stated in the sequence of N, P, and K.
- Grade 28-28-0 indicates that 100 kg of fertilizer material contains 28 kg of N, 28 kg of P and no potash.

10.6.1. Different grades of fertilizers available in India

- 28-28-0
- 20-20-0
- 14-35-14
- 17-17-17
- 14-28-14

10.7. Fertilizer ratio

- Ratio of the percentage of N, P_2O_5 and K_2O in the fertilizer mixture.
- Eg. Fertilizer grade 12-6-6 has a fertilizer ratio of 2:1:1.

10.8. Conditioners

- Low grade organic materials like peat soil, paddy husk, groundnut hulls.
- For reducing hygroscopicity and to improve their physical condition.

10.9. Filler

- Weight making materials like sand, soil, coal powder etc,.

10.10. Neutralizers of residual acidity

- Materials like dolomite, lime stone etc, added in fertilizer mixtures to counteract the acidity of nitrogenous fertilizers.

10.11. Complex fertilizers

10.11.1. DAP $(NH_4)_2HPO_4$:

- Nutrient: 18 % N, 46 % P_2O_5.
- Black colour and uniform size granules.
- Recommended for basal application of fertilizers due to slow releasing nature of DAP.

10.11.2. Ammonium phosphate sulphate $(NH_4)_2(H_2PO_4)(HSO_4)$:

- 20 % nitrogen and 20 % P_2O_5.
- 13 % sulphur.
- Light grey in colour.
- Not readily absorb moisture from the air.

- Good storage properties.
- Greater mobility in the soil.

10.12. Nutrient content of liquid fertilizers

10.12.1. Urea ammonium nitrate:

- 32 % N.
- 7.7 % ammonical nitrogen, 7.7 % nitrate nitrogen and 16.6 % urea nitrogen.

10.12.2. Superphosphoric acid

- 70 % P_2O_5.

10.12.3. Ammonium polyphosphate:

- 10 % N, 34 % P_2O_5.

11

Organic Manures

- Word manure derived from French 'Manoeuvrer, means to manipulate, to work, to produce crop.

11.1. Sources of organic wastes

- Cattle shed wastes, dung, urine and slurry from biogas plants.
- Human habitation wastes-night soil, human urine, town refuse, sewage, sludge and sullage.
- Poultry jitter, droppings of sheep and goat.
- Slaughterhouse wastes-bone meal, meat meal, blood meal, horn and hoof meal, fish wastes.
- Byproducts of agro industries-oil cakes, bagasse and press mud, fruit and vegetable processing wastes etc.
- Crop wastes-sugarcane trash, stubbles and other related material.
- Water hyacinth, weeds and tank silt.
- Green manure crops and green leaf manuring material.

11.2. Nutrient content of organic manures

	N (%)	P (%)	K (%)
Farmyard manure	0.5	0.2	0.5
Sheep and goat manure	3.0	1.0	2.0
Poultry manure	3.03	2.63	1.4

11.2.1. Bulky organic manures

- They supply plant nutrients including micronutrients.
- They improve soil physical properties like structure, water holding capacity etc.,
- They increase the availability of nutrients.
- Carbon dioxide released during decomposition acts as a CO_2 fertilizer.

- Plant parasitic nematodes and fungi are controlled to some extent by altering the balance of microorganisms in the soil.

11.2.2. Concentrated organic manures

1. Average nutrient content of oil cakes

Oil cakes	Nutrient content (%)		
	N	P_2O_5	K_2O
Non edible oil cakes			
Castor cake	4.3	1.8	1.3
Cotton seed cake (Undecorticated)	3.9	1.8	1.6
Karanj cake	3.9	0.9	1.2
Mahua cake	2.5	0.8	1.2
Safflower cake (Undecorticated)	4.9	1.4	1.2
Edible oil cakes			
Coconut cake	3.0	1.9	1.8
Cotton seed cake (Decorticated)	6.4	2.9	2.2
Groundnut cake	7.3	1.5	1.3
Linseed cake	4.9	1.4	1.3
Niger cake	4.7	1.8	1.3
Rape seed cake	5.2	1.8	1.2
Safflower cake (Decorticated)	7.9	2.2	1.9
Sesamum cake	6.2	2.0	1.2

2. Average nutrient content of animal based concentrated organic manures

Organic manures	Nutrient content (%)		
	N	P_2O_5	K_2O
Blood meal	10 - 12	1 - 2	1.0
Meat meal	10.5	2.5	0.5
Fish meal	4 - 10	3 - 9	0.3 - 1.5
Horn and hoof meal	13	-	-
Raw bone meal	3 - 4	20 - 25	-
Steamed bone meal	1 - 2	25 - 30	-

11.3. Green manures

1. **Green manure:** Growing the plants *in-situ* and incorporation in the field.

Biomass production and N accumulation of green manure crops

Crop	Age (Days)	Dry matter (t/ha)	N accumulated
Sesbania aculeata	60	23.2	133
Sunnhemp	60	30.6	134
Cowpea	60	23.2	74
Pillipesara	60	25.0	102
Cluster bean	50	3.2	91
Sesbania rostrata	50	5.0	96

Nutrient content of green manure crops

Plant	Scientific name	Nutrient content (%) on air dry basis		
		N	P_2O_5	K
Sunhemp	*Crotalaria juncea*	2.30	0.50	1.80
Daincha	*Sesbania aculeata*	3.50	0.60	1.20
Sesbania	*Sesbania speciosa*	2.71	0.53	2.21

2. **Green leaf manure:** Addition of green or plant tissues obtained from elsewhere *viz.*, trees, herbs, shrubs prunings and unwanted weeds.

Nutrient content of green leaf manure

Plant	Scientific name	Nutrient content (%) on air dry basis		
		N	P_2O_5	K
Gliricidia	*Gliricidia sepium*	2.76	0.28	4.60
Pongania	*Pongamia glabra*	3.31	0.44	2.39
Neem	*Azadirachta indica*	2.83	0.28	0.35
Gulmohur	*Delonix regia*	2.76	0.46	0.50
Peltophorum	*Peltophorum ferrugenum*	2.63	0.37	0.50
Weeds				
Parthenium	*Parthenium hysterophorus*	2.68	0.68	1.45
Water hyacinth	*Eichhornia crassipes*	3.01	0.90	0.15
Trianthema	*Trianthema portulacastrum*	2.64	0.43	1.30
Ipomoea	*Ipomoea*	2.01	0.33	0.40
Calotrophis	*Calotropis gigantea*	2.06	0.54	0.31
Cassia	*Cassia fistula*	1.60	0.24	1.20

11.3.1. Benefits

- Addition of organic matter.
- Adds macro, secondary and micro nutrients.

- Improves physical condition of the soil.
- Acts as a soil amendment to reclamation of problem soils.
- Acts as a cover or catch crop to prevent soil erosion, conserve moisture, prevent nutrient leaching.
- Leguminous crops fix the atmosphere N and improve N status of the soil.

12

Problem Soils and Their Management

12.1. Problem soils

- **Chemical constraints:** Salinity, sodicity, acidity and nutrient toxicities
- **Physical constraints :** High or low permeability, sub soil hard pan, surface crusting, fluffy paddy soils, sandy soils etc.

12.1.1. Characteristics of saline, saline - sodic and sodic soils

Soil	pH	EC (dSm^{-1})	ESP
Saline	<8.5	>4	<15
Saline-sodic	<8.5	>4	>15
Sodic	>8.5	<4	>15

12.1.2. Saline soils

- Higher amount of water soluble salts like chlorides and sulphates.
- **Reclamation:** Provision of lateral and main drainage channels of 60 cm deep and 45 cm wide and leaching of salts.
- Application of farm yard manure at 5 t /ha at 10 - 15 days before transplanting in paddy and before sowing in garden land crops.

12.1.3. Sodic soils

- Sodic soils are characterised by the predominance of sodium in the complex with the exchangeable sodium percentage exceeding 15 % and the pH more than 8.5.
- **Reclamation:** Plough the soil at optimum soil moisture regime.
- Gypsum ($CaSO_4$ $2H_2O$) application at 5 tons/ha.
- Impound water.
- Provision of drainage for leaching of soluble salts.
- *In situ* incorporation of green manure at 5 t/ ha.

12.1.4. Acid soils

- Low pH (< 6.0).
- Predominance of H^+ and Al_3^+ cause acidity resulting in deficiency of P, K, Ca, Mg , Mo and B.
- Present in hilly tracts of Ooty, Kodaikkanal and Yercaud and laterite soils of Pudukkottai, Kanyakumari etc.
- **Reclamation:** Application of lime ($Ca(OH)_2$) @ 200 to 400kg/ha.
- Application of alternate amendments like dolomite, basic slag, flue dust, wood ash, pulp mill lime.

12.1.5. Iron and aluminum toxicity

- Higher concentration of Fe_2^+ and Al_3^+, more specifically in flooded soils.
- Present in Kanyakumari and Pudukkottai districts (pH 4 - 6).
- **Reclamation:** Application of lime as per the lime requirement along with the recommended dose of NPK.
- Application of organic manure.
- Growing suitable rice varieties.
- Highly susceptible: ADT36.
- Moderately susceptible: ADT42, IR50, CORH1.
- Less susceptible: TPS 1, ASD 16 & 18, IR 64, JJ 92, TKM 9, CO 37 and CO 41.

12.1.6. Fluffy paddy soils

- Low bulk density of the top soil.
- Sinking of farm animals and labourers.
- Poor anchorage to paddy seedlings.
- **Reclamation:** Passing of 400 kg stone roller or oil drum with sand inside eight times.
- Addition of lime @ 2t/ ha once in three years.

12.1.7. Sandy soils

- Predominant amounts of sand resulting in higher percolation rates and nutrient losses.

- **Reclamation:** Compacting the soil with 400 kg stone roller or oil drum with stones inside eight times.
- Addition of tank silt.

12.1.8. Hard pan soils

- Occur in red soil areas below 15 cm depth due to the movement of clay and iron hydroxides and settling at shallow depth.
- Preventing the root proliferation.
- **Reclamation:** Chiselling the soils with chisel plough at 0.5 m interval first in one direction and then in the direction perpendicular to the previous one, once in three years.
- Application of FYM or composted coir pith at 12.5 t/ha.

12.1.9. Surface crusted soils

- Weakly aggregated soil structure is easily broken by the impact of rain drops resulting in the formation of clay crust at the soil surface.
- Prevents the emerging seedlings and arrest the free exchange of gases between the soil and atmosphere.
- **Reclamation:** Easily broken by harrowing or cultivator ploughing.
- Application of lime or gypsum at 2 t/ha and FYM at 12.5 t/ha.
- Retaining crop residues on the surface as a protective cover.

12.1.10. Heavy textured clay soils

- Containing major amounts of clay fraction resulting in the poor permeability and nutrient fixation.
- **Reclamation:** Addition of river sand at 100 t/ha.
- Deep ploughing the field with mould board plough or disc plough during summer to enhance the infiltration and percolation.

12.1.11. Low permeable black soils

- **Reclamation:** Application of 100 cart loads of red loam soil.
- Deep ploughing the field with mould board plough or disc plough during summer to enhance the infiltration and percolation.
- Application of FYM, composted coir pith or pressmud at 25 t/ ha per year.

12.1.12. High permeable red soils

- **Reclamation:** Application of tank silt or black soil at 25 t/ ha per year along with FYM , composted coir pith or pressmud at 25 t/ha.
- Deep ploughing the field with mould board plough or disc plough during summer to improve the water holding capacity of the soil.

12.2. Package of practices for chisel technology

- Occurrence of hard pans at shallow depth is the most prevalent soil physical constraint in soils.
- Characterised by high bulk density (1.8 mg m^{-3}), which in turn lowers infiltration, water storage capacity, available water and movement of air and nutrients, with concomitant adverse effect on the yield of crops.
- Predominantly present in six districts of Tamil Nadu *viz.*, Coimbatore, Erode, Dharmapuri, Tiruchirappalli, Madurai and Salem particularly under rainfed farming affecting a total of 3.8 lakh hectares of land.
- **Technology:**
- Plough the field with chisel plough using tractor at 50 cm interval in both the directions viz., horizontally and vertically.
- Chisel plough is a heavy iron plough which goes up to 45 cm depth, thereby shatters the hard pans.
- Spread 12.5 t/ha of FYM/pressmud/composted coir pith evenly on the surface.
- Give two ploughings using a country plough for incorporating the added manures.
- Broken hard pan and incorporation of manures make the soil to conserve more moisture.

13

Soil Survey

13.1. Soil survey and soil mapping

- Refers to study and mapping of soils in the fields.
- Process of classifying soil types and other soil properties in a given area and geo-encoding such information.
- Draws information on geomorphology, theories of soil formation, physical geography, analysis of vegetation and land use patterns.
- Primary data for the soil survey are acquired by field sampling and by remote sensing.
- Remote sensing principally uses aerial photography (Illuminating the object/target with pulsed laser light and measuring the reflected pulses with sensor), but LiDAR (Light Detection and Ranging) and other digital techniques.

13.2. Main objectives of soil survey

- To describe and classify soils giving uniform system of classification with uniform nomenclature in order to correlate the soils of different area.
- To show distribution of different soils in the field (Soil mapping).
- To provide data for making interpretations on adaptability of particular soils for agricultural purpose and also for many other purposes.

13.3. Types of soil surveys

- Detailed
- Reconnaissance
- Detailed reconnaissance

13.3.1. Detailed soil survey

- Soil boundaries are plotted accurately on maps on the basis of observations made throughout the surveyed area.
- Provide information needed for planning land use and management and formulating agricultural research and extension programmes.

13.3.2. Reconnaissance survey

- Soil boundaries are plotted from the observations made at intervals.

13.3.3. Detailed reconnaissance survey

- A part of surveyed area is plotted on the map by detailed method and remainder by reconnaissance method.
- In mapping of soils, either aerial photographs are taken or good topographic maps are made.
- A scale of four inches to a mile is used.
- Detailed or special soil maps are made on scales of six inches, eight inches or twelve inches to a mile.
- Some surveys of limited areas have been conducted for specific objectives, such as fertility surveys, surveys for soil classification, survey from geological point of view, surveys for physicochemical properties of surface sample and surveys for genetic classification.
- In India, a few generalized soil maps have been prepared under the auspices of International Society of Soil Science, Geological Survey of India and IARI (Indian Agricultural Research Institute), New Delhi.

13.4. Aerial photo-interpretation for soil surveys

- Scales recommended for detailed survey are 1: 10,000 or 1: 25,000.
- Scale recommended for semi - detailed reconnaissance surveys are 1: 25,000 to 1: 1, 00,000 and 1: 2, 50,000.
- Most favored method adopted by soil scientists involves delineation of the soil boundaries by air photos which are sub-divisions of physiographic units.

13.4.1. Aerial photography

- Science of obtaining photographs from the air using aircraft for studying the surface of the earth.
- Pictures taken by camera by flying aircraft over the predetermined height, depending on the scale of aerial photography and focal length of camera.
- Gives a bird's eye view of large areas.
- Sun provides the source of energy (Electromagnetic radiations) and the photosensitive film acts as a sensor to record the images.
- Variations in the grey tones of the various images in a photograph indicate different amounts of energy reflected from the objects as recorded on the film.

- Photographs contain 50 to 65% overlap which is essential for stereoscopic viewing and analysis of stereo pairs.
- Aerial photographs ranging in scale from 1:8000 to 1: 60,000 are used in different types of soil surveys.
- In soil survey, mostly use panchromatic black and white air photos taken with a black and white film.
- Black and white air photos can indicate a lot of information about land forms, vegetation, human interference as well as soils.
- Natural colour and infrared film are also used for aerial photography, especially for forest areas to discreminate forest types/ species.
- Many landforms like terraces, flood plains, sand dunes, coastal plains, plateaus, paleochannels, hills, valleys and mountain can be recognized on the photographs based on their shapes, relative heights and slopes.
- Difference in tones or colour may also reflect soil differences.
- Individual objects like trees, houses, roads, foot paths, field boundaries, lakes, rivers etc, are imaged clearly depending on the scale of photographs.

13.5. Classification of aerial photography

13.5.1. On the basis of scale

- **Large scale :** Between 1:5,000 and 1: 20,000
- **Medium scale:** Between 1: 20,000 and 1:50,000
- **Small scale :** Smaller than 1:50,000

13.5.2. On the basis of tilt

- **Vertical:** When the tilt is within 3 (Nearly vertical).
- **Oblique:** Low oblique (Horizon does not appear, but tilt is more than 30).
- **Horizontal or terrestial:** Camera axis is kept horizontal.

13.5.3. On the basis of angular coverage

- **Narrow angle :** Angle of coverage less than 500
- **Normal angle :** Angle of coverage less than 600
- **Wide angle :** Angle of coverage less than 900
- **Super-wide angle :** Angle of coverage less than 1200

13.5.4. On the basis of film

- Black and white panchromatic.
- Black and white infra red.
- Colour.
- Colour infra red/false colour.

14

Soil Fertility Survey and Evaluation

14.1. Soil fertility

- Ability of the soil to supply essential plant nutrients during growth period of the plants
- Crop producing capacity of the soil

14.1.2. History of soil fertility

- Francis bacon (1591-1624)
- Principle nourishment of plants was water.
- Purpose of soil is to keep plants erect and to protect from heat and cold.
- Jan Baptiste Van Helmont (1577-1644) – Water was sole nutrient of plants
- Robert Boyle (1627-1691) – Plant synthesis salts, spirits and oil etc., from H_2O
- Priestly (1800) – Essentiality of O_2 for the plant growth
- J.B.Boussingault (1802-1882) – Father of field experiments
- Justus von Liebig (1835) – Father of agricultural chemistry
- J.B. Lawes and J.H.Gilbert (1843) – Established permanent manurial experiment at Rothemsted agricultural experiment station, England.
- D.J.Nicholas coined the term "functional or metabolic nutrient".
- Carbon: Priestly (1800)
- Nitrogen: Theodore De Saussure (1804)
- Ca, Mg, K, S: Carl Sprengel (1839)
- Phosphorus: Von Liebig (1844)
- Iron (Fe): E.Greiss (1844)
- Manganese (Mn): J.S.Hargue (1922)
- Zinc (Zn): Sommer and Lipan (1926)

- Copper (Cu): Sommer, Lipman and Mc Kenny (1931)
- Molybdenum (Mo): Arnon and Stout (1939)
- Sodium (Na): Brownell and Wood (1957)
- Cobalt (Co): Ahamed and Evans (1959)
- Boron (B): Warrington (1923)
- Chlorine (Cl): Broyer (1954)
- Nickel: Brown *et al.* (1987)

14.2. Soil fertility survey

- Aims at taxonomical classification of soils in well defined units.
- Properties studied in the survey are ultimately used to plot the extent of boundaries on a map.
- Fertility surveys are carried out in profile studies.
- From samples collected at different depthwise layer, soil physical, chemical and biological properties are thoroughly studied.
- Parameter on the depth of soil, rooting depth, bulk density, particle density, porosity, water holding capacity, possible profile moisture storage, soil reaction, salinity, alkalinity, total and available nutrients, and presence of hard pans.
- By interpretive analysis the fertility, data are grouped and used for mapping.
- Division of Soil Survey and Land Use Planning of State Department of Agriculture, which is operating in selected districts of the Tamil Nadu, compiles the information and publish them as a report.
- Reports are available on cost basis to needy persons and institutions.
- Results of survey are further classified and mapped by National Bureau of Soil Survey and Land Use Planning located to Nagpur.
- Soil maps on agro-climatic regions, fertility status, irrigability, soil depth, crop suitability, salinity, etc are available.

14.2.1. Village fertility indices and mapping

- One of the major functions of soil testing laboratories is to analyze the soil samples collected from the farmers for available N, P and K status.
- For every year few taluks are adopted.
- After analyzing the soil samples for available NPK, they are grouped as low, medium and high based on the soil test value.

- After grouping soil as low, medium or high, village fertility index (VFI) is worked out for every revenue village.

$$VFI = \frac{N_L + 2N_M + 3N_H}{N_L + N_M + N_M}$$

Where N_L, N_M, N_H are the number of soil samples falling under the category low, medium and high which are given weight of 1, 2, 3 respectively.

- Index below 1.5 is low, between 1.5-2.5 is medium and above 2.5 is high.
- Using these fertility indices, the current area wise fertilizer recommendation for each crop can be modified.
- A soil fertility map may be prepared in any outline map of block/taluk by plotting the index values within the boundary of villages.

14.3. Soil fertility evaluation

14.3.1. Soil testing research in India

Year	Landmark
1953	Soil fertility and fertilizer use project.
1955	Establishment of Soil Testing Laboratories.
1956	All India Coordinated Agronomic Research Project.
1960	Establishment of Agricultural Universities.
1967	All India Coordinated Research Project on Soil Test Crop Response (STCR).
1967	All India Coordinated Research Project on Micronutrients in Soils and Plants.
1970	All India Coordinated Research Project on Long term Fertilizer Experiments.
1980	Emphasis on fertilizer prescription for whole cropping system based on initial soil tests.

- Stewart (1947) attempted to relate the knowledge of soils to the judicious use of fertilizers.
- Model Agronomic Experiments on Experimental Farms and Simple fertilizer trials on cultivators fields was started in 1957.
- Soil Test Crop Response (STCR) correlation work carried out at Indian Agricultural Research Institute, New Delhi had resulted in the selection of soil test methods and categorizing the tests into low, medium, and high soil fertility classes.

14.3.2. Soil fertility evaluation

- In order to interpret the data, the results from each analytical procedure must be correlated with the plant response obtained in field experiments by applying that fertilizer nutrient.

- By this calibration, the requirement of fertilizer is calculated for achieving specific yield target.
- Soil fertility evaluation preferably employs a particular method of calibration as per the utility of the outcome.

1. Crop response data

- In nutrient rating experiments, soil test data is correlated with the response of crop.
- Response is measured in terms of 'percent yield' or 'percent yield increase'.
- Both represent the ratio of the yield obtained in unfertilized soil (Nutrient limiting deficient soil) to the yield in fertilizer nutrient applied soil (Non-limiting or nutrient sufficient soil).
- Yield in non-limiting soil is otherwise known as maximum attainable yield or yield of standard treatment with all nutrients applied.
- While comparing native fertility, in experiments that are conducted at different locations, % yield is used.

$$\text{Per cent yield} = \frac{\text{Yield without fertilizer nutrient}}{\text{Yield in standard treatment}} \times 100$$

- While conducting multi location trials or a pot experiment in a single location with soils brought from different locations, percent yield increase is worked out for every level of fertilizer application.
- Highest yield obtained in the experiment with all fertilizer nutrient applied is taken as maximum attainable yield.

Percent yield increase =

$$\frac{\text{Yield at fertilizer nutrient level - Yield without fertilizer nutrient}}{\text{Maximum attainable yield}} \times 100$$

2. Agronomic approach

- Based on fertilizer rate experiments (Recording yield at increasing nutrient levels) conducted at many locations.
- Level at which yields are high are recorded and averaged.
- From these results, the optimum dose of fertilizer is recommended for a crop at given agro-climatic region.
- Eg. A blanket dose of 120-50-50 kg/ha of N, P_2O_5, K_2O respectively is recommended for rice.

3. Critical limit approach

- Waugh and Fitts (1965) developed this technique which is largely meant for less mobile nutrients like P, K and micronutrients.
- Part of applied P is fixed in soil and not readily available to plants.
- Method includes incubation study.
- For P, soil is incubated for 72 hours with graded doses of soluble P in the form of monocalcium phosphate.
- Amount of phosphates released (Extracted) with an extracting reagent (Olsen or Bray) will be determined.
- Extracted P versus the amount of P applied is plotted.
- If the relationship is unique, then for high P fixing soils, a larger amount of fertilizer P application is needed.

Part-2 Dryland Agriculture

1

Dry Farming and Rainfed Agriculture

1.1. Dry farming (Dryland farming)

- Cultivation of crops in a region where the annual rainfall is more than 750 mm.
- Semi arid crops with growing period of 75-120 days.
- Crop failure is less frequent.
- Moisture conservation practices are necessary.

1.2. Rainfed farming

- Growing crops in the region where annual rainfall is more than 1150 mm.
- Crops are not subjected to moisture stress.
- Humid regions with growing period of more than 120 days.
- Rainfed agriculture contributes 40% food grain production and supports 2/3rd of livestock populations.
- In rainfed agriculture, 85% coarse cereals, 83% pulses, 42% rice, 70% oilseeds, 65% cotton are still cultivated.

1.3. Dryland status in India (2019 statistics)

- Total geographical area is 329 million ha.
- Gross cropped area – 195 million ha.
- Net sown area – 141 million ha.
- Rainfed area – 80 million ha.
- Net area irrigated – 65.3 million ha.
- **National organization: CRIDA** – Central Research Institute for Dryland Agriculture, established in 1985 at **Hyderabad**.

1.4. Problems

- Limited rainfall (Less than 800 mm).
- Less crop growing season.
- Low availability of moisture.
- High wind and water erosion.
- Monocropping.

1.5. Dry land soil types

- Five types of soil.

1.5.1. Black soil (Clay to clay loam)

- Deeper soil.
- Low permeability.
- High organic matter content.
- High CEC, water holding capacity and low infiltration rate.
- Fallow soil prone to soil erosion hazards.
- **Crops:** Cotton, sunflower.

1.5.2. Red soil

- Light textured, shallow to medium in depth with compact soil.
- Low water holding capacity.
- Prone to erosion and surface crusting.
- **Crops:** Finger millet, pearl millet.

1.5.3. Alluvial soil

- Deep and light to medium in texture.
- Good permeability.
- **Crops:** Rice, vegetables.

1.5.4. Sierozemic soils

- Deep sandy loams - desert ecosystem.
- Low moisture storage.
- Severe wind erosion.

- Poor soil fertility.
- Formation of surface crust after light showers.
- Mostly available in Dantiwada (Hissar), Jodhpur (Rajasthan).

1.5.5. Sub montane soil

- Dry sub humid environment.
- Hosiarpur in Punjab, Rakhdhiansar in Jammu and Kashmir, humid tract of Dehradun.
- Loamy sand to sandy loam.
- Silty loam and clay loam also present.
- Soil crusting occurs.

1.6. Constraints

- Poor fertility and low productivity.
- Irregular topography with high erodability.
- Difficulty in working especially in vertisols (Black soils).
- Low moisture storage particularly in alfisols (Red soils).
- Presence of dissolved harmful salts and the poor conservations of soil and water.
- High movement of sand and soil particles (Soil erosion and high wind).
- Occurrence of high surface crusting leads to poor crop stand.

1.7. Management of dryland agriculture

1.7.1. Main principle

- Reduction in moisture loss due to evaporation and transpiration.
- Growing early maturing crop with deep root system.
- Reduced number, size and horizontal orientation of leaves.
- Maintaining optimum plant population.
- Pre monsoon sowing.
- Adoption of mixed or intercropping.
- Free from harmful weeds.
- Increasing organic matter content of soil.

- Growing cover crops including mulches.
- Use of antitranspirants, plant modifiers, growth retardants, defoliants, crop ripeners and anti evaporants.

1.7.2. Use of cropping system

- In rainfall areas ranging from 500-700 mm, intercropping can be adopted, to replace monocropping (Growing of traditional varieties).
- Intercropping like cereals + legumes, legumes + legumes, cereals + oilseeds can be recommended.
- Eg. Sorghum + pigeon pea, pearl millet + pigeon pea, sorghum + green gram, sorghum + soyabean, groundnut + red gram.
- Rainfall more than 750 mm, go for sequential cropping.
- Eg. Pulses and oil seeds, maize and chickpea, sorghum and safflower, sorghum and chickpea, maize and chickpea.

1.7.3. Uses of mechanical methods

1. Contour bunding

- Bund section is 1.61 m^2 in vertisols (Black soils)
- Area occupied by the bund is 5% which is lost from cultivation.
- In vertisols, area occupied by the bund is 10-15%.

2. Graded bunding (Steep slope)

- Graded bunds have 0.8 m^2 cross section in vertisols.
- Area lost is 35%.

3. Tie ridges

- Adjacent ridges are joint at regular intervals.
- Act as barriers for facilitating water infiltration and preventing runoff.

4. Bench terracing

- Practiced in steep sloping fields and length of terraces is 100 m with longitudinal grades in the range of 0.2-0.8 %.

5. Ploughing

- Ploughing across the slope and grow low value crops in catchment areas.

- Ploughing deep soil once in 3-4 years.
- Adoption of zero tillage or minimum tillage.

1.8. Reclaiming problem soils in dryland

- Addition of liberal quantity of organic matter.
- Saline soil
- Leaching (Rich in sodium)
- Growing green manure crops
- Apply FYM
- Sodic soil – Gypsum (2 tons/ha) application.
- *In situ* incorporation of green manures – 5 t/ha.
- Acid soil – Lime application (200 - 400 kg/ha).

1.9. Maintenance of soil fertility in dryland agriculture

- Main issues – Loss of plant nutrients due to erosion of surface fertile soil leads to decline in soil fertility and low yield.
- Combined use of FYM, green manure and inorganic fertilizers (Integrated nutrient management).
- Use of crop residues to maintain soil fertility.
- Crop rotation with cereals and legumes or oilseeds.
- Use of biofertilizer to improve microbial activities.
- Use mulching to check loss of moisture and erosion.
- Develop small water shed for collection of runoff water.
- Adopt alternate land use planning for sustainable agriculture, following IFS (Integrated farming system) including agroforestry to get food, fibre, fuel, fire wood, fruits, furniture timber, fodder and farm animals.

1.10. TNAU dryland technologies

1.10. 1. Reducing evaporation loss

- Moisture is lost by evaporation from soil surface and transpiration from plant surface.
- 99% of water absorbed by the plants is lost by transpiration.
- About 60-75% rainfall is lost through evaporation.

1. **Soil mulch or dust mulch**

- Loosening of soil surface is called soil mulch.
- Intercultivation in growing crops creates soil mulch.

2. **Stubble mulch**

- Crop residues like wheat straw or cotton straw are left on the soil surface as stubble mulch.
- Protects soil from erosion and reduction of evaporation loss.

3. **Straw mulch**

- Paddy straw and leaf trashes spread on the soil surface.

4. **Plastic mulching**

- Polyethylene or polyvinyl chloride sheets are used as mulching material.

5. **Vertical mulching**

- Digging narrow trenches across the slope at regular intervals and placing straw and crop residues in these trenches.

1.10.2. Managing transpiration loss

- To reduce transpiration loss, antitranspirants are used.

1. **Stomatal closing**

- PMA spraying (Phenyl mercuric acetate) (Fungicide).
- Spraying of atrazine at low concentration.

2. **Flim forming types**

- Mobileaf, silicone, hexadeconol (Antitranspirants).

3. **Reflectant type**

- 5% kaolin spray, diatomaceous earth product (Celite).

4. **Growth retardants**

- ABA, cycocel (CCC).
- Main disadvantage of antitranspirants is that they reduce photosynthesis

1.10.3. Wind break and shelter belts

- **Wind break:** Obstruct wind blow and reduce wind speed
- Trees used are eucalyptus, silver oak, casuarina, teak, agathi.
- **Shelter belt:** Row of trees planted across the direction of wind to protect crop against wind damage.

- Reduces wind speed and evaporation loss.
- Wind erosion is also reduced.

1.10.4. Weed control

- Weed is the competitor of a crop for water, nutrients and sunlight.
- Eradication of weeds reduces transpirational loss and increase the availability of soil moisture.

1.10.5. Spraying nutrient solution

- Spraying of 2% urea or DAP is useful for quicker regeneration and release from moisture stress in legumes.

2

Erosion and Its Conservation

- Major 2 types of erosion (i) soil erosion and (ii) wind erosion

2.1. Soil erosion

- Detachment of soil particles from the surface
- Transportation by water, wind, ice or gravity.

2.1.1.Types of soil erosion due to water

1. **Sheet erosion**

- Removal of thin layer of surface soil.
- Very dangerous.
- Takes place in gentle slope of soil.

2. **Rill erosion**

- Tiny finger like channels are formed.
- Sheet becomes rill erosion.

3. **Gully erosion**

- Next stage of rill erosion, rills becomes wide and deep.
- Severe type of erosion, heavy loss of soil.

4. **Stream erosion**

- Caused by streams and rivers.
- Velocity of water flow is very high when compared to sheet erosion.

5. **Wave erosion**

- Combination of water and wind to form waves by beating action
- Soil is taken into the water reservoirs.

2.1.2. Loss due to water erosion

- Loss of rain water.
- Loss of fertile top soil.

- Loss of plant nutrients.
- Siltation of reservoirs which reduces the storage capacity of reservoirs and results in desiltation.
- Uncontrolled runoff leading to flood and causes loss of crops, animals and human life.

2.2. Wind erosion

- By action of wind and common in semi arid region. Major problem in land devoid of vegetation.
- **Consists of 3 phases**
- Initiation of movement
- Transposition
- Deposition.
- In India 33 million ha land is affected by wind erosion; 23.46 ha of desert land and 6.5 mha of coastal area.

2.2.1. Losses due to wind erosion

- Fertile in gangetic alluvial plains have been converted in to sandy desert .
- Loss of fertile top soil.
- Conversion of fertile soils into unproductive sandy soils.
- Loss of yield.

2.3. Soil and water conservation measures

2.3.1. Agronomic measures

1. Contour cultivation

- Cultivation of crops across the slope.
- To prevent surface runoff.
- Method is useful up to one and half per cent slope.

2. Mulching

- Use of natural and artificial mulches.
- Reduces evaporation, increase water in filtration, keeps down weed growth and improves soil structure.

3. Contour strip cropping

- Growing 2 or more crops in strips across the slope.

- Eg. Groundnut and cowpea are efficient in checking soil erosion.
- Groundnut and foxtail millet
- Wheat and pulses.

2.3.2. Mechanical measures

1. Contour bunding

- Done in areas where rainfall is less than 600 mm having light textured soil.
- Constructing bunds across the slopes of land.
- Not suitable for soil with low permeability and infiltration.

2. Graded bunding

- Used in areas where annual rainfall is more than 800 mm.
- Formation of broad based graded terraces across the slope.

3. Terracing

- Adopted on the hill slopy sides.
- Formation of series of levelled strips across the slope.
- Drains are provided at suitable distance.
- Also known as **bench terracing**.

4. Gully plucking

- Blocking gullies with dams or rough stones or wood with cut branches.

2.3.3. Agrostological measures

- Raising grasses to prevent soil erosion by rainfall interception.
- **Advantages:** Binding power of soil particles and improving soil structure.
- Grass legume combination is followed which fixes atmospheric nitrogen and also fodder for animals.
- **Grasses:** *Cenchrus ciliaris, Cynodon dactylon, Panicum antidotale, Panicum virgatum, Andropogon sp, Dichanthium annulatum, Heteropogon contortus.*
- **Legumes :** *Stylosanthes, Centrosema.*

3

Coventional Farming and Alternate Land Use in Dryland Farming

3.1. Conventional farming

- Refers to the use of synthetic chemical fertilizers, pesticides, herbicides and other continual inputs, genetically modified organisms, heavy irrigation, intensive tillage and concentrated monoculture production.
- Highly resource and energy intensive, but also highly productive.
- Conventional farming typically involves monocropping, but very expensive.

3.2. Alternate land use in dryland farming

- Refers to use alternative production system in lands to achieve more sustainable biological and economic productivity on long term.
- Classical example is to use of agroforestry (AF) system.
- To cope up increasing population of both human and livestock and rising demand for food, fodder and fibre, a diversified land use system need to be adopted as an alternative to conventional cropping systems.
- Through this approach, the biological productivity and quality of resource base degraded ecosystems can be significantly enhanced.

3.2.1. Agroforestry systems

- Land use systems in which woody perennials (Trees, shrubs, palms, bamboos etc.) are deliberately grown in association with herbaceous plants (Crops, pastures) or livestock in a spatial arrangement, a rotation, or both.

1. **Agrisilviculture system:** Growing trees and agriculture crops.
2. **Alley cropping:** Planting rows of trees in wide spacing with a companion crops grows in the alley ways between rows.
3. **Agrihorticulture system:** Growing agricultural crops along with horticultural crops.
4. **Silvipasture system:** Growing trees and grasses.
5. **Hortipasture system:** Growing fruit trees horticultural crops and grasses.

6. **Agri silvipasture system:** Growing trees field crops and grasses.
7. **Agri silvi horticulture system:** Growing trees along with field and horticultural crops.

3.2.2.Tree based farming

- Marginal lands or the lands, which are not capable of supporting the field crops can be best converted into the tree based farming.

3.2.3. Live fence rows

- Species of legume (*Leucaena leucocephala* or *Sesbania grandiflora*) in wet areas and *Zyziphus*, *Hibiscus*, *Glyricidia* etc. species in dry areas are planted.
- Source for human food and animal feed with leaf protein approaching 6 %.
- Economic trees such as mango or kapok are grown
- Useful for erosion control as well as to separate fields for demarcation.

3.2.4. Trees as hedges

- Trees that are taken as live hedges by the farmers.

3.2.5. Slope agriculture land technology

- Growing seasonal and perennial crops in 3m to5m wide bands between contour rows of N fixing trees.
- N fixing trees are thickly planted in double rows to make hedge rows.
- When hedge is 1.5 to 2m tall, it is lopped to about 40cm height and placed in alleys to serve as mulch cum manure.

1. Trees used for hedges

Common name	Scientific name	Uses
Molucca bean	*Caesalpinia crista*	Reduces soil erosion
Coral tree	*Erythrina variegata*	Non-browsable, soil and water conservation.
Thor	Euphorbia sp	Non-browsable of medicinal value
Jatropha	*Jatropha curcas*	Non-browsable, soil and water conservation
Lantana	*Lantana camara*	Non-browsable, resistant to adverse conditions, improves fertility of exhausted areas and rocky gravelly or hard infertile soil, used in soil conservation, adds humus, green manure, convertible to compost.
Kewada	*Pandanus odoratissimus*	Non-browsable, soil binder, controls erosion from wind & water, shelter belt to sand drifts of food value, used in making paper, ropes, cordage, hats, baskets, umbrellas and fancy articles, roots for brush an basket making.
Horse bean	*Parkinsonia aculeata*	Used as fuel, fodder, paper making of medicinal value.
Vilayati babul	*Prosopis juliflora*	Immunity to grazing, yields firewood, fodder.

Part-3 Watershed and Wasteland Management

1

Watershed Development

1.1. Watershed development

- Refers to the conservation, regeneration and judicious utilization of all the resources viz., land, water, vegetation, animal and human within a particular watershed in integrated manner.
- Water drainage area on earth's surface from which runoff, due to precipitation, flows into a single to a larger stream, a river, a lake or the ocean.

1.2. Concept of watershed management

- Planning and design of soil and water conservation structures such as bunds, waterways, overflow structures, water harvesting structure.

1.3. Principles of watershed management

- Utilizing the land based on its capability.
- Protecting fertile top soil.
- Minimizing silting up of tanks, reservoirs etc.
- Protecting vegetative cover throughout the year.
- *In situ* conservation of rainwater.
- Safe diversion of surface runoff to storage structures through grassed waterways.
- Stabilization of gullies and construction of check dams for increasing ground water recharge.
- Increasing crop intensity through intercropping and sequence cropping.
- Alternate land use systems for efficient use of marginal lands.
- Water harvesting for supplemental irrigation.
- Maximizing farm income through agricultural related activities such as dairy, poultry, sheep and goat farming.
- Improving infrastructural facilities for storage, transport and agricultural marketing.

- Setting up of small scale agro-industries.
- Improving socio-economic status of farmers.

1.4. Classification of watershed

1.4.1. Watershed

- Macro watershed : 400 to 2000 ha.
- Micro watershed : Less than 400 ha.

1.4.2. Agricultural watersheds

- Sub watershed : 10,000 to 50,000 ha.
- Multi watershed : 1000 to 10,000 ha.
- Micro watershed : 100 to 1000 ha.
- Mini watershed : 1 to 100 ha.

1.5. Components of watershed development

1.5.1. Soil and land management

1. **Interceptor drains (Curtain drain)**
   - Trenches made along the contour of a slope to capture surface and sub surface water within permeable soil.
2. **Graded bunding (Graded terraces)**
   - Bunds laid along a pre-determined longitudinal grade very near the contour, but not exactly along contour.
3. **Bench terracing**
   - Benches laid on sloping land with deep soil to retain water and control erosion.
4. **Interbund vegetative barriers**
   - Narrow, parallel strips of stiff, erect, dence grasses close to the contour.
5. **Grass waterways**
   - A 2-48 metre wide native grassland strip of green belt along a valley or water coarse.
6. **Improvement of ill drained soils**
   - Building raised bed or adding organic matter to improve drainage.
7. **Nala training / improvement**
   - Strengthening of nala bunds.

1.5.2. Water harvesting structures

1. **Nala bunding**

- Embankment constructed across nalas for checking runoff velocity for increasing water.

2. **Farm pond**

- Construction of small tank for harvesting and storing surface runoff water.

3. **Percolation tanks**

- Earthen dams with masonary structure in a highly permeable land to improve percolation of surface runoff water.

4. **Minor irrigation tanks**

- Small irrigation reservoirs for ground water recharge.

5. **Stop dams in nalas**

- An artificial lake in hilly areas to harvest water runoff from slopes.

6. **Underground diaphragms**

- Type of retaining wall to resist lateral pressure due to water.

1.5.3. Afforestation cum pasture development for rural energy and forage for animals

- On private marginal and culturable waste lands.
- On community and Government forest lands.

1.5.4. Agricultural development

- Selection of crops and their varieties suitable for local soil and climatic situation.
- Adoption of appropriate cropping system.
- Contour farming.
- Strip cropping.
- Mulching and crop residue management.
- Adoption of alternate land use system depending on land capability such as alley cropping, agri-horticulture, silvi pastural management, dryland horticulture, tree farming, and pasture management.

1.5.5. Farm pond

- Annual rainfall, probable runoff and area of catchment decide size of the farm pond.

- Should be located in lower patches of the field to facilitate better storage and seepage losses.
- Size of the farm pond should be decided by the quantum of water to be stored and nature of the soil strata.
- Hard, rocky field to be provided with shallow pond.
- In clay soil, pond depth may be increased.

2

Wasteland Development

2.1. Definition

- Land which is presently lying unused or which is not being used to its optimum potential due to some constraints.

2.2. Classification

- As per National Wastelands Development Board established in 1985, classified waste land into two groups.
  - Cultivable wastelands
  - Uncultivable wastelands

2.2.1. Cultivable wastelands

- Gullied and/or ravenous lands
- Undulating land without shrubs
- Surface water logging and marsh land
- Salt affected land
- Shifting cultivation area
- Degraded forest land
- Degraded pasture / grazing land
- Degraded forest plantations
- Strip lands
- Sand dunes
- Mining / industrial wastelands

2.2.2. Uncultivable wastelands

- Brown rocky / stony / shut of rocks
- Steep sloppy areas
- Snow covered and / or glacier lands

2.3. Categories of waste land for identification (Based on causative agents)

2.3.1. Water

- Sheet erosion
- Rill erosion
- Gully erosion
- Ravenous land
- Saline soil
- Marshy land
- Water logged
- Alkali soil

2.3.2. Wind

- Sand dunes
- Sand bar
- Coastal
- Sand

2.3.3. Man

- Mine spolls
- Shifting cultivation
- Industrial
- Waste land

2.3.4. Others

- Land slides
- Shallow soils

2.4. Causes of wasteland formation

- Deforestation
- Over cultivation
- Over grazing
- Unskilled irrigation
- Improper developmental activities such as dumping of wastes, mine spoils.

2.5. Choice of species for problem soils

- In Tamil Nadu alone, 0.04 lakhs ha of land is affected by saline and alkaline condition.

2.5.1. Saline soil

- Have high content of soluble salt usually more than 0.2%.
- The soil pH value is generally between 7.3 and 8.5.
- **Suitable species:**

Prosopis juliflora	*Salvadora persica*	*S.oleoides*
Tamarix articulata	*Tamarix troupii*	*Tamarix ericoides*
Acacia farnesiana	*Parkinsonia aculeata*	*Salsola baryosma*
Atriplex spp.	*Suaeda* spp.	*Kochia indica*

2.5.2. Non-saline alkaline soils

- Also called as sodic soils.
- Have more than 15% of their ion exchange sites occupied by Na + ions.
- pH ranges from 8.5 – 10.0.
- **Suitable species:**

Prosopis juliflora	*Tamarix articulata* (Red tamarisk)
Prosopis spiagera	*Tamarix aphylla*
Acacia nilotica (karuvel)	*Casuarina equisetifolia* (Savukku)
Butea monosperma	*Leucaena leucocephala* (Subabul)

2.5.3. Alkali soils

- Have sufficient exchangeable sodium to interface with the growth of crops
- Mostly associated with kankar pan in sub soil.
- **Suitable species:**

Acer sp. (Maple tree)	*Pseudotsuga menziesii* (Douglas fir)
Pinus nigra (Austrian pine)	*Quercus macrocarpa* (Bur oak)
Gleditsia triacanthos (Honey locust)	

2.5.4. Laterite and lateritic soils

- Dominated by hydrous oxides of Fe and Al.
- Humus is absent.

- Poor in N, P, K, Ca and other nutrients.
- About 12 m ha of area in eastern, central and southern states of India are covered by laterite of lateritic soils.
- Annual rainfall varies from 750 – 3750 mm.
- **Suitable species**

Acacia auriculiformis (Kathi karuvel)

Albizia lebbeck (Womens tongue tree)

Bambusa arundinacea (Spiny bamboo)

Holoptelea integrifolia (Indian elm/Jungle cork tree)

Pterocarpus marsupium (Indian kino tree)

Dendrocalamus strictus (Male bamboo)

Madhuca longifolia (Mahua tree)

Shorea robusta (Sal tree)

Dalbergia latifolia (Rose wood)

2.5.5. Afforestation for sand dune stabilization

- Dune means hilly topographical feature.
- Formed by wind
- 58% of western Rajasthan (Thar desert) is covered by moving of semi stabilised sand dunes.
- Distributed almost all the parts of desert, coastal lands and inland riverine areas.
- In India, 11.996 m ha lands of Rajasthan and 1.47 m ha of coastal regions are facing serious wind erosion problems.

1. Classification of sand dunes

Sand dunes of old system: Almost stable sand dunes or partially stabilised sand dunes.

Sand dunes of new system: Barchan and shrub coppiece dunes are the most dominant formation in this new system.

2. Afforestation of sand dunes

Protecting of shifting sand dunes against biotic interference

- **Suitable tree species according to rainfall pattern.**

a) **Rainfall 150-300mm**: *Acacia tortilis* (Umbrella thorn), *A. senegal* (Gum Arabic tree), *Prosopis chilensis* (Mosquito tree), *Prosopis cineraria* (Jammi tree), *Tecomella undulata* (Honey or desert teak).

b) **Rainfall 300-400mm:** *Acacia tortilis, A. senegal, Prosopis chilensis, Prosopis cineraria,Tecomella undulat, Acacia nilotica, Eucalyptus camaldulensis* (River red gum), *Parkinsonia aculeata, Zizyphus sp*

c) **Rainfall 400mm and above:** *Albizia lebbeck, Ailanthus excelsa, Azadirachta indica, Acacia nilotica*

- Erection of micro windbreaks for treating dunes.
- Forestation of sand dunes by direct sowing or transplanting.
- Planting of grass skips or seeds of grasses, castor seeds duly treated with sodium alginate to protect.
- Continuous and proper management.

3. Criteria for choice of species

- Should be highly drought resistant.
- Should be fast growing and fibrous root system.
- Capable of deep vertical penetration to reach lower moisture regime of the soil
- Resistant to high wind velocity.
- Capable of multiple benefit.

2.5.6. Different vegetative methods for soil and water conservation

1. Strip planting

- Erosion permitting and erosion resisting crops are alternatively raised at right angle to the slope of the land to retard the velocity of rain water.

2. Rotational cropping

- Either grain crops, grasses or legumes along with trees are planted in the field.

3. Cover cropping

- Trees and grasses are grown to cover the earth's surface.
- Eg. *Acacia nilotica, Azadirachta indica, Eucalyptus tereticornis*, silk cotton, teak and casuarinas can be planted to arrest erosion.
- *Agave americana* for stabilization of gullies.
- *Acacia nilotica* and *Azadirachta indica* can be planted on the banks of rivers, percolation ponds, lakes to strengthen the bunds and to prevent erosion.

2.5.7. Mine burden reclamation

- Soil is turned over from the bottom to upper horizons during mining called as mine spoils.
- Trees planted: *Picea glauca* (White spruce), *Pinus banksiana* (Jack pine), *Populus tremuloides* (Trembling aspen).

2.5.8. Afforestation in coastal and hilly areas

- Occupy an area of about 8.4 m ha in India in the form of a narrow strip along the eastern and western coasts of the country.
- Soil is generally alkaline in reaction and Na salt content is high.
- Poor in nutrients and low water holding capacity.
- Water table is high and texture is mostly sandy.
- Lime deposition is also common.
- Continuous occurrence of cyclones suggested that 1 to 2 kilometre wide forest belt is necessary for moderating the effects of cyclones.
- Eg. *Casuarina equisetifolia, Cocos nucifera, Borassus flabellifer* (Asian palmyra), *Pongamia pinnata, Avicennia officinalis* (Indian mangrove), *Acacia auriculiformis, Eucalyptus tereticornis* (Forest red gum), *Salvadora persica* (Mustard tree), *Salvadora oleoides* (Pilu tree), *Simarouba* glauca (Paradise tree),*Prosopis sp.*

2.5.9. Afforestation in hilly areas

- **Suitable tree species for temperate hills region:** *Pinus roxburghii* (Long leaf Indian pine), *Pinus wallichiana* (Bhutan pine), *Cedrus deodara* (Deodar tree), *Acacia dealbata* (Silver wattle), *Ailanthus altissima* (Tree of heaven), *Cupressus torulosa* (Himalayan torulosa), *Morus sp, Juglans regia* (Walnut tree), *Robinia pseudoacacia* (Black locust), *Populus spp.*
- **Suitable tree species for subtropical hills region:** *Eucalyptus globulus, Eucalyptus grandis* (Flooded gum), *Eucalyptus tereticornis, Acacia mearnsii* (Black wattle), *Acacia decurrens* (Green wattle), *Acacia dealbata* (Silver wattle), *Acacia melanoxylon* (Australian black wood), *Pinus roxburghii.*

3

Land Use, Suitability and Irrigability Classification

3.1. Land use classification

3.1.1. Classes

- Capability class is the broadest category in the land capability classification system.
- Class codes 1, 2, 3, 4, 5, 6, 7, and 8 are used to represent both irrigated and non-irrigated land capability classes.
- **Class 1**: Soils have slight limitations that restrict their use.
- **Class 2**: Soils have moderate limitations that reduce the choice of plants or require moderate conservation practices.
- **Class 3**: Soils have severe limitations that reduce the choice of plants or require special conservation practices or both.
- **Class 4**: Soils have very severe limitations that restrict the choice of plants or require very careful management or both.
- **Class 5**: Soils have little or no hazard of erosion, but have other limitations, impractical to remove, that limit their use mainly to pasture, range, forestland, or wildlife food and cover.
- **Class 6**: Soils have severe limitations that make them generally unsuited to cultivation and that limit their use mainly to pasture, range, forestland or wildlife food and cover.
- **Class 7**: Soils have very severe limitations that make them unsuited to cultivation and that restrict their use mainly to grazing, forestland, or wildlife.
- **Class 8**: Soils and miscellaneous areas have limitations that preclude their use for commercial plant production and limit their use to recreation, wildlife, or water supply or for esthetic purposes.

3.1.2. Capability subclass

- Second category in the land capability classification system. Class codes **e, w, s** and **c** are used for land capability subclasses.
- **Subclass e:** Soils for which the susceptibility to erosion is the dominant problem or hazard affecting their use. Erosion susceptibility and past erosion damage are the major soil factors that affect soils in this subclass.
- **Subclass w:** Soils for which excess water is the dominant hazard or limitation affecting their use. Poor soil drainage, wetness, a high water table and overflow are the factors that affect soils in this subclass.
- **Subclass s:** Soils that have soil limitations within the rooting zone, such as shallowness of the rooting zone, stones, low moisture holding capacity, low fertility that is difficult to correct and salinity or sodium content.
- **Subclass c**: Soils for which the climate (The temperature or lack of moisture) is the major hazard or limitation affecting their use.
- Subclasses are not assigned to soils of miscellaneous areas in capability classes 1 and 8.

3.2. Land suitability classes

- Fitness of a given type of land for a defined use.
- Land is assessed as suitable or not suitable for the use under consideration.
- Two orders represented by the symbols S and N.

1. **Order S suitable:** Land on which sustained use of the kind under consideration is expected to yield benefits which justify the inputs, without unacceptable risk of damage to land resources.
2. **Order N not suitable:** Land which has qualities that appear to preclude sustained use of the kind under consideration.

- **Three classes are recognized within the order suitable**
- **Class S1 highly suitable:** Land having no significant limitations to sustained application of a given use.
- Not significantly reduce productivity or benefits and will not raise inputs above an acceptable level.
- **Class S2 moderately suitable:** Land having limitations which in aggregate are moderately severe for sustained application of a given use.
- Reduce productivity or benefits.
- Increase in required inputs.

- Appreciably inferior to that expected on class S1 land.
- **Class 3 marginally suitable:** Land having limitations which in aggregate are severe for sustained application of a given use.
- Reduce productivity or benefits
- Increase in required inputs that this expenditure will be only marginally justified.
- **Order not suitable, has two classes**
- **Class N1 currently not suitable:** Land having limitations which may be surmountable in time, but which cannot be corrected with existing knowledge at currently acceptable cost.
- Limitations are so severe as to preclude successful sustained use of the land in the given manner.
- **Class N2 permanently not suitable:** Land having limitations which appear so severe as to preclude any possibilities of successful sustained use of the land in the given manner.

3.3. Land irrigability classification

- Depends on quality and quantity of irrigation water and socio economic factors like land development costs, provision of drainage facilities and production costs of individual crops.
- Six land irrigability classes (Class 1 to 6) have been recognized based on limitation of soil (s), topography (t) and drainage (d).

1. **Class 1:** Few limitation of soil (s), topography (t) and drainage (d) for sustained use under irrigation.
2. **Class 2:** Moderate limitation of soil (s), topography (t) and drainage (d) for sustained use under irrigation.
3. **Class 3:** Severe limitation of soil (s), topography (t) and drainage (d) for sustained use under irrigation.
4. **Class 4:** Lands that are marginal for sustained use under irrigation because of very severe limitation of soil (s), topography (t) and drainage (d).
5. **Class 5:** Lands that are temporarily not suitable for sustained use under irrigation.
6. **Class 6:** Lands that are not suitable for sustained use under irrigation.

4

Size and Distribution of Land Holding

4.1. Highlights of the report on number and area of operational holdings (2015-16)

- Total number of operational holdings in the country has increased from 138 million ha in 2010-11 to 146 million ha 2015-16 i.e. an increase of 5.33%.
- Number of operational holders in different states
- Uttar Pradesh (23.82 million ha)
- Bihar (16.41 million ha)
- Maharashtra (14.71 million ha)
- Madhya Pradesh (10.00 million ha)
- Karnataka (8.68 million ha)
- Andhra Pradesh (8.52 million ha)
- Tamil Nadu (7.94 million ha)
- Rajasthan (7.65 million ha)
- West Bengal (7.24 million ha)
- Five kinds of land holdings in India, depending on various sizes
- **Marginal holdings:** Size 1 hectare or less
- **Small holdings:** Size 1 to 2 hectares
- **Semi-medium holdings:** Size 2 to 4 hectares
- **Medium holdings:** Size 4 to 10 hectares
- **Large holdings:** Size above 10 hectares
- Maximum number of operational land holdings in India is **marginal holdings**.

5

Types and Systems of Farming

5.1. Types of farming

- Arable: Crops.
- Pastoral: Animals.
- Mixed: Crops and animals.
- Subsistence: Grown just for the farmer and his family
- Commercial: Grown to sell.
- Intensive: High inputs of labour or capital usually small.
- Extensive: Low inputs of labour or capital.
- Sedentary: Permanently in one place.
- Nomadic: The farmers move around to find new areas to farm.

5.2. Systems of farming

5.2. 1. Traditional farming

- Generally prevalent in a backward segment of agriculture.
- Characteristics of an overall backward economy.
- Population pressure resulted in perpetual sub-division of holdings leads to size of the farm is very small.
- No marketable surplus on such small farms.
- Also called **subsistence farming**..

5.2. 2. Commercial farming

- Private ownership of a farm by a single farmer.
- Decision making power generally in the hands of employed managers.
- Hired labourers operate the farms.
- Eg: Tea and coffee plantations.

- Also known as **estate farming** or **corporate farming** or **capitalistic farming.**
- Production is meant for the market.

5.2.3. State farming

- State itself is the owner of the farm.
- Decision making power by managers.
- No financial problems.
- Necessary improvements in the land can be made.
- Improved agricultural practices can be adopted.
- Productive assets for efficient production can be procured.
- Well can be dugged, tube wells can be installed, necessary buildings and roads on the farm can be built.
- Market surplus of food grain and raw materials required by the industrial sector are produced.

5.2. 4. Collective farming

- Introduced from U.S.S.R.
- Land is owned by whole village community.
- Community itself can take decisions about production.
- Land and other production assets are held jointly by the village society.
- An executive board manages the farm with nominees of the Government.
- Member of the village community work as labourers.

5.2. 5. Co-operative farming

- Higher intensity of cropping, higher employment level and higher productivity per acre.
- Difficulties like use of improved crop practices and implementation of developmental works like fencing, digging of a well, weak bargaining power in the market etc., are often met.

Part-4 Water Management

1

Water Resource Development and Management

1.1. Status of irrigation in Tamil Nadu

- Geographical area of 13 million hectares is ranked eleventh in size among the Indian states.
- Net area sown is about 6 million hectares.
- 3 million hectares or 50% get irrigation facilities.

1.2. Irrigation sources

- Government canals: 0.95 million hectares.
- Tanks: 0.90 million hectares.
- Wells, Tube wells etc.,: 1.15 million hectares.
- There are 57 P.W.D. dams and 39,202 tanks in the state out of 39,202 tanks.
- P.W.D. tanks: 8,939
- Rainfed : 5242+ System tanks: 3,697
- Eg. Zamin tanks: 9,850.
- Panchayat union tanks: 20,413

1.3. Development of irrigation in Tamil Nadu

- In first five year plan, there were 23 major and medium irrigation projects benefiting 1.10 million hectares.
- In 1997-98, the area under irrigation has risen to 1.476 million hectares.

1.4. Organization set up

- Water Resources Organization has been divided into 4 regions: Chennai, Tiruchi, Pollachi and Madurai.
- Each region is headed by Chief Engineer.

1.4.1. Rivers of Chennai region, Chennai

- Araniyar
- Kosasthalaiyar
- Cooum
- Adyar
- Palar
- Ongur
- Varahanadhi
- Malattaru
- Penniyar
- Gadilam
- Vellar (Upto Coleroon)

1.4.2. Rivers of Trichy region, Trichy

- Cauvery
- Agniar
- Ambuliyar
- Vellar
- Koluvanar
- Pambar

1.4.3. Rivers of Pollachi region, Pollachi

Tributaries of Cauvery such as:

- Amarvathi
- Bhavani
- Noyyal
- Korangam pallam
- Palar
- Periyapallam
- West flowing rivers

1.4.4. Rivers of Madurai region, Madurai

- Manimuthar
- Kottakkariyar
- Vaigai
- Uthrakosamangayar
- Gundar
- Vembaru
- Vaippar
- Kallar
- Korampallamaru
- Thamirabarani
- Karumeniyar
- Nambiyar
- Hanumanadhi
- Palayaru
- Valliyar
- Kodayar

1.4.5. Activities of W.R.O

- Medium irrigation schemes.
- Minor irrigation schemes.
- Special minor irrigation programmes
- Water resources consolidation projects.
- Modernization of Periyar Vaigai irrigation system
- Tank modernization scheme with European Economic Commission assistance.
- Rehabilitation of minor irrigation tanks for rural development for Outer Bank Community Foundation (OBCF) (Japan) assistance.
- Rehabilitation of minor irrigation tanks for rural development with World Bank assistance.
- State tank irrigation projects.
- Desilting of tanks

- Integrated Tribal Development Programme.
- Hill Area Development Programme.
- Western Ghats Development Programme.
- Scheme with NABARD assistance.
- Modernisation and desilting of Cauvery delta.
- Krishna water supply project.
- World Bank aided hydrology project.
- Dam safety assurances and rehabilitation project.
- Rehabilitation and reclamation of Chennai city waterways.
- Anti-sea erosion works.
- Diversion of west flowing rivers of Kerala to Tamil Nadu.

1.5. Rivers which flow in southernmost region of the Indian peninsula

- **Bhavani river:** Fed, mostly by the southwest monsoon and one of the main tributaries of the river Kaveri.
- **Cheyyar river**: Tributary of river Palar is a major seasonal river that flows through the district of Tiruvannamalai.
- **Chittar river**: Main river originates from the Courtallam hills of the Shencottah and Tenkasi taluks in the district of Tirunelveli. It flows through the state along with its 5 tributaries.
- **Ponniyar river**: Flowing across the borders in between the taluks of Villupuram and Cuddalore and finally drains into the Bay of Bengal.
- **Thamirabarani river**: Originates from the peaks named Agasthyamalai, Aduppukkal mottai and Cherumunji mottai in the Tirunelveli district.
- **Vaigai river**: While flowing towards the Palk Strait, it changes its course towards the south east near Sholavandan and passes through the town of Madurai.
- **Gundar river**: Mainly flows through the districts of Tirunelveli and Virudhunagar in Tamil Nadu.
- **Noyyal river**: Tributary of Kaveri flows through Dharapuram and Palladam taluks in Tiruppur district and Coimbatore district, respectively.
- **Suruli river**: Originates from the Suruli Waterfall and major tourist attractions in the Theni district.
- **Vaipar river**: Originates from the bordering hills of the Kerala state, the river runs through the Virudhunagar and Theni districts.

2

Command Area Development

- To achieve maximum possible irrigation efficiencies, there are two approaches.
- Modernization of conveyance system down below the reservoirs upto government controlled outlet. Involves mainly the construction and maintenance of head sluices, main canals, branch canals, and distributaries (Modernization of supplier's side or system level development works)
- Modernization below the government controlled outlets upto the drains (Modernization of user's side or farm level development works). Also known as On-Farm Development Works (OFD).

2.1. Conveyance and distribution system

- Reservoir
- Main canal
- Branch canal
- Minor
- Distributor
- Sluice / outlet
- Field channel
- Distribution boxes
- Turnout
- Checks

2.2. On-Farm Development Works (OFD)

- Include lining of field irrigation channels and infrastructural facilities like bed regulators, diversion and distribution boxes, turnouts and drop structures to regulate and convey the irrigation water from government controlled outlets to individual land holdings.
- To reduce conveyance and application losses, minimize water logging condition and conserve water.

2.2.1. Execution problems

- OFD works are to be executed in the farmer's fields.
- Number of farmers involved is more.
- Influence of socio-economic constraints.
- Undertaken by the State Agricultural Engineering Department.

2.2.2. The OFD strategy

- 10 ha block outlets are the last government outlets having regulating shutter arrangements only at the sluices of branch canals.
- Each sluice serves 1 to 12 blocks through the lined distributory.

1. Problems

- Absence of adequate field channel network causing wastage of irrigation in field.
- Interfering with the distributory (Carrying water down to other 10 ha blocks) by adjoining head reach farmers in each block.
- Leakage and lateral seepage of water from earthen channels with erosive slopes.
- Difficulties in irrigating higher level fields through earthen channels at zero gradient.
- Structural deficiencies essential structures such as channel crossing, small culverts, road crossings with siphon arrangements etc., are to be constructed.

2. OFT measures to overcome the problems

- Provision of proper earthen field channel net work from source upto each land holdings.
- Provision of higher level field channels (Mostly lined to zero point) parallel to the distributory in the upper part of each block.
- Construction of diversion boxes with leading channels in all the required directions.
- Bed dams are constructed to stabilize the slope.
- Drop structures are constructed at the point of sudden drop in bed levels.

3. Rotational water supply (RWS)

- Techniques in irrigation water distribution management.
- Equi-distribution of irrigation water irrespective of location of the land in the command area by enforcing irrigation time schedules.
- Each 10 ha block is divided into 3 to 4 sub units to evolve time schedule.

3

Ground Water Development and Conjunctive Use

3.1. Ground water development

3.1.1. Rain water harvesting

- Collecting and using precipitation water from a catchments surface.
- Extensive rain water harvesting apparatus existed 4000 years ago in the Palestine and Greece.
- In ancient Rome, residences were built with individual cisterns and paved courtyards to capture rain water to augment water from city's aqueducts.
- As early as the third millennium BC, farming communities in Baluchistan and Kutch impounded rain water and used it for irrigation dams.

3.1.2. Artificial recharge to ground water

- Refer to any man made scheme or facility that adds water to an aquifer.

3.1.3. Rain water harvesting

1. **Purpose**

- Surface water is inadequate to meet our demand which demands use of ground water.
- Due to rapid urbanization, there is a decrease in infiltration of rain water into the subsoil and recharging of ground water.

2. **Techniques for ground water recharge**

- Storage of rainwater on surface for future use in underground tanks, ponds, check dams, weirs etc.
- **Pits:** Construction of pits of 1 to 2 m, wide and 2 to 3m deep and back filled with boulders, gravels, coarse sand as filter media.
- **Trenches:** Trench of 0.5 to 1 m wide, 1 to 1.5m deep and 10 to 20 m long.
- Back filled with filter materials.

- **Dug wells:** Existing dug wells are utilized.
- Water should pass through filter media before putting into dug well.
- **Hand pumps**: The existing hand pumps may be used for recharging.
- Water should pass through filter media before diverting it into hand pumps.
- **Recharge wells:** Recharge wells of 100 to 300 mm diameter are generally constructed.
- **Recharge shafts:** For recharging the shallow aquifer which are located below clayey surface, recharge shafts of 0.5 to 3 m diameter and 10 to 15 m deep are constructed and back filled with boulders, gravels and coarse sand.
- **Lateral shafts with bore wells:** For recharging the upper as well as deeper aquifers, lateral shafts of 1.5 to 2 m wide and 10 to 30 m long are constructed.
- Filled with boulders, gravels and coarse sand.
- **Spreading techniques:** Spread the water in streams/nalas by making check dams, nala bunds, cement plugs, gabion structures or a percolation pond may be constructed.

3.1.4. Urbanisation effects on ground water hydrology

- Increase in water demand
- More dependence on ground water use
- Over exploitation of ground water
- Increase in run-off, decline in well yields and fall in water levels
- Reduction in open soil surface area
- Reduction in infiltration and deterioration in water quality

3.1.5. Methods of artificial recharge in urban areas

- Water spreading.
- Recharge through pits, trenches, wells, shafts.
- Roof top collection of rainwater.
- Road top collection of rainwater.
- Induced recharge from surface water bodies.

3.1.6. Computation of artificial recharge from roof top rainwater collection

- Roof top area 100 sq.m. for individual house and 500 sq.m. for multi-storied building.

- Average annual monsoon rainfall - 780 mm.
- Effective annual rainfall contributing to recharge 70% - 550 mm.

3.2. Conjunctive use

- Combined use of surface water and groundwater resources in a unified way to optimize resource use and minimize adverse effects of using a single source.

3.2.1. Purpose of conjunctive use

- **Surface water**
- Lower delivery and extraction costs
- Variability in supply
- Water logging
- **Groundwater**
- Reliable supply
- Expensive to pump
- Decline in ground water table

3.2.2. Advantages

- Aquifer is used to store surface water when there is an excess.
- Conjunctive use can reduce abstraction from rivers.
- Sustainable water management.

3.2.3. Limitations

- Increased energy consumption for pumping from wells and for coping with reduction in pump efficiency due to large fluctuations of water levels.
- Appropriate management plans to be developed.
- Construction of appropriate ground water recharge structures.
- Administrative difficulties in defining acceptable and equitable ground water rates, when surface water is available.
- People participation is essential.

4

Water Use Efficiency (WUE)

4.1. Water Use Efficiency (WUE)

- An efficient irrigation system implies effective transfer of water from the source to the field with minimum possible loss.
- To identify the nature of water loss and to decide the type of improvements in the system.
- Evaluation of performance in terms of efficiency is prerequisite for proper use of irrigation water.

4.1.1. Irrigation efficiency

$$Ei = \frac{Wc}{Wr} \times 100$$

where,

Ei = irrigation efficiency (%)

Wc = irrigation water consumed by crop during its growth period in an irrigation project.

Wr = water delivered from canals during the growth period of crops.

- In most irrigation projects, the irrigation efficiency ranges between 12 to 34%.

4.1.2. Water conveyance efficiency

$$Ec = \frac{Wt}{Wf} \times 100$$

where,

Ec = water conveyance efficiency in per cent

Wf = water delivered to the farm by conveyance system (at field supply channel)

Wt = water introduced into the conveyance system from the point of diversion

- Water conveyance efficiency is generally low; about 21% loss occurs in earthen watercourses.

4.1.3. Water application efficiency

$$Ea = \frac{Ws}{Wf} \times 100$$

where,

Ea = water application efficiency in per cent

Ws = irrigation water stored in the root zone of farm soil

Wf = irrigation water delivered to the farm (at field supply channel)

- Water losses due to inefficient application of water in the field vary from 28 to 50%.
- Sources for loss of irrigation water during application are:

  Rf = surface runoff from the farm

  Df = deep percolation below the root zone soil
- Neglecting evaporation losses during application, we have

  Wf = Ws + Df + Rf

$$Ec = Wf \frac{(Df + Rf)}{Wf} \times 100$$

4.1.4. Water use efficiency

$$Eu = \frac{Wu}{Wd} \times 100$$

where,

Eu = water use efficiency per cent

Wu = water beneficially used

Wd = water delivered

- Water use efficiency is also defined as **crop water use efficiency** and **field water efficiency**.

1. **Crop water use efficiency**: Ratio of yield of crop (Y) to the amount of water depleted by crop in evapotranspiration (ET).

$$CWUE = \frac{Y}{ET}$$

where,

CWUE = Crop water use efficiency

Y = Crop yield

ET = Evapotranspiration

- CWUE is otherwise called **consumptive water use efficiency**. It is the ratio of crop yield (Y) to the sum of the amount of water taken up and used for crop growth (G), evaporated directly from the soil surface (E) and transpired through foliage (T) or consumptive use (Cu)

$$CWUE = \frac{Y}{G+E+T}$$

where,

(G + E + T) = Cu

- In other words ET is Cu, since water used for crop growth is negligible.

$$CWUE = \frac{Y}{CU}$$

- It is expressed in kg/ha/mm or kg/ha/cm.

2. Field water use efficiency:

$$FWUE = \frac{Y}{WR}$$

where,

FWUE = field water use efficiency

WR = water requirement

- Ratio of crop yield to the amount of water used in the field (WR) including growth (G), direct evaporation from the soil surface (E), transpiration (T) and deep percolation loss (D).

$$FWUE = \frac{Y}{G+E+T+D}$$

G + E + T + D = WR

- It is expressed in kg/ha/mm (or) kg/ha/cm

4.1.5. Water storage efficiency

$$Es = \frac{Ws}{Ww} \times 100$$

where,

Es = water storage efficiency in per cent

Ws = water stored in the root zone during the irrigation

Ww = water needed in the root zone prior to irrigation, i.e., field capacity, available moisture.

4.1.6. Water distribution efficiency

$$Ed = \frac{(1-d)}{D} \times 100 = \frac{(1-\text{Average deviation})}{\text{Average depth applied}} \times 100$$

where,

Ed = water distribution efficiency in per cent

d = average numerical deviation in depth of water stored from average depth stored during irrigation

D = average depth of water stored along the run during irrigation

- A water distribution efficiency of 80% means that 10% of water is applied in excess and consequently 10% is deficient.

4.1.7. Consumptive use efficiency

$$Ecu = \frac{Wcu}{Wd} \times 100$$

where,

Ecu = consumptive use efficiency per cent

Wcu = normal consumptive use of water

Wd = net amount of water depleted from root zone soil

- Consumptive use efficiency is useful in explaining the difference in crop response from different methods of irrigation.

5

Moisture Stress and Rain Gun

- Term moisture stress is generally applied to the stomata opening and transpiration increases with time until they close due to high temperature.

5.1. Causes of moisture stress in plants

- Main reason is extent of transpiration which is affected by leaf size and composition, size and distribution of stomata on leaf, atmospheric humidity, temperature, wind velocity and day length.
- High atmospheric temperature due to intense sun and increased transpiration causes closure of stomata and wilting of leaves even, if soil moisture content is not limiting.
- This deficit is made up during night due to decreased transpiration.
- In plants, moisture content decreases either due to increased transpiration or reduction in absorption or both.

5.2. Available water

- Water held by soil between field capacity and wilting point and at a tension between 0.33 and 15 atmosphere is available to plants.
- Comprises the greater part of capillary water.

5.3. Unavailable water

- Two situations at which soil water is not available plants.
- When the soil water content falls below the permanent wilting point and is held at a tension of 15 atmosphere and above.
- When the soil water is above the field capacity and is held at a tension between 0 and 1/3 atmosphere.

5.4. Soil water deficit and plant stress condition

- Plant water stress may be severe when the soil water potential is low and environment or plant factors interfere severely with the absorption of water.

5.4.1. Classification of water stress based on relative water content (RWC) and plant water potential

	Mild stress	Moderate stress	Severe stress
Relative water content	8-10%	10-20%	> 20 %
Plant water potential	-5 to -6 bars	-12 to -15 bars	> 15 bars

5.4.2. Classification based on period of plant water stress

- Stress occurring during 24 hour period of day and night is referred to as **diurnal stress**.
- Lag between absorption and transpiration is minimum at early morning and maximum at 2.00 pm, known as **temporary wilting** and also known as **incipient wilting** and **mid-day depression**.
- Stress that occurs gradually, and increases gradually and progressively with advance of time after irrigation till the next irrigation is referred to as **critical water stress**.

5.5. Effect of moisture stress on crop growth

- Water stress affects plant growth by modifying, anatomy, morphology, physiology and biochemistry.
- Loss of turgidity leading to cell enlargement and stunted growth.
- Decrease in photosynthesis due to decreased diffusion of CO_2 with the closure of stomata.
- Increase in respiration resulting in decreased assimilation of photosynthates.
- Break down of RNA, DNA and proteins.
- Inhibition of synthesis and translocation of growth regulators.
- Hydrolysis of carbohydrates and proteins leading to increase in soluble sugars and nitrogen compounds.
- Affects germination, cell expansion, cell division, growth of leaves, stems, fruits and root development.
- Finally dictates the economic yield.

5.6. Rain guns

5.6.1. Mini rain guns

1. Inlet size : 1.25 inch
2. Pressure : 1.5 to 3 kg/cm^2

3. Discharge : 100 to 280 LPM
4. Suitable For : Sugarcane, groundnut, fodder crops and vegetables
5. Working method : Half circle and full circle
6. Area coverage : 75 feet radius at 3 kg/cm^2
7. Nozzle size : 8mm,10mm,12mm,14mm
8. 5 HP motor runs : 1 or 2 rainguns based on pressure and discharge

5.6.2. Medium range rain guns

1. Inlet size : 1.5 inch
2. Pressure : 2.5 to 4 kg/cm^2
3. Discharge : 180 to 420 LPM
4. Suitable for : Sugarcane, groundnut, fodder crops, coffee, tea, vegetables
5. Working method : Half circle and full circle
6. Area coverage : 90 feet radius at 3.5 kg/cm^2
7. Nozzle size : 10mm,12mm,14mm and 16mm
8. 7.5 HP motor runs : 1 to 3 rain guns based on pressure and discharge

6

Irrigation Water Quality Testing and Advisory

6.1. Quality of irrigation water

- Asia per capita water availability has fallen by around 80 per cent during last five decades.
- Suitability of waters for a specific purpose depends on the types and amounts of dissolved salts.
- Some of the dissolved ions such as NO_3 are useful for crops.

6.2. Parameters for water quality assessment

- Higher salt content (Saline), high EC and SAR (Sodium Adsorption Ratio) (Saline-sodic) (or) high RSC (Residual Sodium Carbonate) affect the physical and chemical properties of soil.
- Important characteristics of irrigation water that have been used in determining its quality are:
  - Salinity hazard
  - Sodicity hazard
  - Carbonate hazard
  - Permeability hazard
  - Concentration of boron and other specific ion toxicity

6.2.1. Salinity hazard

- Accumulation of soluble salts in the soil.
- Restricting the availability of soil water to plants.
- Concentration of soluble salts in irrigation water is determined by electrical conductivity and expressed as dSm^{-1}.
- Total salt concentration can also be measured as TDS (Total Dissolved Salts) by the following equation:
- TDS (ppm) = EC in dSm^{-1} x 640.

1. Classification of irrigation water based on electrical conductivity

EC (dSm^{-1})	Salinity class	Remarks
< 0.25	Low	Safe and no salinity problem.
0.25-0.75	Medium	Need moderate leaching.
0.75-2.25	High	Not to be used on soils with inadequate drainage. Crop with moderate salt tolerance to be grown.
2.25-5.0	Very high	Not to be used on soils with inadequate drainage.Crop with high salt tolerance to be grown.

6.2.2. Sodicity hazard

- Higher proportion of sodium to other cations lead to the problem of sodicity.
- Evaluated by sodium Adsorption Ratio (SAR).
- Relative proportion of sodium to other cations viz., Ca and Mg in water.

$$SAR = \frac{N^{a++}}{\sqrt{(Ca^{++} + Mg^{++})/2}}$$

6.2.3. Carbonate hazard

- Carbonates and bicarbonate ions are having the tendency to precipitate as Ca and Mg.
- Expressed in terms of Residual Sodium Carbonate (RSC) me L^{-1}.

  RSC = ($CO3^{2-}$ + HCO^{3-}) - (Ca^{2+} + Mg^{2+})

1. Classification of irrigation water based on SAR

Class	Classification	Remarks
Water suitability		
< 10	Safe	
10 – 20	Moderate	
> 20	Unsafe	
Crop suitability		
< 5	Non-sodic	All soils and crops
5-10	Normal water	All soils and crops except sodium sensitive crops
10-20	Low sodic water	Semi-tolerant to tolerant crops. Eg: Beans, soybean, maize, field peas.
20 - 30	Medium sodic water	Tolerant crops Eg: Barley, rye, sunflower, sugar beat, wheat, sorghum, rice, potato.
30 - 40	High sodic water	Not suitable
> 40	Very high sodic water	Not suitable

2. Classification of irrigation water based on RSC (me L^{-1})

		Classification	Remarks
Water suitability			
< 1.25		Safe	
1.25 - 2.50		Moderate	
> 2.50		Unsafe	
Crop suitability			
A_o	-ve	Non alkaline water	For all soils and crops
A_1	0	Normal water	For all soils and crops (even crops sensitive to carbonate and bicarbonate)
A_2	0-0.25	Low alkalinity	Semi-tolerant to tolerant crops.
A_3	2.5-5.00	Medium alkalinity	Tolerant crops with little management.
A_4	5.0-10.0	High alkalinity	Can be used in soils with good drainage.
A_5	>10.0	Very high alkalinity	Not suitable

6.2.4. Permeability hazard

- High sodium in irrigation water results in severe soil permeability problem.
- Also affected by to carbonate and bicarbonate content in irrigation water.
- Expressed in me L^{-1}.

$$\text{Permeability Index (PI)} = \frac{\sqrt{\text{Na+HCO}_3}}{\text{Ca+Mg+Na}} \times 100$$

6.2.5. Specific ion toxicity

- Ground water having toxic ions such as B, Cl, F, NO_3, Se.
- Also become problematic for irrigating crops and have consequence of entering human food chain.

1. Degree of problems in relation to concentration of ions

Specification	Degree of problem		
	No problem	Increasing problem	Severe problem
Sodium (adj-SAR)	<3	3.0-9.0	>9.0
Chloride (mg L^{-1})	<4	4.0-10.0	>10.0
Boron (mg L^{-1})	<0.75	0.75-2.0	>2.0
NO_3-M (mg L^{-1})	<5.0	5.0-30.0	>30.0
HCO_3-N (mg L^{-1})	<1.5	1.5-8.5	>8.5
Fluoride (mg L^{-1})	<1.0	1.0-15.0	>15.0

6.3. Management of poor quality water

6.3.1. Management of saline water for irrigation

- Application of organic manures.
- Application of gypsum to the irrigation water to increase the proportion of calcium.
- Combined use of poor quality water with good quality water.
- Raising green manure crops and *in situ* incorporation.
- Scheduling irrigation with small quantity of water at more frequent intervals.
- Optimum use of manures and fertilizers.
- Providing better drainage facilities.
- Adopting ridges and furrows and drip irrigation.
- Deep ploughing to break the impervious layer.
- Mulching with locally available plant materials.
- All soil management practices that improve the infiltration rate and maintain favourable soil structure to reduce salinity hazard.
- Sludge sewage water management.
- Besides salinity and alkalinity hazard of water, some industrial effluents and sewage water are also problem water that can be reused by proper treatment.
- **Sewage:** Mixture of both liquid and disintegrated solid waste without giving any treatment.
- **Effluent:** Filtered liquid portion of the sewage.
- **Sludge:** Anaerobically decomposed organic matter present in sewage water and used as a source of organic manure.

UNIT-IV

CROP MANAGEMENT AND ALLIED AGRICULTURAL ACTIVITIES

Part-1 Crop Management

1

Cropping System and Integrated Farming System

1.1. Cropping pattern

- Yearly sequence and spatial arrangement of crops or crops and fallow on a given area.

1.2. Cropping seasons in India

1.2.1. Kharif season: Southwest monsoon

- Crops - Rice, cotton, jute, jowar (sorghum), bajra (pearl millet) and redgram (Pigeon pea).

1.2.2. Rabi season

- Onset of winter in October-November ends in March-April.
- Crops - Wheat, barley, oats, bengalgram (chickpea), mustard, sorghum, linseed, lentil and peas.

1.2.3. Zaid season

- Short duration summer cropping.
- Crops - Watermelon, muskmelon, cumumber, vegetables and fodder crops.

1.3. Cropping system

- **System**-set of components that are interrelated and interact among themselves.
- **Cropping system**-set of crop systems, making up the cropping activities of a farm system.
- Comprises all components required for the production of a particular crop and their interrelationships with environment.
- Combination of crops in time when crops occupy different growing period and space.

- Combinations in space occur when crops are inter planted.
- Combination of crops within a given year.
- Cropping patterns used on a farm and their interaction with farm resources, other farm enterprises, and available technology which determine their make up.

1.3.1. Types of cropping system in India

1. Mono cropping or monoculture

- Only one crop is grown on farm land year after year.

2. Intensive cropping

- Period between one crop and another is minimised through modified land preparation, when the resources are available in plenty.
- Cropping intensity is higher.
- Crop intensification techniques are intercropping, relay cropping, sequential cropping, ratoon cropping which comes under multiple cropping.
- **Need for intensive cropping**
  - For efficient use of available natural resources.
  - To increase productivity per unit area, unit time and unit resource.
  - To provide enough food for the family, fodder for cattle and generate sufficient cash income for domestic and cultivation expenses.
- **Intensive cropping -** Growing number of crops on the same piece of land during the given period of time.
- **Cropping intensity -** In Punjab and Tamil Nadu, the cropping intensity is more than 100%. In Rajasthan, the cropping intensity is less.

3. Multiple cropping

- Intensification of cropping in time and space dimensions.
- Growing two or more crops on the same field in a year.

Forms of multiple cropping

- **Intercropping -** Growing two or more crops simultaneously on the same field.
- **Mixed intercropping -** Growing two or more crops simultaneously with no distinct row arrangement. Sorghum, pearl millet and cowpea are mixed and broadcasted in rainfed conditions.

- **Row intercropping** - Growing two or more crops simultaneously where one or more crops are planted in rows. Maize + greengram (1:1), maize + blackgram (1:1), groundnut + redgram (6:1).
- **Strip intercropping** - Growing two or more crops simultaneously in strips wide enough to permit independent cultivation, but narrow enough for the crops to interact agronomically. Ex. Groundnut + redgram (6:4 strip).
- **Relay intercropping** - Growing two or more crops simultaneously during the part of the life cycle of each. Second crop is planted after the first crop at reproductive stage of growth, but before it is ready for harvest. Rice-rice fallow pulse, rice-rice fallow cotton or gingelly.

4. Sequential cropping

- Growing two or more crops in sequence on the same field in a farming year.
- Succeeding crop is planted after the preceding crop has been harvested.
- There is no intercrop competition.
- **Double, triple and quadruple cropping**: Growing two, three and four crops, respectively, on the same land in a year in sequence.
- Eg. **Double cropping:** rice: cotton; **triple cropping:** rice: rice: pulses; **quadruple cropping:** tomato: ridge gourd: *Amaranthus* greens: baby corn
- **Ratoon cropping**: Cultivation of crop regrowth after harvest, although not necessarily for grain.
- Ex. sugarcane: ratoon; sorghum: ratoon (for fodder).

1.4. Integrated farming system (IFS)

- Integration of two or more appropriate combination of enterprises like crop, dairy, piggery, fishery, poultry, bee keeping etc.,

1.4.1. Possible enterprises

Wetland based farming system

- Crop + fish + poultry/pigeon
- Crop + fish + mushroom

Gardenland based farming system

- Crop + dairy + biogas
- Crop + dairy + biogas + sericulture
- Crop + dairy + biogas + mushroom + sylviculture

Dry land based farming system

- Crop + goat + agroforestry
- Crop + goat + agroforestry + horticulture

1.4.2. Benefits of IFS

- Higher productivity.
- Profitability
- Sustainability
- Balanced food
- Recycling reduces pollution.
- Money round the year.
- Employment generation.
- Increase input efficiency.
- Increased standard of living of the farmer.
- Better utilisation of land, labour, time and resources.

1.4.3. Components of IFS

1. Crops, livestock, birds and trees are major components of any IFS.
2. Crop: monocrop, mixed/intercrop, multi-tier crops of cereals, legumes (pulses), oilseeds, forage etc.
3. Livestock: milch cow, goat, sheep, poultry, bees.
4. Tree: timber, fuel, fodder and fruit trees.

1. Suitable grain crops for different soil types

Black soil

Cereals : Maize

Millets : Sorghum, bajra

Pulses : Greengram, blackgram, redgram, chickpea, soybean, horse gram

Oilseeds : Sunflower, safflower

FIbre : Cotton

Other crops : Coriander, chillies,

Red soil

Millets : Sorghum

Minor Millets : Ragi, tenai, samai, pani varagu, varagu

Pulses : Lab - lab, greengram, redgram, soybean, horse gram, cowpea

Oilseeds : Groundnut, castor, sesame

2. Suitable forage crops

Black soils

- Fodder sorghum, fodder pearl millet, fodder cowpea, desmanthus, rhodes grass, mayil kondai pul (*Chloris barbata*), *Eleusine sp.*, thomson grass.

Red soils

- Fodder sorghum, fodder pearl millet, neelakolukattai pillu (Blue buffel grass-*Dichanthium annulatum*), fodder finger millet, sanku pushpam (Conch flower creeper - *Clitoria ternatea*), fodder cowpea, muyal masal (*Stylosanthes hamata*), siratro (*Macroptilium atropurpureum*), marvel grass, spear grass, vettiver.

3. Suitable tree species

Red gravelly/sandy red loam soils

- Tamarind, *simarouba glauca*, vagai (Ladies tongue-*Albizia lebbeck*), arappu (*Albizia amara*), kodai vel *(Acacia tortilis),* maan kathu vel (*A. mellifera*), neem, *Hardwickia binata,* ber, Indian gooseberry, *Casuarina,* silk cotton etc.

Black soils

- *Karu vel(Acacia nilotica), A.tortilis, A.albida, Azadirachta indica, Albizia lebbeck, Holoptelia integrifolia, Hibiscus tilifolia, Gmelina arborea, Casuarina* equisetifolia, *Leucaena leucocephala* and *Adina cordifolia.*

4. Suitable livestock and birds

- Goat, sheep, white cattle, black cattle, pigeon, rabbit, quail and poultry.

1.4.4. IFS for various agro-climatic zones of Tamil Nadu

1. Western zone

Wetland

- Cropping (0.90 ha) + fishery (0.10 ha) + poultry (50 layers) + 5 kg oyster mushroom production /day.

- Rice-gingelly-maize and rice-soybean-sunflower in 0.90 ha + polyculture fish rearing (0.10 ha), pigeon (100 pairs) and 5kg mushroom production per day.
- Goat (20 female + one male) + fish (400 numbers of polyculture) + improved cropping system for wetlands.

Rainfed land

- Integration of grain crop cultivation + fodder production + silvipastoral system involving trees like subabul (*Leucaena leucocephala*), *Acacia* spp., and thornless *Prosopis* interplanted with *Cenchrus* grass and rearing of 20 female + one male of tellicherry goat.

2. North western zone

- Integration of crops with 3 milch cows, 6 layers of poultry in 0.80 ha land.
- Integration of cropping with 2 milch animals, 6 goats in 1.25 ha rainfed land, out of which 0.25 ha with mulberry cultivation for sericulture.

3. Hilly zone

- Integration of crop cultivation with 2 milch cows, 6 poultry layers and 9 broilers.

4. Cauvery delta zone

- Integration of rice based cropping with 2 milch cows.
- Crop cultivation along with goat rearing (6 nos.).
- Cropping with duck and fish rearing.

5. Southern zone

- Rice based cropping + fish rearing + poultry in one ha land area in Periyar-Vaigai command area.
- Milch cow + fish rearing + rice based cropping system in wetlands of Tirunelveli district.
- In rainfed black clay soil, cropping + fruit tree cultivation + goat rearing.

2

Organic Manures and Green Manures

- Word manure derived from French word 'Manoeuvrer, means to manipulate, to work, to produce crop.

2.1. Sources of organic wastes

- Cattle shed wastes-dung, urine and slurry from biogas plants.
- Human habitation wastes-night soil, human urine, town refuse, sewage, sludge and sullage.
- Poultry litter, droppings of sheep and goat.
- Slaughter house wastes-bone meal, meat meal, blood meal, horn and hoof meal, fish wastes.
- Byproducts of agroindustries-oil cakes, bagasse and press mud, fruit and vegetable processing wastes.
- Crop wastes-sugarcane trash, stubbles and straw.
- Water hyacinth (*Eichhornia crassipes*), weeds and tank silt.
- Green manure crops and green leaf manuring materials.

2.1.1. Nutrient content of organic manures

Bulky organic manures	N (%)	P_2O_5 (%)	K_2O (%)
Farmyard manure	0.5	0.2	0.5
Sheep and goat manure	3	1	2
Poultry manure	3.03	2.63	1.4

- Supply plant nutrients including micronutrients.
- Improve soil physical properties like structure, water holding capacity etc.,
- Increase the availability of nutrients.
- Carbon dioxide released during decomposition acts as a CO_2 fertilizer.
- Plant parasitic nematodes and fungi are controlled to some extent by altering the balance of microorganisms in the soil.

2.1.2. Concentrated organic manures

Oilcakes	Nutrient content (%)		
	N	P_2O_5	K_2O
Non edible oilcakes			
Castor cake	4.3	1.8	1.3
Cotton seed cake (undecorticated)	3.9	1.8	1.6
Karanj cake	3.9	0.9	1.2
Mahua cake	2.5	0.8	1.2
Safflower cake (undecorticated)	4.9	1.4	1.2
Edible oil-cakes			
Coconut cake	3	1.9	1.8
Cotton seed cake (decorticated)	6.4	2.9	2.2
Groundnut cake	7.3	1.5	1.3
Linseed cake	4.9	1.4	1.3
Niger cake	4.7	1.8	1.3
Rape seed cake	5.2	1.8	1.2
Safflower cake (decorticated)	7.9	2.2	1.9
Sesamum cake	6.2	2	1.2

2.1.3. Average nutrient content of animal based concentrated organic manures

Organic manures	Nutrient content (%)		
	N	P_2O_5	K_2O
Blood meal	12-Oct	2-Jan	1
Meat meal	10.5	2.5	0.5
Fish meal	10-Apr	9-Mar	0.3 - 1.5
Horn and hoof meal	13	-	-
Raw bone meal	4-Mar	20 - 25	-
Steamed bone meal	2-Jan	25 - 30	-

2.2. Green manures

- Growing the plants and *in situ* incorporation in the field.

2.2.1. Biomass production and N accumulation of green manure crops

Crop	Age (days)	Dry matter (t/ha)	N accumulated
Sesbania aculeata	60	23.2	133
Sunnhemp	60	30.6	134
Cowpea	60	23.2	74
Pillipesara	60	25	102
Cluster bean	50	3.2	91
Sesbania rostrata	50	5	96

2.2.2. Nutrient content of green manure crops

Plant	Scientific name	Nutrient content (%) on air dry basis		
		N	P_2O_5	K
Sunnhemp	*Crotalaria juncea*	2.3	0.5	1.8
Dhaincha	*Sesbania aculeata*	3.5	0.6	1.2
Sesbania	*Sesbania speciosa*	2.71	0.53	2.21

2.3. Green leaf manure

- Addition of green or plant tissues obtained from elsewhere *viz.*, trees, herbs, shrubs prunings and unwanted weeds.

2.3.1 Nutrient content of green leaf manure

Plant	Scientific name	Nutrient content (%)		
		N	P_2O_5	K
Gliricidia	*Gliricidia sepium*	2.76	0.28	4.6
Pungam	*Pongamia glabra*	3.31	0.44	2.39
Neem	*Azadirachta indica*	2.83	0.28	0.35
Gulmohur	*Delonix regia*	2.76	0.46	0.5
Peltophorum	*Peltophorum ferrugineum*	2.63	0.37	0.5
Weeds				
Parthenium	*Parthenium hysterophorus*	2.68	0.68	1.45
Water hyacinth	*Eichhornia crassipes*	3.01	0.9	0.15
Trianthema	*Trianthema portulacastrum*	2.64	0.43	1.3
Ipomoea	*Ipomoea*	2.01	0.33	0.4
Calotrophis	*Calotropis gigantea*	2.06	0.54	0.31
Cassia	*Cassia fistula*	1.6	0.24	1.2

2.3.2. Benefits

- Addition of organic matter.
- Adds macro, secondary and micro nutrients.
- Improves physical condition of the soil.
- Acts as a soil amendment to reclamation of problem soils.
- Acts as a cover or catch crop to prevent soil erosion, conserve moisture, prevent nutrient leaching.
- Leguminous crops fix the atmosphere N and improve N status of the soil.

3

Integrated Nutrient Management

3.1. Integrated nutrient management

- Maintenance of soil fertility and plant nutrient supply at an optimum level for sustaining the desired productivity through optimization of all possible sources of organic, inorganic and biological components in an integrated manner.

3.2. Concepts

- Regulated nutrient supply for optimum crop growth and higher productivity.
- Improvement and maintenance of soil fertility.
- Zero adverse impact on agro – ecosystem quality by balanced fertilization of organic manures, inorganic fertilizers and bioinoculants in an integrated manner.

3.3. Determinants

- Nutrient requirement of cropping system as a whole.
- Soil fertility status and special management needs to overcome soil problems, if any
- Local availability of nutrients resources (organic, inorganic and biological sources)
- Economic conditions of farmers and profitability of proposed INM option.
- Social acceptability.
- Ecological considerations.
- Impact on the environment.

3.4. Advantages

1. Enhances the availability of applied as well as native soil nutrients
2. Synchronizes the nutrient demand of the crop with nutrient supply from native and applied sources.

3. Provides balanced nutrition to crops and minimizes the antagonistic effects resulting from hidden deficiencies and nutrient imbalance.
4. Improves and sustains the physical, chemical and biological functioning of soil.
5. To protect ecosystem and environment, minimizes the deterioration of soil, water and ecosystem by promoting carbon sequestration, reducing nutrient losses to ground and surface water bodies and also to atmosphere.

3.5. Components

3.5.1. Soil source

- Mobilizing unavailable nutrients and to use appropriate crop varieties, cultural practices and cropping system.

3.5.2. Mineral fertilizer

- Super granules, coated urea, direct use of locally available rock phosphates in acid soils, single super phosphate (SSP), muriate of potash (MOP) and micronutrient fertilizers.

3.5.3. Organic sources

- By products of farming and allied industries. FYM, droppings, crop wastes, residues, sewage, sludge, industrial wastes.

3.5.4. Biological sources

- Microbial inoculants substitute 15 - 40 kg N/ha.

4

Physiological Disorders in Plants and Their Management

4.1. Physiological disorders in field crops

Identification of physiological disorders in agricultural crops

Crop	Malady	Corrective measure
Rice	Severe chlorosis of leaves	1%super phosphate and 0.5% ferrous sulphate
	Irregular flowering and chaffiness Multiple deficiency of nutrients	1% super phosphate and magnesium sulphate.
	Tip drying and marginal scorching and browning	1%super phosphate and 0.5% zinc sulphate.
Maize	Chlorosis	0.5% ferrous sulphate and 0.5% urea.
	White bud yellowing in the bud leaves only	0.5% zinc sulphate spray with 1% urea.
	Tip drying and marginal scorching Pinkish colouration of lower leaves	1% super phosphate and 0.5% zinc sulphate.
	Marginal scorching and yellowing. Irregular drying of tips and margins	0.5% ferrous sulphate and 1%urea. 25 kg of zinc sulphate / ha as soil application.
Sorghum	Chlorosis of younger leaves	0.5% ferrous sulphate with 0.5% urea and 0.5% ammonium sulphate
Cowpea	Water soaked necrotic spots on leaf surface. Root growth very much restricted in 10-12 days old seedling	Spray sulphate containing chemical and 0.1% zinc sulphate and 0.1% urea.
Groundnut	Chlorosis of terminal leaves	0.5% ferrous sulphate and urea 1%

4.2. Disorders associated with low temperature

4.2.1. Leaf chlorosis and frost banding

- Caused by a disruption of chloroplasts due to winter cold.
- Green chlorophyll pigments are often converted in to yellow pigment namely chlorophyllins.

- Leaf may appear with distinct bleached bands across the blade of young plants called frost banding.
- Eg: Sugarcane, wheat and barley.

4.2.2. Leaf necrosis and malformation

- Spring frost cupping, crinkling finishing and curling of leaves of apple trees and stone fruits.
- Distortion is caused by death of the developed tissues before the expansion of leaves.

4.2.3. Stem disorders

- Frost cracks develop when tree trunk or limps lost their heat too rapidly.
- Outer layer of bark and wood cool most rapidly and subjected to appreciable tension causing marked shrinkage and cracking following a sudden temperature drop.
- Affected timber is of poor quality.

4.3. Disorders associated with high temperature

4.3.1. Leaf scorch

- High temperature causes leaf scorch directly or indirectly by stimulating excessive evaporation and transpiration.
- Eg: Tip burn of potato.

4.3.2. Sunscald

- In leaf vegetable crops, when leaves on the top of the head are exposed to intense heat, water soaked lesions or blistered appearance occur.
- These irregular shaped areas become bleached and parched later.
- Eg: Lettuce, cabbage.

4.3.3. Water core

- In tomato, exposure to high temperature causes death of the outer cells of fruit skin.
- Subsequently, corky tissue occurs beneath the skin, with watery appearance of the flesh near the core of the fruits faster.
- Often light stress is coupled with heat stress.
- Eg: Sun scald of bean, sun burning of soybean and cowpea.

- In chrysanthemum, increase in light intensity affects flower bud formation.
- Reproduction phase does not commence and modified into leaf like bracts.

4.4. Physiological disorders caused by light stress

- Adverse light intensity causes impaired growth and reduced vigour.
- Subsequently, leaves gradually loose green colour, turning pale green to yellow, stems may dieback little every year.
- Insufficient light limits photosynthesis, causing food reserves to be depleted.

4.5. Hen and chicken disorder

- Caused by the deficiency of boron resulting into more number of undeveloped round or oblate shot berries with few berries attaining the normal shape characteristic of the variety in grapes.

5

Irrigation Management of Different Crops

5.1. Irrigation requirement of field crops

Crop	Duration (Days)	Total water requirement (mm)	Number of irrigations
Rice 110	125	18	
Sugarcane	360	2200	24
Groundnut	105	510	10
Sorghum	105	500	6
Maize	100	500	8
Finger millet	95	310	6
Cotton	165	600	11
Blackgram	65	280	4
Soybean	85	320	4
Sesame	85	150	-
Sunflower	110	450	-
Pearl millet	110	300-400	3

5.2. Water requirement of horticultural crops

Crop	Total water requirement (mm)
Tomato	600-800
Potato	500-700
Pea	350-500
Onion	350-550
Bean	300-500
Cabbage	380-500
Banana	1200-2200
Citrus	900-1200
Grapes	500-1200
Pineapple	700-1000
Coconut	80-100 (lit/plant/day)
Guava	22-30 (lit/plant/day)
Mango	30-40 (lit/plant/day)
Banana	8-12 (lit/plant/day)

5.3. Critical stages for irrigation

Crop	Critical stages / Sensitive stages
Rice	Panicle initiation, heading and flowering
Sorghum	Flowering and grain formation
Maize	Just prior to tasseling and grain filling
Pearl millet	Heading and flowering
Finger millet	Primordial initiation and flowering
Wheat	Crown root initiation, tillering and booting
Groundnut	Flowering, peg initiation and penetration and pod development
Sesame	Blooming to maturity
Sunflower	Two weeks before and after flowering
Soybean	Blooming and seed formation
Safflower	From rosette to flowering
Castor	Full growing period
Cotton	Flowering and boll formation
Sugarcane	Maximum vegetative stage
Tobacco	Immediately after transplanting
Onion	Bulb formation to maturity
Tomato	Flowering and fruit setting
Chilli	Flowering
Cabbage	Head formation to maturity
Alfalfa	Immediately after cutting for hay crop and flowering for seed crop
Beans	Flowering and pod setting
Peas	Flowering and pod formation
Coconut	Nursery stage and root enlargement
Potato	Tuber initiation and maturity
Banana	Throughout the growth
Citrus	Flowering, fruit setting and enlargement
Mango	Flowering and fruiting to maturity
Coffee	Flowering and fruit development
Pine apple	Vegetative growth
Grapes	Vegetative growth. Frequent irrigation during vegetative stage may cause rotting of fruits
Guava	Period of fruit growth
Ber	A drought resistant plant. Irrigation is required during fruit growth

5.4. Irrigation management in rice

- Total water requirement is 1100-1250 mm.

5.4.1. Systems of rice cultivation in Tamil Nadu

- Moist the soil in the early period of 10 days.
- Maximum depth of 2.5cm after development of hairline cracks.
- Increasing irrigation depth to 5.0cm after panicle initiation.

5.4.2. Transplanted puddled lowland rice/transplanted rice

1. Nursery

- Drain the water 18 to 24 hrs after sowing
- Care must be taken to avoid stagnation of water in any part of the seedbed.
- Allow enough water to saturate the soil from 3rd to 5th day. From 5th day onwards, increase the water depth to 1.5cm depending on the height of the seedlings.
- Thereafter, maintain 2.5cm depth of water.

2. Main field management

- Puddling and leveling minimizes the water requirement
- Plough with tractor drawn cage wheel to reduce percolation losses and to save water requirement up to 20%.
- Maintain 2.5cm water over the puddle and allow the green manure to decompose for a minimum of 7 days in the case of less fibrous plants like sunnhemp and 15 days for more fibrous green manure plants like kolinchi (*Tephrosia purpurea*).
- At the time of transplanting, a shallow depth of 2cm of water is adequate since high depth of water will lead to deep planting resulting in reduction of tillering.
- Maintain 2 cm of water up to seven days of transplanting.
- After the establishment stage, cyclic submergence of water is the best practice for rice crop. This cyclic 5cm submergence has to be continued throughout the crop period.

5.4.3. Wet seeded puddled lowland rice

- During first one week just wet the soil by thin film of water.
- Depth of irrigation may be increased to 2.5cm progressively along the crop age.
- Afterwards, follow the schedule as given to transplanted rice.

5.4.4. Dry seeded irrigated un-puddled lowland rice

- As that of irrigated rice when canal water is used for irrigation.
- Possibility of subsequent conversion towards deep water situation is seen in this tract, then variety should be specific for those areas.

5.5. Crop water requirement

- Water required by the plants for its survival, growth, development and to produce economic parts.
- The crop water requirement includes all losses like:
- Transpiration loss through leaves (T)
- Evaporation loss through soil surface in cropped area (E)
- Amount of water used by plants (WP) for its metabolic activities which is estimated as less than 1% of the total water absorption.
- ET loss is taken as crop water use or crop water consumptive use.
- Other application losses are conveyance loss, percolation loss, runoff loss, etc., (WL).
- The water required for special purposes (WSP) like puddling, ploughing, land preparation, leaching, requirement for weeding, for dissolving fertilizer and chemical, etc.

WR = T + E + WP + WL + WSP

Combined loss of evaporation and transpiration from a cropped field is termed as evapotranspiration.

ET is also known as consumptive use.

CU = E + T + WP

Therefore,

WR = CU + WL + WSP

1. Irrigation requirement

- Water requirement of crops exclusive of effective rainfall and contribution from soil profile.

IR - WR – (ER + S)

IR - Irrigation requirement

WR - Water requirement

ER - Effective rainfall

S - Soil moisture contribution

2. Net irrigation requirement

- Actual quantity of water required in terms of depth to bring the soil to field capacity level to meet the ET demand of the crop.
- Water applied by irrigation alone in terms of depth to bring the field to field capacity level.
- Difference between the F.C and the soil moisture content in the root zone before starting irrigation.

$$d = \sum_{i}^{n} \frac{Mfci - Mbi}{100} \times Ai \times Di$$

d = Net irrigation water to be applied (cm)

Mfci = FC in i^{th} layer (%)

Mbi = Moisture content before irrigation in i^{th} layer (%)

Ai = Bulk density (g/cc)

Di = Depth (cm)

n = Number of soil layer

3. Gross irrigation requirement

- Total quantity of water used for irrigation.
- Includes net irrigation requirement and losses in water application and other losses.

$$\text{Gross irrigation} = \frac{\text{Net irrigation requirement}}{\text{Field efficiency of system}} \times 100$$

4. Irrigation frequency

- Interval between two consecutive irrigations during crop periods.
- Number of days between irrigation during crop periods without rainfall.
- Irrigation should be given at about 50 per cent not over 60 per cent depletion of the available moisture from the effective root zone in which most of the roots are concentrated.

$$\text{Design frequency (days)} = \frac{\text{F C - moisture content of the root zone prior to starting irrigation}}{\text{Peak period consumptive use rate of crop}}$$

5. Irrigation period

- Number of days that can be allowed for applying one irrigation to a given design area during peak consumptive use period of the crop

$$\text{Irrigation period} = \frac{\text{Net amount of moisture in soil at start of irrigation (FC-PWP)}}{\text{Peak period consumptive use of the crop}}$$

- At critical stages, favorable water level should be ensured through timely irrigations.

Part-2 Agricultural Microbiology

1

Recycling of Agricultural Wastes

1.1. FYM compost

- Decomposed mixture of dung and urine of farm animals along with litter and left over materials from roughages or fodder fed to the cattle.
- Farmyard manure contains 0.5% N, 0.2% P_2O_5 and 0.5% K_2O.
- Urine contains one per cent nitrogen and 1.35% potassium.
- 10 to 20 t/ha is applied, but more than 20 t/ha is applied to fodder grasses and vegetables like potato, tomato, sweet potato, carrot, radish, onion etc., respond well to the farmyard manure.
- Crops like sugarcane, rice, oranges, banana, mango and coconut respond good to FYM.
- About 30% of nitrogen, 60 to 70% of phosphorus and 70 per cent of potassium are available to the first crop.
- Compost is a rich source of organic matter which improves physico-chemical and biological properties of soil and fertility.

1.1.1. Composting methods

- Natural process of rotting or decomposition of organic matter by microorganisms under controlled conditions.
- Raw organic materials - crop residues, animal wastes, food garbage, municipal wastes and suitable industrial wastes.
- **Farm compost:** refers to compost made from farm wastes like sugarcane trash, paddy straw, weeds and other plants/wastes.
- Farm compost contains 0.5% N, 0.15% P_2O_5 and 0.5% K_2O.
- Composting increased by application of superphosphate or rock phosphate at 10 to 15 kg/ton of raw materials at the initial stage of filling.
- **Town compost:** compost made from town refuse like night soil, street sweepings and dustbin refuse. It contains 1.4 per cent N, 1.0 per cent P_2O_5 and 1.4 per cent K_2O.

1. Coimbatore method

- A layer of wastes is first laid in the pit.
- Moistened with 5-10 kg cow dung dissolved in 2.5 to 5.0 litres of water and sprinkle 0.5 to 1.0 kg fine bone meal.
- Layers are laid one over the other till the material rises 0.75 m above the ground level.
- Plaster with wet mud and left undisturbed for 8 to 10 weeks.
- Plaster is then removed, moistened with water, give a turning and made into a rectangular heap under a shade.

2. Indore method

- Organic wastes are spread in the cattle shed to serve as bedding.
- Urine soaked material along with dung is removed every day and formed into a layer of about 15 cm thick at suitable sites.
- Urine soaked earth scraped from cattle sheds is mixed with water and sprinkled over the layer of wastes twice or thrice a day.
- A thin layer of well decomposed compost (acts as a inoculum) is sprinkled over top and heap is turned and reformed.
- Heap is left undisturbed for about a month.
- Moistened thoroughly and give turning.
- Compost is ready in another one month.

3. Bangalore method

- Dry waste materials of 25 cm thick are spread in a pit and a thick suspension of cow dung slurry is sprinkled over for moistening.
- A thin layer of dry waste is laid over the moistened layer.
- Pit is filled alternately with dry layers of material and cow dung suspension till it rises 0.5 m above ground level.
- Left exposed without covering for 15 days.
- Give a turning, plastered with wet mud and left undisturbed for about 5 months or till required.
- In Coimbatore method, there is anaerobic decomposition to start with, followed by aerobic fermentation.
- It is the reverse in Bangalore method.
- Bangalore compost is not so thoroughly decomposed as the Indore compost.

1.1.2. Nutritive value of animal solid and liquid excreta

Animal	Dung (mg/g)			Urine (%)		
	N	P	K	N	P	K
Cattle	20-45	4-10	7-25	1.21	0.01	1.35
Sheep and goat	20-45	4-11	20-29	1.47	0.05	1.96
Pig	20-45	6-12	15-48	0.38	0.10	0.99
Poultry	28-62	9-26	8-29	-	-	-

1.1.3. Enrichment of FYM

- P content of FYM is low (0.4-0.8%).
- Addition of superphosphate is recommended to P insufficiency.
- This process is called **reinforcing** or **enriching** and resultant material is called enriched farmyard manure.
- Single super phosphate and/or rock phosphate is sprinkled either in the cattle shed or on the manure heap.
- Application of 1 kg superphosphate, bone meal or phosphate rock is applied over each layer of dung to increase the P_2O_5 content.
- **Use of animal bones:** Brokened bones are boiled with wood ash leachate or lime, decanted and residues are applied to the pits.
- Wood ash waste can also be added to increase the K content of compost.
- Addition of N-fixing and P-solubilizing cultures
- Inoculation of N fixers like *Azotobacter*, *Azospirillum lipoferum*, and *Azospirillum brasilense* and P-solubilizers like *Bacillus megaterium* or *Pseudomonas* sp.
- This will increase N content of straw compost by 2 percent.

1.2. Coirpith compost

- By products of coconut is coconut husk from which coir fibre is extracted.
- Extraction process generates a large quantity of dusty material called coir dust or coir pith.
- 7.5 million tons of coir wastes are generated from coir industry in India and 5 lakh tons in Tamil Nadu.
- Has a wider carbon and nitrogen ratio and lower biodegradability due to high lignin content
- Composting reduces its bulkiness and converts plant nutrients to the available form.

1.2.1. Coir pith composting technology

1. Collection of raw material

- Collected from the coir industry without any fibre.
- Fibrous materials are removed by sieving which hinders the composting process.

2. Site selection

- Elevated and shady areas are better.
- Shady area conserves the moisture in the composting material.
- Levelled earthen floor is made to hard by hard pressing and also by applying cow dung slurry.
- Presence of roof over the composting material is advantageous, since it protects the material from rain and severe sunshine.
- Coir pith is an aerobic composting.
- Coir pith should be spread to the length of 4 feet and breadth of 3 feet.
- Put coir pith upto 3 inch height and moistened.
- Apply 5 kg urea for one ton of coir pith at 1 kg per layer of material.
- Alternatively, apply 200 kg of fresh poultry litter for one ton of coir pith.
- After adding the nitrogen source, the microbial inoculums such as *Pleurotus* and TNAU biomineralizer (2%) are added over the material.
- Heap upto 4 feet height to retain the temperature generated during composting.

3. Turning of material

- Turn once in 10 days to allow the stale air trapped inside the compost material to go out and fresh air will get in.
- Organism performing the composting requires oxygen for its metabolic activity.
- Alternate method is to insert perforated PVC pipes both vertically and horizontally.

4. Moisture maintenance

- Maintain 60 % moisture.

5. Compost maturity

- Sixty days (60 days).
- Volume reduction of waste material.

- When the waste material is composted, the compost heap height will be reduced by 30%.
- Waste materials are turned to black in colour.
- Waste particle size is reduced.
- Composted material emits earthy odour.
- Well composted coir pith has C:N ratio of 20:1, less oxygen uptake, less number of microorganism, more amount of available nutrients and high cation exchange capacity.

6. Nutritive value of raw and composted coir pith

Parameters	Raw coir pith	Composted coir pith
Lignin (%)	30.00	4.80
Cellulose (%)	26.52	10.10
Carbon (%)	26.00	24.00
Nitrogen (%)	0.26	1.24
Phosphorous (%)	0.01	0.06
Potassium (%)	0.78	1.20
Calcium (%)	0.40	0.50
Magnesium (%)	0.36	0.48
Iron (ppm) (%)	0.07	0.09
Manganese(ppm)	12.50	25.00
Zinc(ppm)	7.50	15.80
Copper(ppm)	3.10	6.20
C:N ratio	112.1	24:1

7. Benefits of composted coir pith

- Improves soil texture, structure and tilth.
- Sandy soil becomes more compact and clayey soil becomes more arable.
- Improves the soil aggregation.
- Improves the water holding capacity (more than 5 times of its dry weight).
- Reduces bulk density soil.
- Improvement in cation exchange capacity of soils.
- Increase in ammonification, nitrification and nitrogen fixation due to improved microbiological activity.

8. Application of coir pith compost

- 5 tons/ha as basal.
- For nursery potting mixture, 20% of composted coir pith can be mixed with the soil and sand before filling it in the polybag.
- Established trees like coconut, mango, banana and other fruits: apply 5 kg composted coir pith per tree.

9. Limitation in using composted coir pith

- Not economical
- Ensure that the material is composted completely and check for quality analysis certificate.
- If immature compost is applied to the soil, it will undergo decomposition inside the soil by taking nutrients from the soil.
- Standing crop is affected.

1.3. Sugarcane trash composting

- Sugarcane produces about 10 to 12 tonnes of dry leaves per hectare per crop.
- Detrashing is done on 5th and 7th month.
- Composted by using the fungi like *Trichurus, Aspergillus, Penicillium* and *Trichoderma.*
- Addition of rock phosphate and gypsum facilitates for quicker decomposition.

1.3.1. Collection of trash

- Detrashed material pooled together and transported to the compost yard.
- Shred the waste into small particles using shredder or chop cutters.
- Reduces the volume of material and increases the surface area of the waste.

1.3.2. Inputs for composting

1. Microbial consortium

- TNAU biomineralizer at 2 kg/ton of trash.

2. Animal dung

- Inoculate with fresh dung at 50 kg/ton of trash in 100 litres of water.
- Apply rock phosphate at 5 kg/ton of waste to increase the phosphorus content.

1.3.3. Making heap formation

- Heap should be formed with a minimum height of 4 feet.

1.3.4. Turning the compost material

- Turned periodically once in 15days.

1.3.5. Moisture content

- 60% moisture should be maintained.

1.3.6. Compost maturity

- Volume reduction, earthy odour, brownish black colour and reduction in particle size are maturity parameters.

1.3.7. Nutritive value of sugarcane trash compost

- 0.5% nitrogen, 0.2% phosphorus and 1.1% potassium, 28.6% organic carbon

1.3.8. Compost application

- 5 tons per hectare as basal application.

2.4. Parthenium compost

- Collect the parthenium biomass before flowering for making compost either by a farmer named Narayan Deotao Pandharipande (NADEP) from Maharashtra or open pit method.
- Pit size: 3x 6x10 feet (depth x width x length).
- Cover the surface and sidewalls of the pit with stone chips.
- 100 kg dung, 10 kg urea or rock phosphate, soil (100-200 kg) and one drum of water near the pit.
- Spread about 50 kg of parthenium on the surface of pit.
- Over this sprinkle 500 gm urea or 3 kg rock phosphate.
- Add *Trichoderma viride or Trichoderma harzianum* (kind of fungi cultured powder) at 50 gm per layer.
- Make several layers till the pit is filled upto 1 feet high from the ground surface.
- If there is no soil with parthenium roots, then add 10-12 kg of loamy soil on each layer.
- After 4-5 months, well decomposed compost is ready.

- 37–45% of compost from 3700-4200 kgs of parthenium biomass.
- Parthenium compost contains 1.05% N, 0.84% P, 1.11% K, 0.9% Ca and 0.55% Mg.

1.5. Composting weeds

- Composting weeds are parthenium, water hyacinth (*Eichhornia crassipes*), cyperus (*Cyperus rotundus*) and cynodon (*Cynodon dactylon*).

1.5.1. Materials required

- 250 g of *Trichoderma viride* and *Pleurotus sajor-caju* consortia, and 5 kg of urea.
- An elevated shaded place is selected or a thatched shed is erected.
- An area of 500 × 150 cm is marked out.
- Spread 100 kg of cut material (10-15 cm in size) over the marked area.
- About 50 g of microbial consortia is sprinkled over this layer.
- Then, spread 100 kg of weeds on this layer.
- One kilogram of urea is sprinkled uniformly over the layer.
- Repeat this until the level rises to 1 m above the ground surface.
- Water is sprinkled as necessary to maintain a moisture level of 50-60%.
- Thereafter, the surface of the heap is covered with a thin layer of soil.
- Do thorough turning on the twenty-first day.
- Compost is ready in about 40 days.

1.6. Farm waste composting

- In Tamil Nadu, 190 lakh tons of crop residues are available.
- Harvest refuses include straws, stubble, stover and haulms of different crops.
- Processing wastes include groundnut shell, oil cakes, rice husks and cobs of maize, sorghum and cumbu.
- Field residues like sorghum, maize, soybean, cotton, sugarcane etc.

1.6.1. Collection and shredding of waste materials

- Collect all crop wastes at compost yard.
- Shred all the crop residues to 2 to 2.5 cm size using shredder machine.

1.6.2. Mixing of green waste and brown waste

- Narrow C: N ratio of 30:1 is ideal for composting.
- To get a narrow C: N ratio, carbon and nitrogen rich materials should be mixed together.
- Green waste materials like glyricidia leaves, parthenium, freshly harvested weeds, sesbania leaves are rich in nitrogen.
- Whereas, brown waste materials like straw, coir dust, dried leaves and dried grasses are rich in carbon.
- Alternate layers of carbon and nitrogen rich material with intermittent layers of animal dung are essential.

1.6.3. Compost heap formation

- Minimum 4 feet height should be maintained for composting.
- Composting area should be elevated one and have sufficient shade.
- While heap formation, all the crop residues should be mixed together to form a heterogeneous material rather than a single homogenous material.
- After heap formation, the material should be thoroughly moistened.

1.6.4. Bioinputs for composting

- Two kg TNAU biomineralizer should be mixed with 20 litres of water and the slurry is added for one ton of wastes.
- Alternatively, 40 kg fresh cow dung is mixed with 100 litres of water and added for a ton of waste.
- Cow dung slurry acts as nitrogen source as well as source of microbial inoculum.

1.6.5. Moisture maintenance

- 60% moisture should be maintained.

1.6.6. Compost maturity

- Volume reduction, black colour, earthy odour, reduction in particle size is all the physical factors to be observed for compost maturity.
- After maturity, compost heap can be disturbed and spread on the floor for curing for one day.
- Composted material is sieved through 4 mm sieve to get uniform composted material.

1.6.7. Compost enrichment

- Beneficial microorganisms like *Azotobacter* or *Azospirillum, Pseudo monas,* Phosphobacteria (0.2%) and rock phosphate (2%) have to be inoculated for one ton of compost having 40% moisture and kept for 20 days incubation.

1.6.8. Nutrient content of biocompost prepared from different crop residues

Biocompost	Nutrient content (%)		
	Nitrogen	Phosphorous	Potash
Animal refuse			
Cattle dung	0.3 - 0.4	0.1 – 0.2	0.1 – 0.3
Horse dung	0.4 – 0.5	0.3 – 0.4	0.3 – 0.4
Sheep dung	0.5 – 0.7	0.4 – 0.6	0.3 - 0.1
Night soil	1.0 – 1.6	0.8 – 1.2	0.2 – 0.6
Poultry manure	1.8 – 2.2	1.4 – 1.8	0.8 – 0.9
Sewage sludge	2.0 – 3.5	—	—
Cattle urine	0.9 – 1.2	Trace	0.5 – 1.0
Horse urine	1.2 – 1.5	Trace	1.3 – 1.5
Sheep urine	1.5 – 1.7	0.1 – 0.2	0.1 – 0.3
Wood ash			
Ash coal	0.73	0.45	0.53
Ash wood	0.1 – 0.2	0.8 – 5.9	1.5 – 3.6
Habitation waste and factory waste			
Rural compost	0.5 – 1.0	0.4 – 0.8	0.8 – 1.2
Urban compost	0.7 – 2.0	0.9 – 3.0	0.3 – 1.9
Farmyard manure	0.4 – 1.5	0.3 – 0.9	2.0 – 7.0
Straw and stalk			
Pearl millet	0.65	0.75	2.50
Cotton	0.44	0.10	0.66
Banana pseudo stem	0.61	0.12	1.00
Sorghum	0.40	0.23	2.17
Maize	0.42	1.57	1.65
Paddy straw	0.36	0.08	0.71
Tobacco	1.12	0.84	0.80
Pigeon pea	1.10	0.58	1.28
Sugarcane trash	0.53	0.10	1.10
Wheat	0.53	0.10	1.10
Tobacco dust	1.10	0.31	0.93

1.6.9. Benefits of biocompost

- Contains both nutrients and beneficial microorganisms.
- Improves the physical, chemical and biological properties of the soil due to regular addition of biocompost.

1.6.10. Compost application

- Maintains soil health and increasing soil organic carbon content.
- For one hectare of land, 5 tons of enriched biocompost is recommended as basal application.

1.7. Press mud compost

- Mix thoroughly the sugar press mud (87.8%), carbon materials (9.5%) such as grass powder, straw powder, germ bran, wheat bran, chaff, sawdust etc., molasses (0.5%), single super phosphate (2.0%), sulfur mud (0.2%), pile the mixed material for 20m length above ground level, 2.3-2.5m in width and 5.6m high in semicircle shape
- Digestion is completed in 14-21 days.
- During digestion, mixture is mixed, turned and watered after every three days to maintain 50-60 % moisture.
- Matured compost is in loose shape, grey colour and no odour.
- **Granulation unit**: Add molasses (0.5% of total raw material) and water to the compost mixture and then sent to granulation unit – new type organic fertilizer granulator.
- Rotary drum dryer is used to form granules at a temperature of 24-25°C and to reduce the moisture content to 10%.
- **Screening unit**: After granulation of compost, it is sent to screening unit (rotary drum screen machine) to get 5mm diameter granules.

1.7.1. Features and functions

- High disease resistance as it contains antibiotics, hormones and other metabolites.
- High fertilizer efficiency.
- Culturing soil fertility and improving the soil.
- Improving crop yield and quality.
- Wide application in agriculture.

2

Vermicomposting

- Vermicomposting is the process of turning organic debris into worm castings by earthworms.
- Earthworms feed on decaying material, pass through alimentary canals and excreted as castings.
- pH of the castings is 7 (neutral) and the castings are odourless.
- Contain high amounts of nitrogen, potassium, phosphorus, calcium, and magnesium.
- Contains 5 times the available nitrogen, 7 times the available potash, and 1 ½ times more calcium and rich in humus.
- Improves aeration, porosity, structure, drainage, and moisture holding capacity.

2.1. Method of vermicomposting

2.1.1. Biodegradable wastes

- Crop residues
- Weed biomass
- Vegetable wastes
- Leaf litter
- Hotel refuse
- Waste from agro industries
- Biodegradable portion of urban and rural wastes.

2.1.2. Phase of vermicomposting

Phase 1 : Collection of wastes, shredding, mechanical separation of the metal, glass and ceramics and storage of organic wastes.

Phase 2 : Predigestion of organic waste for 20 days by heaping the material along with dried cattle dung slurry. This process partially digests the material and fit for earthworm consumption. Wet dung should not be used.

Phase 3 : Preparation of earthworm bed with concrete base. Loose soil allows the worms to go into soil and also leaching of nutrients.

Phase 4 : Collection of earthworm after vermicompost collection. Sieving the composted material to separate fully composted material. The partially composted material will be again put into vermicompost bed.

Phase 5 : Storing the vermicompost in proper place to maintain moisture and allow the beneficial microorganisms to grow.

2.1.3. The five essentials of compost worms

1. An hospitable living environment usually called bedding.
2. A food source
3. Adequate moisture (greater than 50% water content by weight)
4. Adequate aeration
5. Protection from temperature extremes

- **Bedding:** Any material that provides the worms with a relatively stable habitat.
- **High absorbency:** Materials should retain moisture fairly well for earthworm survival. If a worm's skin dries out, it dies.
- **Good bulking potential**.
- **Low protein and/or nitrogen content (high carbon: nitrogen ratio).**

2.1.4. Vermicompost production methodology

1. Selection of suitable earthworm

- Surface dwelling earthworm alone should be used.
- African earthworm (*Eudrilus engeniae),* red worms (*Eisenia foetida)* and composting worm (*Perionyx excavatus)*.
- African worm (*Eudrilus eugeniae)* is preferred over other two types, because it produces higher production of vermicompost in short period of time and more young ones in the composting period.

2. Selection of site for vermicompost production

- Any place with shade, high humidity and cool.
- Thatched roof may be provided to protect the process from direct sunlight and rain.

3. Containers for vermicompost production

- A cement tub may be constructed to a height of 2½ feet and a breadth of 3 feet.
- Also be prepared in wooden boxes, plastic buckets or in any containers with a drain hole at the bottom.

4. Vermiculture bed

- Placing saw dust or husk or coir waste or sugarcane trash in the bottom of tub / container.
- A layer of fine sand (3 cm) should be spread over the culture bed followed by a layer of garden soil (3 cm).
- **Common bedding materials**

Bedding Material	Absorbency	Bulking Pot.	C:N Ratio
Horse Manure	Medium-Good	Good	22 - 56
Peat Moss	Good	Medium	58
Corn Silage	Medium-Good	Medium	38 - 43
Hay – general	Poor	Medium	15 - 32
Straw – general	Poor	Medium-Good	48 - 150
Straw – oat	Poor	Medium	48 - 98
Straw – wheat	Poor	Medium-Good	100 - 150
Paper from municipal waste stream	Medium-Good	Medium	127 - 178
Newspaper	Good	Medium	170
Bark – hardwoods	Poor	Good	116 - 436
Bark — softwoods	Poor	Good	131 - 1285
Corrugated cardboard	Good	Medium	563
Lumber mill waste — chipped	Poor	Good	170
Paper fibre sludge	Medium-Good	Medium	250
Paper mill sludge	Good	Medium	54
Sawdust	Poor-Medium	Poor-Medium	142 - 750
Shrub trimmings	Poor	Good	53
Hardwood chips, shavings	Poor	Good	451 - 819
Softwood chips, shavings	Poor	Good	212 - 1313
Leaves (dry, loose)	Poor-Medium	Poor-Medium	40 - 80
Corn stalks	Poor	Good	60 - 73
Corn cobs	Poor-Medium	Good	56 - 123

5. Worm food

- Able to consume in excess of their body weight each day.
- General rule-of-thumb is ½ of their body weight per day.
- Manures are the most commonly used worm feedstock, with dairy and beef manures generally considered the best natural food for *Eisenia*.

6. Selection for vermicompost production

- Cattle dung (except pig, poultry and goat), farm wastes, crop residues, vegetable market waste, flower market wastes, agroindustrial wastes, fruit market wastes and all other bio degradable wastes are suitable for vermicompost production.

7. Putting the waste in the container

- Predigested waste material should be mud with 30% cattle dung either by weight or volume.
- Moisture level should be maintained at 60%.
- For one metre length, one metre breadth and 0.5 metre height, 1 kg of worm (1000 Nos.) is required.

8. Harvesting vermicompost

- Castings formed on the top layer are collected periodically once in a week.

9. Harvesting earthworm

- In the vermibed, before harvesting the compost, small, fresh cow dung balls are made and inserted inside the bed in five or six places. After 24 hours, the cow dung ball is removed.
- All the worms will be adhered into the ball. Putting the cow dung ball in a bucket of water will separate this adhered worm.

Manual method

- Used by hobbyists and smaller-scale growers, particularly those who sell worms to the home-vermicomposting or bait market.
- Involves hand-sorting or picking the worms directly from the compost by hand.

Self-harvesting (Migration) methods

- Based on the worms tendency to migrate to new regions, either to find new food or to avoid undesirable conditions, such as dryness or light.

Screen or onion bag method

- Screen bottom box having a mesh of ¼ (1.8" also used) is used.
- Two approaches namely down ward and upward migration.
- In downward migration, worms migrate downward due to strong light into a container filled with moist peat mass and they are removed.

- In upward migration, box with mesh bottom or onion hag filled with damp peat mass sprinkled with food attractive like chicken mass, coffee grounds or fresh cattle manure and kept on the vermibed without disturbing it.

10. Nutritive value of vermicompost

Organic carbon	: 9.5 – 17.98%
Nitrogen	: 0.5 – 1.50%
Phosphorous	: 0.1 – 0.30%
Potassium	: 0.15 – 0.56%
Sodium	: 0.06 – 0.30%
Calcium and magnesium	: 22.67 to 47.60 meq/100g
Copper	: 2 – 9.50 mg kg^{-1}
Iron	: 2 – 9.30 mg kg^{-1}
Zinc	: 5.70 – 11.50 mg kg^{-1}
Sulphur	: 128 – 548 mg kg^{-1}

11. Storing and packing of vermicompost

- Stored in dark and cool place.
- Minimum 40% moisture.
- Vermicompost can be stored in laminated bag for one year without loss of its quality.
- If stored in open, sprinkle water to maintain moisture.

2.2. Advantages of vermicompost

- Rich in all essential plant nutrients.
- Provides excellent effect on overall plant growth, encourages the growth of new shoots / leaves and improves the quality and shelf life of the produce.
- Free flowing, easy to apply, handle and store, and does not have bad odour.
- Improves soil structure, texture, aeration, and water holding capacity and prevents soil erosion.
- Rich in beneficial microflora such as a N fixers, P solubilizers, cellulose decomposing microflora etc.
- Contains earthworm cocoons and increases the population and activity of earthworm in the soil.
- Neutralizes the soil reactions.

- Prevents nutrient losses and increases the use efficiency of chemical fertilizers.
- Free from pathogens, toxic elements, weed seeds etc.
- Minimizes the incidence of pest and diseases.
- Enhances the decomposition of organic matter in soil.
- Contains valuable vitamins, enzymes and hormones like auxins, gibberellins etc.

2.3. Pests and diseases of vermicompost

- Compost worms are not subjected to disease.
- But they are subjected to predation by certain animals and insects (red mites are the worst).
- Disease known as sour crop is caused by environmental conditions.

2.4. Types of earthworms

1. Epigeic

- Surface living worms.
- Reddish brown in colour.
- Enhance the rate of organic manure production through biodegradation or mineralization.
- Eg. *Lampito mauritii, Octochaetona serrata, Perionyx excavatus, Dendrobaena octaedra, D.attemsi, Dendrodrilus rubidus.*

2. Endogeic

- Burrowing worms.
- Burrow and mix the soil from different horizons in the profile.
- Eg. *Allolobophora chlorotica, Aporrectodea caliginosa, A.rosea, Murchieona muldali, Octolasion cyaneum, O.lacteum.*

2.5. Types of vermibed

1. Bed method

- Pucca / kachcha floor by making bed size of 6x2x2 feet.

2. Pit method

- Cemented pits of size 5x5x3 feet covered with thatched roof.
- Not preferred due to poor aeration, water logging at bottom, and more cost of production.

2.6. Dosages for different crops

- Field crops: 5-6 t/ha
- Vegetables: 10-12 t/ha
- Flower plants: 100-200 g/sq ft
- Fruit trees: 5-10 kg/tree

2.7. Favourable conditions for vermicomposting

- pH: 6.5 to7.5
- Moisture: 60-70% moisture. Below and above this range mortality of worms taking place
- Aeration: 50% aeration from the total pore space
- Temperature: 18°C to 35°C.

2.8. Transport of live worms

- Live earthworm-s is packed with moist feed substrate (Roughly 0.5-1.5 g/ individual for 24 hours of transportation journey) in a cardboard/plastic container with aeration. .
- If worms do not acclimatize to new environment, cocoons are helpful for population build up.

3

Biofertilizers

- Preparations containing living cells or latent cells of efficient strains of microorganisms that help crop plants to uptake of nutrients by their interactions in the rhizosphere when applied through seed or soil.
- Accelerates microbial processes in the soil.

S.No.	Groups	Examples
N_2 fixing biofertilizers		
1.	Free-living	*Azotobacter, Beijerinckia, Clostridium, Anabaena, Nostoc*
2.	Symbiotic	*Rhizobium, Frankia, Anabaena azollae*
3.	Associative symbiotic	*Azospirillum*
P solubilizing biofertilizers		
1.	Bacteria	*Bacillus megaterium var., phosphaticum, Bacillus subtilis, Bacillus circulans, Pseudomonas striata*
2.	Fungi	*Penicillium* sp, *Aspergillus awamori*
P mobilizing biofertilizers		
1.	*Arbuscular mycorrhiza*	*Glomus* sp., *Gigaspora* sp., *Acaulospora* sp., *Scutellospora* sp., *and Sclerocystis sp.*
2.	Ectomycorrhiza	*Laccaria* sp., *Pisolithus* sp., *Boletus* sp., *Amanita* sp.
3.	Ericoid mycorrhiza	*Pezizella ericae*
4.	*Orchid mycorrhiza*	*Rhizoctonia solani*
Biofertilizers for micro nutrients		
1.	Silicate and zinc solubilizers	*Bacillus* sp.
Plant growth promoting rhizobacteria		*Bradyrhizobium sp, Pseudomonas, Azospirillum, Azotobacter, Bacillus, Rhizobium, Enterobacter, Erwinia, Mycobacterium, Mesorhizobium, Flavobacterium.*

3.1. Different types of biofertilizers

- Carrier based biofertilizers or inoculants.

3.1.1. *Rhizobium*

- Soil habitat bacterium able to colonize the legume roots and fixes the atmospheric nitrogen symbiotically.

- They have seven genera and highly specific to form nodule in legumes, referred as cross inoculation group.
- First made in USA and commercialized by private enterprise in 1930.
- Chronicled by Fred (1932).

3.1.2. *Azotobacter*

- *A.chroococcum* - Dominant inhabitant in arable soils capable of fixing N_2 (2-15 mg N_2 fixed/g of carbon source) in culture media.
- Produce abundant slime that helps in soil aggregation.

3.1.3. *Azospirillum*

- *Azospirillum lipoferum and A.brasilense* (*Spirillum lipoferum* in earlier literature) are primary inhabitans.
- Associative symbiotic relation with the poaeceous plants.
- Gram negative.
- *Vibrio* or *Spirillum* having abundant accumulation of polybutadroxybutrate (70%) in cytoplasm.
- Five species of *Azospirillum – A.brasilense, A.lipoferum, A.amazonense, A.halopraeferens and A.irakense*.
- Produces IAA that helps in disease resistance and drought tolerance.

3.1.4. Cyanobacteria

- Free-living as well as symbiotic cyanobacteria (blue green algae) found in rice ecosystem.
- Composite culture of BGA having heterocystous *Nostoc*, *Anabaena*, *Aulosira* is given as primary inoculum in trays and mass multiplied in the field for application as soil based flakes to rice at 10 kg/ha.

3.1.5. *Azolla*

- Free-floating water fern.
- Fixes atmospheric nitrogen in association with nitrogen fixing blue green algae (*Anabaena azollae*).
- *Azolla* fronds consist of sporophyte with a floating rhizome and small overlapping bilobed leaves and roots.
- Fixes and contributes 40-60 kg N/ha per rice crop.

3.1.6. Phosphate solubilizing microorganisms (PSM)

- Phosphobacteria, *Pseudomonas, Bacillus, Penicillium, Aspergillus* etc., secrete organic acids and lower the pH that solubilize fixed phosphates in the soil.
- Increased yields of wheat and potato due to inoculation of peat based cultures of *Bacillus polymyxa* and *Pseudomonas striata.*

3.1.7. AM fungi

- *Glomus, Gigaspora, Acaulospora, Sclerocytes* and *Endogone.*
- Posses vesicles for storage of nutrients and arbuscles for funnelling these nutrients into the root system.
- Transfer of P, Zn and S from soil to root through these fungi.

3.1.8. Silicate solubilizing bacteria (SSB)

- Capable of degrading silicates and aluminium silicates.
- Supply H^+ ions to the medium and promote hydrolysis and the organic acids like citric, oxalic, keto and hydroxy carbolic acids which form complexes with cations, promote their removal and retention in the medium in a dissolved state.

3.1.9. Plant growth promoting rhizobateria (PGPR)

- Group of bacteria that colonize roots or rhizospere soil beneficial to crops.
- Suppression of plant disease (termed bioprotectants), improved nutrient acquisition (termed biofertilizers) or phytohormone productions (termed biostimulants). Eg. *Pseudomonas and Bacillus.*
- Produces biostimulants and phytohormones like IAA, cytokinins, gibberellins etc that give resistance to pests and diseases and drought.

3.1.10. Yield increase and N fixed by biofertilizers

Name	Crops suited	Benefits usually seen	Remarks
Rhizobium strains	Legumes like pulses, groundnut, soybean	10-35% yield increase, 50-200 kg N/ha.	Fodders give better results. Leaves residual N in the soil.
Azotobacter	Soil treatment for non- legume crops including dry land crops	10-15% yield increase adds 20-25 kg N/ha	Also controls certain diseases.
Azospirillum	Non-legumes like maize, barley, oats, sorghum, millet, Sugarcane, rice etc	10-20% yield increase	Fodders give higher/enriches fodder response. Produces growth promoting substances. It can be applied to legumes as co-inoculant
Phosphate solubilizers (there are 2 bacterial and 2 fungal species in this group)	Soil application for all crops	5-30% yield increase	Can be mixed with rock phosphate.
Blue-green algae and *Azolla*	Rice/wet lands	20 -30 kg N/ha, Azolla can give biomass upto 40-50 tonnes and fix 30-100 kg N/ha	Reduces soil alkalinity, can be used for fishes as feed. They have growth promoting hormonal effects.
Microhizae (VAM)	Many trees, some crops, and some ornamental plants	30-50% yield increase, enhances uptake of P. Zn, S and Water.	Usually inoculated to seedlings.

3.2. Liquid biofertilizers

- *Rhizobium, Azosprillum* and phosphobacteria provide nitrogen and phosphorous nutrients to crop plants through nitrogen fixation and phosphorus solubilization processes.
- Effectively utilized for rice, pulses, millets, cotton, sugarcane, vegetables and other horticulture crops.
- Enhances the crop growth and yield.
- Improves the soil health and sustain soil fertility.

3.2.1. Benefits

- Longer shelf life 12-24 months.
- No contamination.
- No loss of properties due to storage upto 45°C.
- Greater potential to fight with native population.
- High population can be maintained i.e., more than 10^{-9} cells/ml in storage.

- Easy identification by typical fermented smell.
- Cost saving on carrier material, pulverization, neutralization, sterilization, packing and transport.
- Quality control protocols are easy and quick.
- Better survival on seeds and soil.
- No need of running biofertilizer production units throughout the year.
- Very much easy to use by the farmer.
- Dosages in 10 time less than carrier based powder biofertilizers.
- High commercial revenues.
- High export potential.
- Very high enzymatic activity since contamination is nil.

3.2.2. Characteristics of different liquid bioferilizers

1. *Rhizobium*

- Symbiotic nitrogen fixation.
- Infects the legume roots and form root nodules within which they reduce molecular nitrogen to ammonia.
- Fixes 40-250 kg N/ha/year in different legumes.
- Dull white in colour.
- No bad smell.
- No foam formation.
- pH 6.8 - 7.5.

2. *Azospirillum*

- Fixes 20 - 40 kg N/ha in the rhizosphere in non leguminous plants such as cereals, millets, oilseeds, cotton, vegetables etc.,
- Three species namely *A.lipoferum A.brasilense* and *A.amazonese* actively denitrify or reduce nitrate to nitrite.
- *A.lipoferum* - present in roots of *Digitaria, Panicum, Brachiaria*, maize, sorghum, wheat and rye.
- Blue or dull white colour.
- Bad odours confirms improper liquid formulation.
- Production of yellow gummy colour materials confirms the quality product.
- Acidic pH confirms that there is no *Azosprillum* bacteria in the liquid.

Production of growth hormones

- Synthesize vitamins, nicotinic acid, indole acetic acids, gibberellins and help the plants in better germination, early emergence, better root development.

Role of liquid *Azospirillum* under field conditions

- Stimulates growth and imparts green colour which is a characteristic of a healthy plant.
- Aids utilization of potash, phosphorous and other nutrients.
- Encourages plympness and succulences of fruits and increase protein percentage.

3. *Azotobacter*

- Free living nitrogen fixing aerobic bacterium.
- Biofertilizer for all non leguminous plants especially vegetables etc.,
- The pigmentation produced by *Azotobacter* in aged culture is melanin due to oxidation of tyrosine by tyrozinase, an enzyme which has copper.
- ***A.chroococcum***: Produces brown black pigmentation.
- ***A.beijerinchii***: Produces yellow light brown pigmentation.
- ***A.vinelandii***: Produces green fluorescent pigmentation.
- ***A.paspali***: Produces green fluorescent pigmentation.
- ***A.macrocytogenes***: Produces pink pigmentation.
- ***A.insignis***: Produces less, gumless, grayish blue pigmentation.
- ***A.agilis***: Produces green fluorescent pigmentation.
- Produces antifungal substances which inhibit the growth of *Aspergillus, Fusarium, Curvularia, Alternaira, Helminthosporium*.

4. *Acetobacter*

- Sacharophillic bacteria and associates with sugarcane, sweet potato and sweet sorghum.
- Fixes 30 kg N/ha/year.
- Commercialized for sugarcane crop and increase yield by 10-20 t/acre and sugar content by about 10-15 percent.
- Have capacity to fix 300 kg N.

3.2.3. Liquid biofertilizer application methodology

1. **Seed treatment-**Seeds are treated with *Rhizobium*, *Azotobacter* and *Azospirillum*.
2. **Root dipping-**Mix the recommended dosage in 5-10 litres of water and root dipped for half – an- hour before transplanting.
3. **Soil application-**Mix recommended dosage with 400 to 600 kg of cow dung FYM along with ½ bag of rock phosphate. Keep over night, maintain 50% moisture and apply to field.

Dosage of liquid biofertilizers in different crops

Crop	Recommended biofertilizer	Application method	Quantity to be used
Pulses: Chickpea, pea, groundnut, soybean, beans, lentil, lucerne, berseem, greengram, blackgram, cowpea, and pigeon pea.	*Rhizobium*	Seed treatment	200 ml/acre
Cereals: Wheat, oat, barley	*Azotobacter* / *Azospirillum*	Seed treatment	200 ml/acre
Rice	*Azospirillum*	Seed treatment	200 ml/acre
Oilseeds: Mustard, sesasum, linseed, sunflower, castor	*Azotobacter*	Seed treatment	200 ml/acre
Millets: Pearl millet, finger millet, kodo millet	*Azotobacter*	Seed treatment	200 ml/acre
Maize and sorghum	*Azospirillum*	Seed treatment	200 ml/acre
Forage crops and grasses: Bermuda grass, sudan grass, naiper grass, para grass, star grass etc.,	*Azotobacter*	Seed treatment	200 ml/acre
Tea, coffee	*Azotobacter*	Seed treatment	400 ml/acre
Rubber, coconuts	*Azotobacter*	Seed treatment	2-3 ml/plant
Agroforestry/fruit plants: All fruit/agroforestry (herb, shrubs, annuals and perennial) plants for fuel wood, fodder, fruits, gum, spice, leaves, flowers, nuts and seed purpose	*Azotobacter*	Seed treatment	2-3 ml/plant at nursery
Leguminous trees	*Rhizobium*	Seed treatment	1-2 ml/plant

3.3. Application of biofertilizers (carrier based)

3.3.1. Seed treatment

Inoculants (1 packet) is mixed with 200 ml of rice cruel or jaggery solution to make a slurry.

- Uniform coating of the inoculants over the seeds and then, shade dried for 30 min.
- One packet of the inoculant (200 g) is sufficient to treat 10 kg of seeds.

3.3.2. Seedling root dip

Two packets (400 g) is mixed with 40 litres of water and roots are dipped for 5-10 min.

3.3.3. Main field application

Four packets (800 g) of the inoculants are mixed with 20 kg of dried and powdered farmyard manure and then, broadcasted in one acre of main field.

- ***Rhizobium:*** For all legumes *Rhizobium* and applied as seed treatment.
- ***Azospirillum/Azotobacter:*** For transplanted crops, *Azospirillum* is inoculated through seed or seedling root dip or soil application methods. For direct sown crops, *Azospirillum* is applied through seed treatment and soil application.
- **Phosphobacteria:** Inoculated through seed, seedling root dip and soil application.
- **Combined application:** Phosphobactera is mixed with *Azospirillum* and *Rhizobium* in equal quantities.

3.3.4. Precautions

- Bacterial inoculants should not be mixed with insecticide, fungicide, herbicide and fertilizers.
- Seed treatment with bacterial inoculants is to be done at last when seeds are treated with fungicides.

Part-3 Mushroom

1

Mushroom

1.1. Mushroom

- Fruiting body of fungus appears above ground and contains spores.
- Fungi are network of fine microscopic threads collectively known as mycelium.
- Fungi are the part of the plant kingdom and don't contain chlorophyll and root system.
- Mushroom comes under sub division Basidiomycotina, order – Agaricales.

Three types of mushroom based on nutritional habits.

i) Saphrophytes – living on dead wood or dead tissue of living tree or dung.

ii) Parasites – attacking living plant or animal tissues.

iii) Mycorrhizae – symbiotic relationship with plants.

1.2. Importance of mushroom

- Delicious food, rich in protein, carbohydrates and vitamins.
- Protein – 3.7%
- CHO – 2.4%
- Fat – 0.4%
- Minerals – 0.6%
- Water content – 91%
- Low calorific value and contains 9 essential amino acids for human growth.
- Excellent source of thiamin (vitamin B_1), riboflavin (B_2), niacin, pantothenic acid, biotic, folic acid and vitamin C, D, A, K which retained after cooking.
- Rich in fibre content, high potassium and sodium ratio, excellent source of potassium that helps in lower blood pressure and diminish risk of stroke.
- Recommended for hypertension.
- Rich in copper and helps in cardio protective properties.
- High fibre and low calorific is suitable for diabetic patient and anti cancer activity.

- White button mushroom reduces breast cancer and prostate cancer.
- Shiitake mushroom is associated with immune system.
- Fighting against AIDS and anti tumour activity.
- Oyster mushroom is used for relaxation.
- Plays a vital role in integrated rural development by self employment for village youths, womenfolks and house wife to make them financially independent.

1.3. Classification of mushroom

- Fungi have 50,000 species and edible ones are 10,000.
- Of which, 180 are mushrooms and 70 are accepted as food.
- Cultivated techniques are developed for 20 mushrooms and about dozens (12) of them are recommended for commercial cultivation.

Only 6 mushrooms are preferred for large scale cultivation.

1. Paddy straw mushroom - *Volvariella sp*
2. Oyster mushroom - *Pleurotus spp*
3. Button mushroom - *Agaricus spp*
4. Shiitake mushroom - *Lentinula spp*
5. Jews ear mushroom - *Auricularia spp*
6. Milky mushroom - *Calocybe spp*

1.4. Morphology

1.4.1. Button mushroom - *Agaricus bisporus*

- Cap: 3-16 cm, convex to broadly convex.
- Gills free from stem.
- Colour-pinkish to pinkish brown at first and become dark brown to blackish.
- Stipe - 2-8 cm long, 1-3 cm thick.
- Spore-basidia-2 spores.

1.4.2. Oyster mushroom-*Pleurotus spp*

- Cap is tongue shaped having 5-15 cm diameter, whitish to grey to blue grey in colour.
- Gills are white with short stem.
- Spore: Basidia (4 haploid spores) (tetrapolar).

1.4.3. Paddy straw mushroom - *Volvariella spp*

- Cap: 5-15 cm, broad egg shaped.
- White initially, become dark tan in colour and to pale tan in age.
- At stem base, there is a encasement called volva (cuplike structure).
- Stem : 4-20 cm long.
- Basidiapolar (4 haploids).

1.5. Mushroom poisoning

- Toxic due to eating certain mushroom as they contains poisonous compounds (toxic substance).
- Symptom: Mild pain and discomfort to death.
- Toxins are secondary metabolites causes allergy, gastrointestinal upset and irritation leads to vomiting and diarrhea.

Mushroom species	Toxin	Symptoms
Amanita ocreata, Conocybe filaris, Lepiota josserandii	Alpha amanitin	Gastrointestinal upset after 6 hours of eating i.e., vomiting and profuse, watery diarrhea. Damage to liver begins after 24 hours of ingestion.
Amanita phalloides	Phallotoxin	Gastrointestinal upset such as vomiting and diarrhoea normally leads to death
Cortinarius orellanus	Orellanine	After 20 days of ingestion, symptoms such as pain in kidney area, thirst, vomiting and headache occurs
Boletus luridus, Omphalotus olivasceus, Inocybe fastigiata	Muscarine	Sweating, salivation, tears, blurred vision, palpitations and respiratory failure.
Gyromitra esculenta, Gyromitra infula	Gyromitrin	It blocks neurotransmitter leading to muscle cramps, loss of coordination, tremors, seizures & gastrointestinal upset such as vomiting & diarrhoea. It also causes red blood cells to break down, leading to jaundice, kidney failure, and signs of anemia.
Coprinus variegata, Coprinus atramentarius	Coprine	It inhibits aldehyde dehydrogenase, which generally causes no harm. If alcohol is ingested, it prevents removal of alcohol in body leading to flushed skin, vomiting, headache, dizziness, weakness, apprehension, confusion, palpitations, and sometimes trouble breathing.
Amanita muscaria, Amanita pantherina	Ibotenic acid	Nausea, vomiting, confusion, euphoria, or sleepiness is possible. Loss of muscular coordination, sweating, and chills are likely.
Amanita gemmata, Amanita pantherina, Amanita muscaria	Muscimol	It alters neuronal activity & displays sedative, hypnotic and dissociative psychoactive effects including dissociation, synesthesia, auditory and visual distortions.
Psilocybe cyanescens, Panaeolus cyanscens	Psilocybin	Euphoria, visual and religious hallucinations, and heightened perception.
Amanita pantherina, Hygrophoropsis aurantica	Arabitol	Gastrointestinal irritation
Boletus satanas	Bolesatine	It is a protein synthesis inhibitor and main symptoms include violent vomiting, which can last up to six hours.

1.5.1. Treatments

- Initial treatment - gastric leavage to remove toxins.
- Use of benzyl penicillin, an anti toxin and silymarin – anti oxidant.

1.6. Preparation of media for raising pure culture

1.6.1. Potato Dextrose Agar (PDA) medium

- Peeled and sliced potato (250 g), dextrose 20 gm, agar agar powder 20gm, water 1000 ml, pH-7.
- Boil peeled potato (25-30 min).
- Filter through muslin cloth.
- Make the volume to 1000 ml.
- Add dextrose and agar agar powder and boil.
- Adjust the pH 7.
- Then, fill the medium in flask.
- Plug with sterilized non absorbent cotton.

1.6.2. Potato Dextrose Yeast Agar (PDYA) medium

- 2 gm of yeast is added and similar to PDA.

1.6.3. Malt Extract Agar (MEA) medium

- Malt extract 25 g, agar agar 20 gm, distilled water 1000 ml, pH -7.

1.6.4. Compost Extract Agar (CEA) medium

- Pasteurized compost (150 gm), agar agar powder (20 gm), water (1000 ml), pH 7.
- Boil the compost in 1.5 - 2 litres of water till the volume is reduced into half.
- Filter through muslin cloth.
- Add all other mixtures and boil.

1.6.5. Malt Peptone Grain Agar (MPGA) medium

- Malt extract - 20 gm, rye or wheat grains 5gm or yeast 2 gm, agar agar 20 gm, peptone 5 gm, pH 7, water 1000 ml.
- Boil rye in water for 1-1/2 hours.
- Filter and mixed with other ingredients.

1.7. Isolation techniques to get pure cultures and their maintenance

- **Two methods** – spore culture, tissue culture

1.7.1. Spore culture

1. Spore print

- Healthy disease free mushroom cap is removed and surface cleaned using alcohol dipped cotton and place on clean sterilized white paper or clean glass plate.
- Cover with a clean glass or glass jar or cup to prevent air flow.
- Spores fall on white paper within 24-48 hours.
- Spore print on paper is preserved for long time by folding into two halves.

2. Spore transfer and germination

- Using sterilized and clean scalpel, scrap some spores from spore print and transfer them on agar medium aseptically.
- Minimum three agar dishes are inoculated in each spore print.
- Incubate at appropriate temperature for culture development - known as **multispore culture**.
- Dilute the spores, transfer to sterile PDA or MEA medium and incubated at $25^{\circ}C \pm 2^{\circ}C$ for 2 weeks to obtain pure culture.

1.7.2. Tissue culture

- Cut a small bit from the pelial region using sterilized blade.
- Wash several times in distilled water repeatedly.
- Dried in clean tissue paper.
- Inoculate aseptically on a petriplate, containing PDA or MEA and incubate at $25^{\circ}C \pm 2^{\circ}C$ for 6-12 days.
- Discard the contaminated cultures and retain pure culture.

1.7.3. Sub culturing

- Pure culture of edible mushroom established through either spore culture or tissue culture is maintained in cool atmosphere or refrigerator.
- Sub culturing is done from time to time aseptically transferring small piece of growing pure culture in pure culture medium on test tube or other suitable medium
- Pure culture is used for mother culture for large scale spawn production.

1.7.4. Spawn production

- Vegetative mycelium network of mushroom developed after germination with one or more fungal spores is grown on convenient medium.

1.8. Preparation of mother spawn

- Mother spawn – mushroom grown on a grain based medium.
- Sorghum grains are the best substrate.

1.8.1. Procedure

- Soak cleaned grains in a clean water.
- Cook the grains for 30 mins.
- Spread the cooked grains evenly on platform to remove excess water.
- At 50% moisture, mix calcium carbonate at 20g/kg, cook and dry.
- Fill the cooked and dried grains in saline bottle upto $3/4^{th}$ height (300-350 g/bottle).
- Plug the mouth with non absorbent cotton.
- Autoclave and sterilize bags under 20lbs pressure for 2hrs.
- Cook the bags and keep then inside culture room under UV light for 20 mins.
- Cut the fungal culture into two halves and transfer them in two bottles.
- Incubate them in a clean room at room temperature for 10 days.
- This can be used for bed spawn.

1.9. Preparation of bed spawn

- Method of preparation of bed spawn is similar to the mother spawn except the use of polypropylene bags.
- Sterilize the polypropylene bags and put them under UV light for 20 minutes.
- Take well grown mother spawn and transfer 10 gram of mother spawn in each sterilized bags.
- Keep the bag in clean room for fungal growth – this is the **first generation bed spawn.**
- After 10 days, bed spawn are used for bed preparation.

1.9.1. Precautions to be observed

- Avoid over cooking time.
- Dry the cooked grains either on a wire mesh tray or hessian cloth.

- Use only recommended dose of calcium carbonate ($CaCO_3$) to reduce fungal growth.
- Avoid further sub culturing of second generation spawn.

1.9.2. Characters of good spawn

- Proper coating of mycelium around every grain.
- Mycelium growth should be strandy, but not cottony or pluffy.
- Fresh spawn is white in colour and changes brown when it grows.
- No slimy growth (bacterial contamination).
- Should not be greenish or blackish spot which indicates mould contamination

1.9.3. Storage of spawn

- Always use fresh spawn.
- Can be stored at 4 to 6°C for one month.

1.10. Cropping/cultivation

1.10.1. Oyster mushroom

1. Cropping room

- Thatched shed, diffused light with good ventilation.
- Transfer the bed spawns into the cropping room for button development.

2. Methods of spawn cropping

- **Open bed method:** Completely remove polythene cover and allow for button development.
- **Closed bed method:** Cover the bed with polythene having holes and the button will emerge through holes.
- **Half open cover open method:** One half of polythene cover is removed for cropping and second half after first harvest.
- **Strip method:** Polythene cover of longitudinal strips of 5cm breadth is kept at 4-5 places in the bed.
- **Tear method:** Polythene cover is teared longitudinally at several places.
- **Round opening method:** Polythene cover is given with round shape openings of 5cm diameter at random.
- **Complete removal of polythene cover is the best method (open bed method).**

3. Procedure

- Allow the bed to dry for a day as fresh bed spawn has more moisture.
- Using atomizer spray water on the bed from the second day onwards.
- 2-3 days after opening, pinheads of mushroom button develop and ready for harvest in another 4 days (6-7 days)
- Harvest the mushroom in early morning.
- Pack neatly in polythene cover having ventilation holes @ 200 gm/bag.
- After first harvest, scrap out the mushroom bed with new comb and then spray water daily.
- Second harvest - 7 to10 days of first harvest.
- Third harvest - 7 to 10 days after second harvest.

Precautions

- Optimum temperature of 23 - 25°C is maintained inside the shed.
- Humidity - 80-85%.
- Avoid excess watering, spray water on the beds only after harvesting.
- Always harvest in morning hours.

1.10.2. Milky mushroom

- Transparent polythene covers of 60 x 30 cm with a thickness of 80 gauge is used for cultivation.
- Wash hands thoroughly with antiseptic lotion.
- Take the polythene cover and tie the bottom end using thread and turn it inwards.
- Mix the dried straw and moisture it uniformly.
- Take well grown bed spawn and divide into two halves.
- Two beds are used for a single spawn bag.
- Fill the straw to a height of 3 inches in the bottom of polythene bag and sprinkle the water and keep a hand full of spawn over the straw layer concentrating more on edges.
- Fill the second layer of straw to the height of 5 inches and spawn it as above.
- Repeat this process to get 5 straw layer with spawn.
- Gently press the bed and tie tightly with a thread.

- Put 6 ventilation holes randomly for ventilation and removal of excess moisture.
- Arrange the beds inside the thatched shed with good ventilation following rack system of hanging.
- Observe daily for contamination; maintain the temperature of 22-25°C and RH of 85-90%.
- To control insect and pest like flies, beetles and mites, spray malathion at 1ml/litre of water.
- Fully spawn run beds are shifted to **blue coloured tent after casing for button initiation**.
- **Casing** – application of thin layer of sterilized soil on the surface of the mushroom bed to induce button.
- Soil or river sand or mixed soil and river sand in equal proportion can be used.
- Mix $CaCO_3$ (calcium carbonate) at 100 g/ha of sand or soil.
- Sterilize the soil in autoclave for 45 min using mud pot or vessels.

1. Casing room

- Casing is the extra process for milky mushroom for button initiation.
- Blue polythene covered tent should be maintained with optimum temperature of 30-35°C.
- RH 80-85% and light intensity of 2000-3000 lux.
- Apply casing soil at height of 1cm and press it gently.
- Spray water to wet the cased soils.
- Place the bed inside the blue tent, observe daily for contamination and spray water to keep them wets.
- First harvest – 7 days.
- Second harvest – 10 days after first harvest.
- Third harvest – 10 days after 2 harvest.
- After each harvest, disturb the top soil and stir the top of the bed and water regularly.
- Mushroom yield – 350-400 gm/250 gm dried weight of straw.

1.11. Post harvest processing

- Browning of mushroom takes place at high temperature due to high respiration.

1.11.1. Storage of fresh mushrooms

1. Refrigeration /instant packing

- Pack the mushroom in 25 gauge polythene bags
- Store it at 5°C which extends the life for 3-5 days.

2. Freeze drying

- Slice the mushroom and immerse in 0.05% sodium meta bisulphate and 2% common salt (NaCl) for 30 minutes.
- Blanch in boiling water for 2 minutes and cooling.
- Frozen for 1 min at -12°C and store it at -20°C for 3-4 months.
- Long distance transport.

3. Dehydration

i) Pre treatment

- Blanch the mushroom with boiling water for 2 min.
- Immerse in cold water for 2 min.
- Dip the mushroom in water containing 0.2% potassium metabisulphate and 1% citric acid.

ii) Drying

- Dry in open sun light, till it reaches 1/10th weight of fresh product.
- Stored for 3 months.
- Artificial drying at 60°C for 6-8 hours to bring the final moisture to 3-5%.

4. Canning

- Only to preserve button mushrooms.
- Harvest the mushroom at an early stage.
- Wash with clean water followed by boiling water for 2 min and cold water for 2 min.
- Fill the mushroom in 1pound cans up to 3/4th capacity (220gm).
- Add salt solution containing 2% common salt + 2% sugar + 0.3% citric acid upto brim (125 ml solution) approximately.
- Seal the can air tight and sterilize in autoclave for 20-25 min.
- Cool it in cold water and can be stored for 12 months.

1.11.2. Value addition

1. Mushroom soup

- Button mushroom and oyster mushroom is preferable.
- Dried and sliced mushroom is finely ground and passed thorough 0.5 mm sieve, mix with milk powder and corn flour and other ingredients.

2. Mushroom biscuit

- Button and oyster mushroom is used.
- Ingredients to be used are maida, sugar, ghee, mushroom powder, coconut powder, baking soda, ammonium bicarbonate and milk powder.
- Using dough kneader, make the mixture homogenous for 20-25 min, then add 500 ml of water and mix them for 10-15 mins.
- Cool the dough for 10 mins and cut in to different sizes of biscuits.

3. Mushroom nuggets

- Mushroom powder is mixed with greengram dhal powder by adding water.
- Prepare a paste and make round balls of 2-4 cm and then sun drying them.
- Deep fry and used as snacks.
- Add in vegetable curry preparation.

4. Mushroom catch up

- Button mushroom is used.
- Cook the button mushroom in 50% of water for 20 min.
- Grind them to make paste.
- Add acetic acid (1.5%) and other ingredients and then cook and cool. Fill in sterilized bottle or jars.

5. Mushroom candy

- Oyster and button mushroom are preferable.
- Cut the mushroom longitudinally into two pieces.
- Blanched in 0.05% KMS (potassium meta bisulphite) for 5 min.
- Drain the water and add sugar at 1.5 kg /kg of mushroom.
- Divide the sugar into the three equal parts, one part is covered over the mushroom and kept for 24 hours.

- Next day, the mushroom is covered with second part of sugar and kept overnight.
- Third day, remove the mushroom from the sugar syrub and boiled with third part of sugar and add 0.1% citric acid.
- Mix the mushroom with sugar syrub and boiled for 10 minutes and dry at 60°C for about 10 hours.

6. Mushroom preserve (Murabba)

- Button mushroom are used.
- Blanching in 0.5% KMS for 10 min.
- Mix with 40% of its weight of sugar daily for 3 days. (45 kg of mushroom mixed with 55 kg of sugar).
- Take out from syrub and treat with 0.1% citric acid, add 40% of sugar in the syrub and cook it and then add mushroom in the syrub.

7. Mushroom pickle

- Both button and oyster mushroom are used.
- Blanch with 0.5% KMS solution for 5 min.
- Subject the mushroom to salt curing process by adding 10% sodium chloride over night.
- Remove the mushroom and prepare pickle using turmeric powder, black mustard seed powder, red chilli, cumin seed powder, fenugreek seed powder, black pepper, fennel seed powder, carom seed and mustard oil.
- Add acetic acid and sodium benzoate (Preservatives).
- Pickle made from oyster mushroom considered as good quality and tasty.

8. Mushroom chips

- Button mushroom are blanched in 2% brine solution.
- Dip the mushroom overnight in 0.5% citric acid and 1.5% sodium chloride + 0.3% red chilli powder solution.
- Drain the water and dry it and fry in the refined oil.
- Add karam masala and other spices and mix it.

1.12. Diseases

1.12.1. Fungal diseases and mould

1. White button mushroom

a) Competitor/indicator/weed mould

i) False truffle

Causes

- High temperatures during spawn run.

Symptoms

- White cream coloured mycelium in compost and casing soil.
- Grows into irregular fungal mass looking like peeled walnut.
- At maturity, becomes pink and disintegrates into powdery mass, emit like chlorine.

Control measures

- Compost should be prepared on concrete floor.
- Avoid temperature above 26-27°C during spawn run.
- Maintain temperature below 18°C during cropping.

ii) Olive green mould: *Chaetomium olivaceum, C.globosum*

Causes

- Lack of aeration during compost preparation.
- Temperature should not exceed 16°C during pasteurization.
- Noticed in spawn bottles.

Symptoms

- The fungus is greyish white fine mycelium in the compost and produce musty odour.

Control measures

- Composting period should not be too short. Spray dithane Z-78 (0.2%).

iii) Brown plaster mould: *Papulaspora byssina*

Symptoms

- Mycelium – light brown to cinnamon brown and becomes rust colour.

Causes

- Use of poor and old straw.
- High moisture compost.
- Use of less gypsum during composting.

Control measures

- Localized application 2% formalin.
- Use sufficient gypsum in composting.

iv) Yellow mould: *Myceliophthora lutea*

Causes

- Use of old poultry manure.
- Cause mat disease.
- Yellow brown corky mycelial layer

Control measures

- Spraying of 0.05% bavistin, 15% calcium hypochlorite, 0.04% blitox.

v) Tikki mould: *Sepedonium spp.*

Symptoms

- Turning yellow or tan at maturity.

Causes

- Use of old poultry manure and improper pasteurization.
- Mould is observed in compost and mycellium shows variegated colours.

Control measures

- Incorporation of 0.5% carbendazim in compost.
- Sterilize the poultry manure with 2% formalin.

vi) Ink caps: *Coprinus spp.*

Causes

- High nitrogen and ammonia in compost.
- High moisture content of compost.
- Use of old straw.

Control measures

- Use fresh straw.
- Ammonia content of compost should be less than 10 ppm.

vii) Cinnamon mould – *Chromelosporium fulva*

Causes

- Casing soil is too wet.

Symptoms

- Secondary disease.
- White to pink to cherry red to dull orange.

Control measures

- Casing soil should be sterilized with formaldyde and spraying of dithanze Z78 0.15%.

viii) Lipstick mould – *Sporendonema purpurescens.*

Causes

- Too wet compost.

Control measures

- Good hygiene is essential.

b) Fungal diseases

i) Wet bubble *Mycogone perniciosa*

Causes

- Due to improper casing.

Symptoms

- Gives fowl odour and deformation of gills.

Control measures

- Benomyl 0.1% spray after casing.
- Incorporation of carbendazim 0.1% in the casing mixture.

ii) Dry bubble – *Verticillium fingicola (Lecanicillium fungicola)*

Symptoms

- Most common and very serious fungal disease of mushroom.
- Onion shaped mushroom.
- Destroys the crops in 2-3 weeks.

Control measures

- 2-3 sprays of dithane Z78 0.15%.

iii) Green mould – *Trichoderma viride, Trichoderma harzianum, Pencillium spp, Aspergillus spp*

Symptoms

- Dense pure white growth of mycelium.

Control measures

- Spray mancozeb 0.2% or bavistin 0.1%.

2. Oyster mushroom (*Pleurotus* spp.)

a) Competitor moulds

- Different fungi like *Aspergillus niger, A. flavus, A. fumigatus, Penicilluim sp, Rhizopus spp, Trichoderma viride, Cephalosporium spp.*
- Loss in yield upto 70%.

Control measures

- 50 ppm benomyl solution or 100 ppm carbendazim.

b) Fungal diseases

i) Cob web - *Cladobotryum apiculatum, C.verticillatum, C.variospermum*

- White cottony growth on substrate.

Symptoms

- Decay of sporophores.
- Emitting fowl smell.

Control measures

- Spray bavistin 0.05%.

ii) Green blotch: *Gliocladium virens*

Symptoms

- Fruit bodies covered with mycelium and green spots.

Control measures

- Spray 0.01% bavistin.

iii) Brown rot: *Arthrobotrys pleuroti*

Symptoms

- Fluffy growth on substrate and fruit bodies.

Control measures

- 0.05% bavistin spray.

iv) Sibirina rot: *Sibirina fungicola*

Symptoms

- Powdery white growth on stipe, gills and primodia.

Control measures

- 50 ppm benomyl

1.12.2. Bacterial disease for all mushrooms

i) **Bacterial blotch:** *Pseudomonas tolaasii.*

Symptoms

- Mostly it affects white button mushroom.
- Brown spots or blotches on the cap.

Control measures

- Application of bleaching powder (0.15%).

1.12.3. Viral diseases

Symptoms

- Infects button, oyster, shiitake and paddy straw mushrooms.

- Most serious in button mushroom.
- Die back.
- Mushroom dense cluster and matures early.

Control measures

- Spray 2% sodium pentachlorophenate and 1% sodium carbonate.
- Spray malathion 0.05% or 0.1% formalin.

1.13. Pests of mushroom

1.13.1. Sciarids – flies (Diptera: Sciaridae)

- Resemble like midges, major pest in mushroom and cause 50% reduction in yield.

1.13.2. Phorids (Diptera: Phoridae)

- Black or light dark brown flies.
- Also known as scuttle flies, manure flies, coffin flies.

Control measures

- Use of light traps and screening of doors and ventilators.
- Spraying of 30 ml nuvan for 100 m^3 or 0.1% malathion or 0.5% pyrethrum at 1cc/cm.

1.13.3. Cecids (flies) (Diptera: Cecidomyiidae)

- Larva of the flies damages the mushroom and gill tissues break down to produce black fluid.

Control measures

- Spraying of fenthion or fenitrothion 1gm a.i./m^2 or diazinon 50 ppm.

1.13.4. Spring tails

- Affects oyster, milky, button mushrooms.
- Feeds on gills and damage the gill linings.

Control measures

- Spraying 0.5% malathion.
- Mixing 30 ppm diazinon in the compost.

1.13.5. Mites: *Tyrophagus spp, Caloghyphus spp, Oppia sp, Pygmephorus sp.*

- Make large cavities on stipe and caps.

Control measures

- Disinfect the mushroom house by spraying 0.1% dicofol.
- Spray diazinon at 2ml/litre of water in the compost.

1.14. Abiotic disorders

- Due to temperature, RH, moisture of the substrate, pH of compost, CO_2 concentration in the room, wind velocity and fumes.

1.14.1. Stroma

- Aggregation of mushroom mycelium on the surface of compost or casing.
- **Reason:** excessive CO_2, high water content in compost, prolonged spawn run period.
- **Control measure :** Removal of large patches of 8-12 inches from the compost.

1.14.2. Weepers (strinkers/leakers)

- **Leakers:** Exudation of small water droplets from stem or cap.
- Stinkers means development of musty odour or putrid odour.
- **Reason:** Low moisture compost, coupled with high moisture casing leading to weepers.

1.14.3. Hollow core and brown pith

- Circular gap is seen in the centre of the stem.
- Extend to the length of stipe.
- **Reason:** Water stress and excess watering.

1.14.4. Purple stem/black leg/storage burn

- Mushroom develops deep purple colour immediately after harvest.
- **Reason:** Excess watering before harvest.

1.14.5. Rose comb

- Pinkish gills form large lumps and grow an abnormal growth on the cap and looks like a comb.
- **Reason:** Smoke or vapours from paints, kerosene, petrol, oil products.

1.14.6. Brown discoloration

- Browning of small pinheads or half grown mushroom.
- **Reason:** Excess or high wind velocity coupled with low RH.

1.15. Varieties

- *Pleurotus* - M 2, CO1, MDU 1, MDU 2, CO 2, APK 1.
- Button mushroom - Ooty 1, Ooty 2.
- Milky mushroom – APK 2, CO 3.
- Pink oyster mushroom – APK 1.
- Pleurotus sajor-caju – M2.

1.16. Production and export

- Directorate of Mushroom Research (National Research Centre for Mushroom) Solan, Himachal Pradesh.
- Button mushroom production - 73%.
- Oyster mushroom production - 16%.
- Major states – Punjab, Uttarkhand, Haryana, Uttar Pradesh, Tamil Nadu and Maharashtra.
- In Tamil Nadu, 20-30% of production in Erode and Coimbatore districts.
- Also cultivated in Krishnagiri, Salem, Dharmapuri, Namakkal districts.
- Mushroom production in Tamil Nadu – 10,000 metric tonnes (2016).
- In India – 1, 29,782 metric tonnes.
- Export and import – Polland is the first largest exporter in mushroom.
- India is second largest exporter.
- Exporting to US, UK, Russia, Netherland, Denmark, Switzerland, Israel and UAE.

Part-4 Bee Keeping (Apiculture)

1

Beekeeping

- Rerd. L.L.Langstroth (Father of modern bee keeping) discovered movable frame hive in 1851.
- In 1882, beekeeping was introduced in India.
- In 1911, Rerd. Newton introduced beekeeping to South India.

1.1. Bee species (Apidae)

a) Apinae

- *Apis dorsata* (The rock bee).
- *Apis cerana indica* (The Indian hive bee).
- *Apis florea* (The little bee).
- *Apis mellifera* (The European or Italian bee).

b. Meliponnae

- *Tetragonula (Melipona) iridipennis* (Dammer bee or mosquito bee or stingless bee)

1.1.1. *Apis dorsata*

- Constructs single comb in open space (6 x 3 ft – Long x Deep).
- Shifts the place of the colony often.
- Ferocious and difficult to rear.
- Produces 25-37 kg honey /hive/year.
- Largest among the bees.

1.1.2. *Apis florea*

- Constructs single small comb in open space on branches of bushes, hedges, buildings, caves, empty cases etc.
- Produces 0.5 to 1 kg honey/year/hive.

- Not rearable as they frequently change their place.
- Smallest among the bees.
- *Apis sp*. described (smaller than Indian bee).
- Distributed only in plains and not in hills above 450MSL.

1.1.3. *Apis cerana indica*

- Makes multiple parallel combs in the cavities and hallows of trees in darkness.
- Larger than *Apis florea* but smaller than *Apis mellifera*.
- Produces about 5kg of honey/hive /year.
- More prone to swarming and absconding.
- Native of India/Asia.

1.1.4. *Apis mellifera*

- Also makes multiple parallel combs in cavities in darkness.
- Larger than Indian bees, but smaller than rock bees.
- Imported from European countries (Italy).
- Produces 35 kg/hive/year.
- Less prone to swarming and absconding.

1.1.5. *Tetragonula* (Melipona or Trigona) *iridipennis* (Dammer bee or stingless bee)

- Only vestigial sting is present.
- Inhabits crevices in walls and hollow taunts of trees.
- Comb is made up of a dark material called **cerumen** which is a mixture of wax and earth or resin.
- Yield only 60 to 180 g/hive/year.

1.1.6. In addition to above species, the following bee species have been added

1. Single comb bees

- **A.*andreniformis* (Black dwarf bee):** This is the sister species of *A.florea*
- **A.*laboriosa* (Himalayan bee):** World's largest wild species.
- ***A.binghami***
- ***A.breviligula***

2. Multiple comb bees

- *A.koschevnikovi* **sp**
- *A.nigrocincta* **sp**
- *A.nulvensis* **sp**

1.2. Apiary site requirements

- Site should be dry without dampness.
- High RH will affect bee flight and ripening of nectar.
- **Water** - Natural source/Artificial provision.
- **Wind breaks** - Trees serve as wind breaks in cooler areas.
- **Shade** - Hives can be kept under shade of trees. Artificial structures can also be constructed.
- **Bee pasturage/Forage** - Plants that yield pollen/nectar to bees are called bee pasturage/forage.

1.3. General apiary management practices

1.3.1. Hive inspection Opening the hive atleast twice a week and inspecting for

- Presence of queen.
- Presence of eggs and brood.
- Honey and pollen storage.
- Hive record to be maintained for each hive.
- Presence of bee enemies like wax moth, mite, disease.

1.3.2. Expanding brood net

- Done by providing comb foundation sheet in empty frame during honey flow period.

1.3.3. Sugar syrup feeding

- Sugar dissolved in water at 1:1 dilution.
- Used to feed bees during dearth period.

1.3.4. Supering (Addition of frames in super chamber)

- This is done when brood chamber is filled with bees on all frames are covered.
- Comb foundation sheet or constructed comb provided in super chamber.

1.3.5. Honey extraction

- Bee escape board.
- Kept between brood and super chamber.
- Bees bushed away using brush.
- Cells uncapped using uncapping knife.
- Honey extracted using honey extractor.
- Combs replaced in hive for reuse.

1.3.6. Swarm management

- Remove brood frames from strong colony and provide to weak colony.
- Pinch off the queen cells during inspection.
- Divide strong colonies into two to three.
- Trap and hive primary swarm.

1.3.7. Uniting bee colonies - Done by newspaper method

- Bring colonies side by side by moving 30 cm/day.
- Remove queen from weak colony.
- Keep a newspaper on top of brood chamber of queen.
- Right colony.
- Make holes on the paper.
- Keep queen less colony on top.
- Close hive entrance (the smell of bees will mix).
- Unite bees to the brood chamber and make it one colony.

1.4. Seasonal management

- Pollen and nectar available only during certain period.
- Honey flow season (surplus food source) and dearth period (scarcity of food).
- Extremes in climate like summer, winter and monsoon – Need specific management tactics.

1.4.1. Honey flow season management (Coincides with spring)

- Provide more space for honey storage by giving comb foundation sheet or built combs.

- Confine queen to brood chamber using queen excluder.
- Prevent swarming.
- Prior to honey flow - Provide sugar syrup and build sufficient population.
- Divide strong colonies into 2-3 new colonies, if colony multiplication is needed.
- Queen rearing technique may be followed to produce new queens for new colonies.

1.4.2. Summer management

- Bees have to survive intense heat and dearth period
- Provide sufficient shade (under trees or artificial structure).
- Increase RH and reduce heat.
- Sprinkle water twice a day on gunny bag or rice straw put on hive.
- Increase ventilation by introducing a splinter between brood and super chamber.
- Provide sugar syrup, pollen supplement/substitute and water.

1.4.3. Winter management

- Maintain strong and disease free colonies.
- Provide new queen to the hives.
- Winter packing in cooler areas (Hilly areas).

1.4.4. Management during dearth period

- Remove empty combs.
- Use dummy division board to confine bees to small area.
- Unite weak colonies.
- Provide sugar syrup, pollen supplement/substitute.

1.4.5. Rainy season/monsoon management

- Avoid dampness in apiary site.
- Provide proper drainage - In rainy season when bees are confined to the hive, provide sugar syrup feeding.

1.5. Bee pasturage/bee forage

- Plants that yield pollen and nectar are collectively called bee pasturage or bee forage.

1.5.1. Good source of nectar

- Tamarind (Rich source) (*Tamarindus indica*)
- Neem (*Azadirachta indica*)
- Soapnut tree (*Sapindus mukorossi* and *Sapindus trifoliatus*)
- Eucalyptus (*Eucalyptus globulus*)
- Pungam (*Millettia pinnata*)
- Moringa (*Moringa oleifera*)
- *Prosopis juliflora*
- Gliricidia (*Glyricidia maculata*)
- *Tribulus terrestris*

1.5.2. Good source of pollen

- Sorghum (Rich source)
- Maize
- Millets like pearl millet
- Roses, varagu, finger millet
- Pomegranate
- Sweet potato
- Tobacco
- Coconut
- Castor
- Date palm

1.5.3. Good source of pollen and nectar

- Banana
- Citrus
- Apple
- Berries
- Pear
- Plum
- Peach
- Guava

- Sunflower
- Safflower
- Mango
- Coconut
- Sesamum
- Mustard
- Bhendi
- Lucerne
- Clover
- Antigonon creeper
- Cruciferous and cucurbitaceous vegetables
- Cotton (Very rich source)

1.6. Foraging

- Refers to collection of nectar and pollen by bees.

1.6.1. Nectar foragers

- Collect nectar from flowers using lapping tongue.
- Passes the nectar to hive bees.
- Hive bees repeatedly pass the nectar between preoral cavity and tongue to ripen honey.
- Later drops into cell.

1.6.2. Pollen foragers

- Collects pollen by passing flower to flower.
- Pollen sticking to body removed using pollen comb.
- Packed using pollen press into corbicula (pollen basket).
- A single bee carries 10-30 mg pollen (25% of bee's wt).
- Dislodge by middle leg into cell.
- Mix with honey and store.

1.6.3. Floral fidelity

- A bee visits same species of plant for pollen/nectar collection until exhausted.
- Bees travel 2-3 km distance to collect pollen/nectar.

1.7. Pollinators and role of bees in crop productivity

1.7.1. Honey bees as pollinators

- Among insects, honey bees play major role.
- Honey bees and flowering plants have coevolved.
- In insect pollinated plants, flowers are large, brightly coloured, distinct fragrance, presence of nectar and sticky pollen.

1.7.2. Role of bees in pollination

- Bees aid in cross pollination in fruits, vegetables, ornamentals, cotton, tobacco, sunflower and many other crops.
- Entomophily refers to cross pollination aided by insects.
- Melitophily refers to pollination by bees.

1. Honeybees as sapromyophily (carrion and dung flies)

- All bee species aid in pollination.
- Value of honey bees in pollination is 15-20 times higher than that of the honey and wax it produces.

2. Carpenter bee, *Xylocopa* sp. (Xylocopinae:Anthophoridae hymenoptera)

- Robust dark bluish bees with hairy body.
- Dorsum of abdomen bare, pollen basket absent.
- Adults are good pollinators.
- Construct galleries in wood and store honey and pollen.

3. Digger bees, *Anthophora* sp. (Anthophorinae, Anthophoridae: Hymenoptera)

- Stout, hairy, pollen collecting bees.
- Abdomen with black and blue bands.

1.7.3. Role of bees in crop productivity

1. Qualities of honeybees which make them good pollinators

- Body covered with hairs and have structural adaptation for carrying nectar and pollen.
- Bees - Not injurious to plants.
- Adult and larva feed on nectar and pollen - Available in plenty.

- Superior pollinators - Since store pollen and nectar for future use.
- No diapause - Need pollen throughout year.
- Body size and proboscis length - Suitable for many crops.
- Pollinate wide variety of crops.
- Forage in extreme conditions also (weather).

2. Effect of bee pollination on crop

- It increases yield (seed yield, fruit yield) in many crops.
- It improves quality of fruits and seeds.
- Bee pollination is a must in some self incompatible crops for seed set.

3. Crops benefited by bee pollination

Fruits and nuts	Vegetable and vegetable seed crops	Oil seed crops	Forage seed crops
Almond	Cabbage	Sunflower	Lucerne
Apple	Cauliflower	Niger	Clover
Apricot	Carrot	Rape seed	
Peach	Coriander	Mustard	
Strawberry	Cucumber, Melon	Safflower	
Citrus	Onion, Pumpkin	Gingelly	
Litchi	Radish, Turnip		

4. Per cent increase in yield due to bee pollination

Crop	Botanical name	Per cent yield increase
Mustard	*Brassica* sp	43
Sunflower	*Helianthus annus*	32– 48
Cotton	*Gossypium* sp.	17-19
Lucerne	*Medicago sativa*	112
Onion	*Allium cepa*	93
Apple	*Malus domestica*	44

5. Scope of bee keeping for pollination in India

- Total area under bee dependent crops - 50 million ha.
- At the rate of 3 colonies/ha - 150 million colonies needed.
- In India, only 1.2 million colonies exist.

6. Management of bees for pollination

- Place hives very near the field (source) - to save bee's energy.
- Migrate colonies near field at 10% flowering.

- Place colonies at 3/ha - Italian bee; 5/ha - Indian honey bee.
- The colonies should have 5-6 frame strength of bees, possess sealed brood, have young mated queen.
- Allow sufficient space for pollen and honey storage.

7. Pollination by bees - Cross studies with selected crops

Sunflower

- It is a cross-pollinated crop by bees because of self incompatibility mechanism.
- Pollen should come from different plants.
- Improves quality and quantity of seeds.
- Requires 5 strong *A. cerana indica* colonies or 3 *A. mellifera* colonies.

Cucurbitaceous vegetables

- Monoecious - Staminate and pistillate flowers in the same plant.
- 30-100% increase in fruit set due to bee pollination.

Alfalfa (Lucerne)

- When bees sit on keel petal, staminal column strikes against standard petal and pollen shatters called as tripping.

Coriander

- Yield increase upto 187% due to bee pollination.

Cardamom

- Important commercial crop depends on bee pollination. Yield increase upto 21-37%

Gingelly

- An oilseed crop, bee pollination causes 25% increase in yield.

Apple

- Seed set occurs due to bee pollination.
- If improper seed set, fruit shape is lopsided (market value decreases).

8. Migratory vs. stationary beekeeping

- Migratory beekeeping is advantageous to beekeeper and farmer.

Part-5 Silkworm Rearing

1

Silkworm Rearing

1.1. Sericulture

- Refer to rearing of silkworms for the production of raw silk.
- Major activities are mulberry cultivation, silkworms spin silk cocoons and reeling the cocoons for unwinding the silk filament for value added benefits.
- *Bombyx mori* is the most widely used. Silk-fiber is a protein produced from the silk-glands (modified labial glands) of silkworms.

1.2. Races

- Multivoltine : Tamil Nadu white, Pure mysore, Hosa mysore, BL 23, BL 24.
- Bivoltine : KA-NB4D2, BL 23xNB4D2, BL 24 x NB4D2.
- Cross breed : PM x NB4D2, BL23 x NB4D2, BL 24 x NB4D2.
- Bivoltine hybrids : KA x NB4D2 (Nandi), NB7 x NB18 (chamundi), CSR 2 x CSR4, CSR3 x CSR 5, CSR 18 x CSR 19

1.3. Mulberry varieties

Irrigated : Kanva 2, MR 2, S 30, S 36, S 54, DD (Viswa), V1

Semi irrigated : Kanva 2, MR 2

Rainfed : S 13, S 34, RFS 135, RFS 175, S 1635

1.3.1. Manures, fertilizers and water requirement

Item	Rainfed	Irrigated	Chawki plot/garden
Manures FYM (tonnes/ha)	10	20	40
Fertilizer NPK (kg/ha/year)	100:50:50	300:120:120	225:150:150
Water requirement (mm)	600-700	1200-1250	1300-1400
Irrigation	-	Once in 10-15 days	Once in 7-10 days

1. Biofertilizers

Nursery

- *Azospirillum* 1 kg for treating the cuttings

Mainfield

- *Azospirillum*
- Phosphobacteria } Each 10 kg/ha/year in two splits at six months interval.

1.3.2. Pest and disease management

1. Pink mealy bug *(Maconellicoccus hirsutus* - responsible for tukra)

Control measures

- **Mechanical-**Collection and destruction of affected branches by burning.
- **Biological-**Releasing Australian lady bird beetle *Cryptolaemus montrouzieri* and indigenous coccinellid *Scymnus coccivora* .
- **Chemical-**Spraying with dichlorvos (2 ml/litre).

2. Root rot-*Macrophomina phaseolina* (*Fusarium moniliforme*)

Control measures

- **Biological-***Trichoderma viride, Bacillus subtilis*.
- **Chemical-**Carbendazim (2 ml/litre).

1.4. Rearing house

- A separate house is ideal for rearing of silkworm.
- Sufficient number of windows to permit cross ventilation.
- Air tight for proper disinfection.
- Optimum temperature of 26-28° C and RH of 60-70% .

1.5. Important principles

1.5.1. Avoid

- Damp condition
- Stagnation of air
- Direct and strong drift of air
- Exposure to bright sun light and radiation

- An equable temperature and humidity
- Good ventilation.

1.5.2. Features

- Rearing area of 2 sq.ft/ dfl for floor rearing and 3 sq. ft/ dfl for shoot rearing
- Main rearing hall, an ante room (8 x 8 ft) and leaf preservation room.
- Maintain a separate chawki room (must for two- plot rearing system; rearing room of size 10' x 14' with a height of 9-10 ft for an acre of garden).
- Face east-west direction.
- 480 sq.ft area is required for rearing 100 dfls.

1.5.3. Preparation of rearing house

- Rearing room: Kept Ready after disinfection atleast 3-4 days in advance.
- Arrange rearing appliances and provision of essential environmental conditions one day in advance.

1.5.4. Preparation for brushing

- Thoroughly wash and disinfect rearing equipments and rearing houses with chlorine dioxide and kept closed for about 24 hours after disinfection.
- 500 ml of chloride dioxide + 50 g of activator + 100 g lime powder are dissolved in 20 litres of water.
- In a tray of 120 cm x 90 cm x 105 cm size, 20 dfls are brushed and reared upto second stage.

1.5.5. Selection of leaves

- Select from the largest glossy leaf, 3rd or 4th from the top.
- Next 6 to 8 leaves are used to rear the young age worm's upto II moult.
- Size of the chopped leaf is 0.5 to 1.0 sq.cm during 2nd age.

1.5.6. Leaf preservation

- Silkworm grows best when fed with succulent leaves which are rich in nutrients and moisture.
- Harvested leaves must be preserved in fresh condition in a wet gunny cloth.
- If the climate is too hot and dry, the leaves are preserved in a leaf chamber which is lined with gunny cloth and kept wet by spraying water at frequent intervals.

1.5.7. Cleaning

- Remove excreta and left over leaves in the rearing bed
- In the first age, one cleaning is just a day before the worms settle for moulting.
- In the second age, two cleanings are given, one after resuming feeding and the other before second moult.
- Net with mesh size of 0.5 x 0.5 cm is spread over the rearing bed and feeding is given.
- Worms crawl through the net and come to fresh leaves.
- Net along with the worms and leaves are transferred to another tray.
- Left over leaves and litter are discarded.

1.5.8. Moulting

- During moult, the rearing bed should be kept thin and dry by applying lime @ 30 – 50 g/m^2 and should have proper aeration.

1.6. Shoot rearing for late age worms

- Silkworm larvae consume 85% of their food requirement during fifth instars. Fifty per cent of the labour input is utilized during the last seven days of rearing.

1.6.1. Rearing house

- Provide separate rearing house for shoot rearing in shady areas. Separate room should be provided for young age worm rearing, leaf storing and hall for late age worm rearing.

1.6.2. Shoot rearing rack

- A rearing rack of 1.2m x 11m size is sufficient to rear 50 dfls.
- Fabricate the rack stand with wood, or steel and the rearing seat with wire mesh/bamboo mat.

1.6.3. Shoot harvesting

- Harvest the shoots at 1 m height from ground level at 60 to 70 days after pruning.
- Store the shoots vertically upwards in dark cooler room.
- Provide thin layer of water (3 cm) in one corner of storage room and place the cut of shoots in the water for moisture retention.

1.6.4. Feeding

- Provide a layer of newspaper in rearing shelf.
- Disinfect the bed; spread the shoot in perpendicular to width of the bed.
- Transfer the third instars larvae to shoots immediately after moulting.
- Watch for feeding rate from 4th day of fourth instar. If 90% of larvae have not settled for moulting, provide one or two extra feedings.
- Provide 3 feedings during rainy/winter months and 4 feedings during summer rearing

1.6.5. Spacing

- 18-36 m^2/100 dfls.

1.6.6. Bed cleaning

- Bed cleaning is done once during second day of fifth instar following rope (or) net method.
- In rope method, spread 2 m length of rope (two numbers) at parallel row leaving 0.5m on other side.
- After 2 to 3 feedings, ends of the ropes are pulled to the centre to make it into a bundle.
- In net cleaning method, spread 1.5 cm^2 sizes net across the bed.
- After 2 or 3 feedings, the nets are lifted and the old bed is cleaned and disinfected.
- Transfer the net to newer shelf, spread the net over the shoots; larvae will migrate to lower layer.

1.7. Pest of silkworm

1.7.1. Uzi fly (*Exorista bombycis*)

- Mature maggot causes reduction in yield of cocoon and cocoon quality.
- Causes death of silkworm larva.

Symptoms

- Presence of creamy white oval eggs on the skin of larvae in the initial stage.
- Presence of black scar on the larval skin
- Silkworm larvae die before they reach the spinning stage (if they are attacked in the early stage).
- In later stage, pierced cocoon is noticed.
- Throughout the year, severity is more in winter months.

Management

- Spray 1% benzoic acid over the larvae to kill the eggs of uzi fly.
- Dissolve the uzicide tablets in the water (2 tablets/l) to attract the adults.
- Release the gregarious, ectopupal hyperparasitoid, ***Nesolynx thymus*** (Eulophidae: Hymenoptera) at 1 lakh adults/100 dfls during night hours. Release the hyperparasitoid in three split doses at 8000, 16,000 and 76,000/100 dfls during fourth and fifth instars and after cocoon harvest.

1.8. Diseases of silkworm

1.8.1. Viral Disease: *Grasserie (Borrelina virus (NPV), Flacherie, Infectious flacherie virus (IFV)*

Symptom

- Sluggish larvae with swollen inter segmantal region.
- Integument of diseases larvae will be fragile and breaks easily.
- On injury, milky fluid containing many polyhedral inclusion bodies oozes out from the larval body.
- Diseased larvae do not settle for moult and shows shining integument
- Larvae appear to be restless.
- Dead larvae hang by hind legs head downward (tree top disease).

Management

- Sun drying of rearing appliances for one/two days.
- Disinfection of rearing room and appliances with 5% bleaching powder.
- Dust the bed disinfectant, vijetha (or) resham keet oushadh or TNAU seri dust on the larvae, after each moult and ½ hr. before resumption of feeding (3 - 4 kg/100 dfl) and for bivol time hybrids 5-6 kg/100 dfls.
- Dose of disinfectant (g/100 dfl) (BV x BV hybrid)
- After I moulting - 50 g
- After II moulting - 150 g
- After III moulting - 900 g
- After IV moulting - 1900 g
- 5th instar – 4th day - 3000 g

  6000 g (or) 6 kg
- Spray 1% of extract of *Psoralea corylifolia* on mulberry leaves, shade dry and feed worms once during third instars.

1.8.2. Bacterial diseases

- Fluctuating temperature, humidity and poor quality mulberry predispose the disease development.
- Diseased larvae will be stunted in growth, dull lethargic soft and appear flaccid.
- Cephalothoracic region may be translucent.
- Larvae vomit gut juice, develop dysentery and excrete chain type faces.
- Larvae on death putrefy, develop different and emit foul smell.

Management

- Avoid spraying commercial Bt insecticides in nearby mulberry field.
- Apply antibiotics like streptomycin/tetracyclin/ampicillin.

1.8.3. Fungal disease: *Muscardine* (or) calcino

- **White muscardine:** *Beauveria bassiana.*
- **Green muscardine:** *Spicaria prasina, Metarhizium anisopliae.*
- Aspergillosis is common in young age silkworms and the infected larvae will be lustrous and die. Dark green (*Aspergillus flavus*) or rusty brown (*Aspergillus tamarii*) mycelial clusters are seen on the dead body.
- Diseased larvae prior to death will be lethargic and on death are flaccid.
- Oil specks may be seen on the surface of larvae.
- They gradually become hard, dry and mummify into a white or green coloured structure.
- The diseased pupae will be hard, lighter and mummified.

Management

- Sun dry the rearing appliances.
- Disinfect the rearing room and utensils with 5% bleaching powder.
- Avoid low temperature and high humidity in the rearing room.
- Keep the rearing bed thin and dry.
- Early diagnosis and rejection of infected lots.
- Apply dithane M45 / vijetha (4 kg/100 dfls) supplement as disinfectant on the larvae.
- Disinfect rearing rooms and trays with 4% pentachlorophenol to control Aspergillosis.

1.8.4. Protozoan disease: Pebrine (*Nosema bombycis*)

- Diseased larvae show slow growth, undersized body and poor appetite.
- Diseased larvae reveal pale and flaccid body.
- Tiny black spots appear on larval integument.
- Dead larvae remain rubbery and do not undergo putrefaction shortly after death.

Management

- Surface disinfects the layings in 2% formalin for 10 minutes before incubation.
- Collect and burn the diseased eggs, larvae, pupae and moths, bed refuses, faecal pellets, etc.

1.9. Other utilization

- Aesthetic materials made from waste cocoons.
- Flower vase.
- Interior decoration.
- Flower pot.
- Greeting cards.
- Animal toys.
- Garland.

1.9.1. Other utilization of silk and mulberry plants

- Used as operational thread in hospital.
- From mulberry fruits, juice can be prepared.
- Used as thread in parachutes and rocket.
- In sports cycle, thread can be used for preparing rubber wheel.
- From cocoon, protein extracted that can be used for cosmetics, medicine and value added products.

Part-6 Agricultural Engineering

1

Energy in Agricultural Production

1.1. Types of energy

1.1.1. Potential energy

- Stored energy and the energy of position (gravitational).

1.1.2. Kinetic energy

- Energy in motion - the motion of waves, electrons, atoms, molecules and substances. It exists in various forms.

1.2. Various forms of energy

1.2.1. Chemical energy

- Energy stored in the bonds of atoms and molecules.
- Eg. Biomass, petroleum, natural gas, propane and coal.

1.2.2. Nuclear energy

- Energy stored in the nucleus of an atom - the energy that holds the nucleus together.
- Eg. Nucleus of a uranium atom.

1.2.3. Stored mechanical energy

- Energy stored in objects by the application of a force.
- Eg. Compressed springs and stretched rubber bands.

1.2.4. Gravitational energy

- Energy of place or position.
- Eg. Water in a reservoir behind a hydropower dam.
- When the water is released to spin turbines, it becomes rotational energy.

1.2.5. Radiant energy

- Electromagnetic energy that travels in transverse waves.
- Includes visible light, x-rays, gamma rays and radio waves.
- Eg. Solar energy

1.2.6. Thermal energy

- Internal energy in substances.
- Vibration and movement of atoms and molecules within substances.
- Eg. Geothermal energy

1.2.7. Electrical energy

- Movement of electrons.
- Eg. Lightning and electricity

1.2.8. Motion

- Movement of objects or substances from one place to another is motion.
- Eg. Wind and hydropower.

1.2.9. Sound

- Movement of energy through substances in longitudinal (compression/rarefaction) waves.

1.2.10. Light energy

- Light energy is a type of wave motion.
- It enables us to see, as objects are only visible when they reflect light into our eyes.

1.3. Commercial primary energy resources

1.3.1. Non-renewable energy or conventional energy

- They are being accumulated in nature for a very long time and can't be replaced, if exhausted.
- Eg. Coal, petroleum, natural gas, thermal power, hydro power and nuclear power.

1.4. Renewable sources of energy

- Energy sources, which are continuously and freely produced in the nature and are not exhaustible.
- Eg: Solar energy, biomass and wood energy, geo thermal energy, wind energy, tidal energy and ocean energy.
- But main attention has to be directed to the following sources of renewable energy namely:
  a) Solar photovoltaic
  b) Wind
  c) Hydrogen fuel cell

Energy resource	Advantages	Disadvantages
Fossil fuels	Provide a large amount of thermal energy per unit of mass. Easy to get and easy to transport. Can be used to generate electrical energy and make products such as plastic.	Nonrenewable. Burning produces smog. Burning coal releases substances that can cause acid precipitation. Risk of oil spills.
Nuclear	Very concentrated form of energy Power plants do not produce smog.	Produces radioactive waste. Nonrenewable.
Solar	Almost limitless source of energy Does not produce air pollution.	Expensive to use for large scale energy production. Only practical in sunny areas.
Water	Does not produce air pollution.	Dams disrupt a river's ecosystem available only in areas that have rivers.
Wind	Renewable. Relatively inexpensive to generate. Does not produce air pollution	Only practical in windy areas.
Geothermal	Almost limitless source of energy Power plant require little land	Only practical areas near hot spots Waste water can damage soil
Biomass	Renewable	Requires large area of farmland Produces smoke

2

Energy Sources: Solar, Wind, Animal, Biomass and Biogas

2.1. Solar energy

- States like Andhra Pradesh, Bihar, Gujarat, Haryana, Madhya Pradesh, Maharashtra, Odisha, Punjab, Rajasthan, and West Bengal have greater potential for tapping solar energy due to their location.

2.1.1. Advantages

- Inexhaustible source of energy.
- Environment friendly and inexpensive.
- Variety of purposes like heating, drying, cooking and electricity.
- Also used in cars, planes, large power boats, satellites, calculators etc.

2.1.2. Disadvantages of solar energy in India

- Cannot generate energy during the night time, cloudy or rainy, with little or no sun radiation.
- Require inverters and storage batteries to convert direct electricity to alternating electricity so as to generate electricity.
- Land space is quite large
- Energy production is quite low.
- Solar panels require considerable maintenance as they are fragile and can be easily damaged.

2.1.3. Solar constant

- Total radiation energy received from the sun per unit of time per unit of area on a theoretical surface perpendicular to the sun's rays and at earth's mean distance from the sun.
- Value of the constant is approximately 1.366 kilowatts per square metre.

- **Constant** is fairly constant, increasing by only 0.2 percent at the peak of each 11 year solar cycle.

2.1.4. Applications

1. Solar water heating

- A solar water heating unit comprises a blackened flat plate metal collector with an associated metal tubing facing the general direction of the sun.
- The plate collector has a transparent glass cover above and a layer of thermal insulation beneath it.
- The metal tubing of the collector is connected by a pipe to an insulated tank that stores hot water during cloudy days.
- The collector absorbs solar radiations and transfers the heat to the water circulating through the tubing either by gravity or by a pump.
- This hot water is supplied to the storage tank via the associated metal tubing.

2. Solar heating of buildings

- Solar energy is admitted directly into the building through large south-facing windows.
- Using separate solar collectors which may heat either water or air or storage devices which can accumulate the collected solar energy for use at night and during inclement days.

3. Solar distillation

- The abundant sunlight can be used for converting saline water into potable distilled water by the method of solar distillation.
- In this method, solar radiation is admitted through a transparent air tight glass cover into a shallow blackened basin containing saline water.
- Solar radiation passes through the covers and is absorbed and converted into heat in the blackened surface causing the water to evaporate from the brine (impure saline water).
- The vapors produced get condensed to form purified water in the cool interior of the roof.

4. Solar pumping

- In solar pumping, the power generated by solar-energy is utilized for pumping water for irrigation purposes.

- The requirement for water pumping is greatest in the hot summer months which coincide with the increased solar radiations during this period and so this method is most appropriate for irrigation purpose.

5. Solar drying of agricultural and animal products

- Agricultural products are dried in a simple cabinet dryer which consists of a box insulated at the base, painted black on the inner side and covered with an inclined transparent sheet of glass.
- At the base and top of the sides ventilation holes are provided to facilitate the flow of air over the drying material which is placed on perforated trays inside the cabinet.
- These perforated trays or racks are carefully designed to provide controlled exposure to solar radiations.

6. Solar furnaces

- High temperature is obtained by concentrating the solar radiations onto a specimen using a number of heliostats (turnable mirrors) arranged on a sloping surface.
- The solar furnace is used for studying the properties of ceramics at extremely high temperatures above the range measurable in laboratories with flames and electric currents.
- An important future application of solar furnaces is the production of nitric acid and fertilizers from air.

7. Solar cooking

- A simple solar cooker is the flat plate box type solar cooker.
- Consists of a well insulated metal or wooden box which is blackened from the innerside. The solar radiations entering the box are of short wavelength.
- As higher wavelength radiations are unable to pass through the glass covers, the re-radiation from the blackened interior to outside the box through the two glass covers is minimised, thereby minimising the heat loss.
- Maintenance cost of the solar cooker is negligible.
- Main disadvantage of the solar cooker is that the food cannot be cooked at night, during cloudy days or at short notice.
- Cooking takes comparatively more time and chapattis cannot be cooked in a solar cooker.

8. Solar electric power generation

- Electric energy or electricity can be produced directly from solar energy by means of photovoltaic cells.
- The photovoltaic cell is an energy conversion device which is used to convert photons of sunlight directly into electricity.
- It is made of semi conductors which absorb the photons received from the sun, creating free electrons with high energies.
- Photovoltaic cells have been used to operate irrigation pumps, rail road crossing warnings, navigational signals, highway emergency call systems, automatic meteorological stations etc.
- Also used for weather monitoring and as portable power sources for televisions, calculators, watches, computer card readers, battery charging and in satellites
- Also used for the energisation of pump sets for irrigation, drinking water supply and for providing electricity in rural areas i.e. street lights.

9. Solar thermal power production

- Conversion of solar energy into electricity through thermal energy.
- Solar energy is first utilised to heat up a working fluid, gas, water or any other volatile liquid.
- This heat energy is then converted into mechanical energy in a turbine.
- Finally a conventional generator coupled to a turbine converts this mechanical energy into electrical energy.

10. Production of power through solar ponds

- A solar pond is a natural or artificial body of water utilised for collecting and absorbing solar radiation and storing it as heat.
- It is very shallow (5-10 cm deep) and has a radiation absorbing (black plastic) bottom.
- It has a curved fibre glass cover over it to permit the entry of solar radiation but reduces losses by radiation and convection (air movement).
- Loss of heat to the ground is minimised by providing a bed of insulating material under the pond.
- Solar ponds is used for space heating, industrial process heating and to generate electricity by driving a turbine powered by evaporating an organic fluid with a low boiling point.

11. Solar green houses

- A green house is a structure covered with transparent material (glass or plastic) that acts as a solar collector and utilises solar radiant energy to grow plants.
- It has heating, cooling and ventilating devices for controlling the temperature inside the green house.
- Solar radiations can pass through the green house glazing but the thermal radiations emitted by the objects within the green house cannot escape through the glazed surface.
- As a result, the radiations get trapped within the green house and result in an increase in temperature.
- As the green house structure has a closed boundary, the air inside the greenhouse gets enriched with CO_2 as there is no mixing of the greenhouse air with the ambient air.
- Further, there is reduced moisture loss due to restricted transpiration.
- All these features help to sustain plant growth throughout the day as well as during the night and all year round.

2.2. Wind energy

The power in the wind is proportional to:

- Area of windmill being swept by the wind
- Cube of the wind speed
- Air density which varies with altitude

Power = density of air x swept area x velocity cubed

$P = \frac{1}{2}.\rho.A.V^3$

Where, P is power in watts (W)

ρ (Rho) is the air density in kilograms per cubic metre (kg/m^3)

A is the swept rotor area in square metres (m^2)

V is the wind speed in metres per second (m/s)

2.2.1. Advantages

- Wind energy is an inexhaustible source of energy and is virtually a limitless resource.
- Energy is generated without polluting environment.
- Wind power taps a natural physical resource.

- Don't emit any emissions that can lead to acid rain or greenhouse effect.
- Used directly as mechanical energy.

2.2.2. Disadvantages

- Wind energy requires expensive storage during peak production time.
- Unreliable energy source as winds are uncertain and unpredictable.
- Visual and aesthetic impact on region.
- Requires large open areas for setting up wind farms.
- Noise pollution problem is usually associated with wind mills.

2.2.3. Principles of wind energy conversion

- Two primary physical principles by which energy can be extracted from the wind; these are through the creation of either lift or drag force or through a combination of the two.
- Basic features that characterize lift and drag are:
- Drag is in the direction of air flow
- Lift is perpendicular to the direction of air flow
- Generation of lift always causes a certain amount of drag to be developed
- With a good aerofoil, the lift produced can be more than thirty times greater than the drag
- Lift devices are generally more efficient than drag devices.

2.2.4. Classification of wind energy conversion systems

- Broadly categorized into two types according to their axis alignment.

1. Horizontal axis wind turbine

- It can be further divided into three types

Dutch type grain grinding wind mills

- Grain grinding windmills that are widely used in Europe since the middle ages were Dutch.
- These are operated on the thrust exerted by the wind.
- Blades, generally four, are inclined at an angle to the plane of rotation.
- Wind being deflected by the blades exerted a force in the direction of rotation.
- Blades were made of sails or wooden slats.

Multiblade water pumping windmill

- Modern water pumping windmills have a large number of blades generally wooden or metallic driving a reciprocating pumps.
- As the mill has to be placed directly over the well, the criterion for site selection concerns about water availability and not windiness.
- Large number of blades gives a high torque, required for driving a centrifugal pump, even at low wind speeds. Hence sometimes these are called as fan mills.
- Blades are made of flat steel plates, working on the thrust of wind.
- These are hinged to a metal ring to ensure structural strength, and the low speed of rotation adds to the reliability.
- Orientation is generally achieved by tail vane.

High speed propeller type wind machines

- Used for electricity generation and not does operate on thrust force.
- They depend mainly on the aerodynamic forces that develop when wind flows around a blade of aerofoil design.
- Windmills working on thrust force are inherently less efficient.
- So all the modern wind turbine blades are designed based on aerofoil section.

2. Vertical axis wind turbines

Three different designs

The savonius rotor

- Extremely simple vertical axis device that works entirely because of the thrust force of wind.
- Basic equipment is a drum, cut in two halves vertically.
- Two parts are attached to the two opposite sides of a vertical shaft. As the wind blowing into the structure meets with two dissimilar surfaces one convex and the other concave the forces exerted on the two surfaces are different, which gives the rotor a torque (force that tends to cause rotation).
- By providing a certain amount of overlap between the two drums, the torque can be increased.
- This is because the wind blowing into the concave surface turn around and give a push to the inner surface of the other drum, partly cancelling the wind thrust on the convex side.

- It has been found that an overlap of about one third the drum diameter gives optimum result.

The darrieus rotor

- Two or more flexible blades are attached to a vertical shaft.
- Blades bow outwards, taking approximately the shape of a parabola and are of symmetrical airfoil section.
- Here the torque is zero when the rotor is stationary.
- It develops a positive torque only when it is already rotating. This means that such a rotor has no starting torque and has to be start using some external means.

Giromill

- A variant of Darrieus wind turbine is the Giromill which uses the same concept.
- Here the blades are straight resulting in simple construction.
- However, in such a case the centrifugal force developed in the blade will produce stress, trying to bend it.
- Blades have to be strong enough in the transverse direction to withstand this stress.
- Moreover the vertical shaft cannot be secured with guywires, and so the coupling at the base has to be strong enough to keep it vertical when subjected to strong winds.
- Also called as H-type windmill because of its shape.

2.3. Animal power

- Power developed by an average pair of bullocks is about 1 hp.
- Other animals like camals, buffaloes, horses, donkeys, mules and elephants are also used.
- Average force a draft animal can exert nearly one tenth of its body weight.

Merits and demerits of animal power

S.No.	Merits	Demerits
1.	Easily available.	Not very efficient.
2.	Used for all types of work.	Seasons and weather affect the efficiency.
3.	Low initial investment.	Cannot work at a stretch.
4.	Supplies manures to the field and fuels to farmers.	Requires full maintenance when not in use.
5.	Lives on farm products.	Creates unhealthy and dirty atmosphere near the residence.Very slow in doing work.

2.4. Biomass

- Refers to organic materials made from plants and animals.
- Contains stored energy from the sun.
- Eg. Fuels like wood, crops, manure, and some garbage.
- Biomass can be converted to energy like methane gas or transportation fuels like ethanol and biodiesel.
- Methane gas is the main ingredient of natural gas.
- Smelly stuff, like rotting garbage, and agricultural and human wastes release methane gas.
- Also called as landfill gas or biogas marsh gas.
- Corn and sugarcane can be fermented to produce ethanol.

2.4.1. Typical biomass supply materials

- Woody forest residue, fuelwood, mill residues, short rotation crops.
- Non-woody agricultural crops, crop residues, processing residues.
- Animal wastes such as manure and municipal sewage and wastes.

2.4.2. Biomass energy – Direct and indirect uses

- **Direct use:** Firewood used by combustion.
- **Indirect use:** Fuels like ethanol and methane.

2.4.3. Biomass energy conversion

- Agricultural wastes or manures undergo thermal and biological process to form gas.

1. Thermal biomass energy conversion

- **Combustion**: Biomass is burned to make electricity.
- Solid municipal waste is also burned to generate electricity.
- When mixed with coal, cofiring process is very efficient.
- **Pyrolysis**: Chemical process of decomposition of biomass materials.
- Heating process without the involvement of oxygen.
- Recycling of used tyres uses pyrolysis.
- **Gasification**: Process that turns parts of solid biomass materials into gas.

  Gasification stages are

Stage I: **Auto-thermal heating of the reaction mixture.**

- Necessary heat for this process is obtained by initial oxidation exothermic reactions by combustion.

Stage II: **Pyrolysis stage**: Being passed through a bed of fuel at high temperature pyrolyzes combustion gases.

- Heavier biomass molecules distillate into medium weight organic molecules and CO_2.
- Tar and char are also produced.

Stage III: Initial products of combustion namely carbon dioxide (CO_2) and (H_2O) are reconverted by reduction reaction to carbon monoxide (CO), hydrogen (H_2) and methane (CH_4).

2. Biological conversion

- **Digestion and fermentation:** Biodegradable wastes in large digester power plants where bacteria convert waste into gas.
- Produced gas drives turbines that generate electricity.
- Solids that are left behind may be used as fertilizer.

3. Biofuels from biomass

- Biogas
- Bioethanol
- Biobutanol and
- Biodiesel

2.4.4. Biomass as a source of energy

- Biomass fuels represent the second largest source of energy used after fossil fuels in the world.
- About 35 per cent of the energy used is from biomass in developing countries.
- About 40 per cent of the total energy consumed in India comes from fuel wood, charcoal and various agricultural residues.
- About half of all the trees cut in the world are used as fuel for cooking and heating.
- These methods are highly inefficient.
- On the other hand, utilization of the residues through gasification becomes economical and promising for thermal and power to rural areas and for small scale agro industries.

2.5. Biogas

- A mixture containing 55-65% methane, 30-40% carbon dioxide (CO_2) and the rest being the impurities (H_2, H_2S, N_2).
- Produced from decomposition of animal, plant and human wastes.
- Calorific value 5000 to 5500 kcal/kg.
- Used directly in cooking, reducing the demand for firewood.
- Used as a fertilizer.
- Popular as "**Gobar gas**" because it is produced from cow dung.
- Produced by digestion, pyrolysis, or hydrogasification.
- Digestion biological process that occurs in the absence of oxygen and in the presence of anaerobic organisms at ambient pressures and temperatures of 35-70°C in a digester.

2.5.1. Biogas production method

1. Nature manure that are used for biogas production

- **Raw manure:** Manure is excreted with a solid content of 8 to 25%.
- **Liquid manure:** Manure is diluted to a solid content of less than 5%.
- **Slurry manure**: Manure is diluted to a solid content of about 5 to 10%.
- **Semi-solid manure:** Manure has a solid content of 10 to 20%.
- **Solid manure**: Manure with a solid content of greater than 20%.

2. Size for biogas plant

- Depends on,
- Daily feed
- Retention time
- Digester volume
- 1 kg of cow dung along with equal quantity of water (1:1) under anaerobic conditions in a day produces 0.04 m^3 or 40 litres of biogas.

3. Selection of construction site

- Plant must be as close as possible to the cattle shed to avoid wastes.
- Slightly higher elevation than the surrounding to avoid water logging.
- Temperature of 20-35°C has to be maintained inside the digester.
- Avoid damp and cool place and sunny site is preferable.
- Well or ground water source should be at least 10 metre away from the biodigester especially the slurry pit to avoid the ground water pollution.
- Plant should not be located further than 5 metres from the field.

4. Biogas plant and its components

- **Mixing tank-**Cow dung is mixed with the water in equal proportion (1:1) to make a homogenous mixture (slurry) in the mixing tank
- **Feed inlet pipe/tank-**Homogenous slurry is let into the digester through this inlet pipe (KVIC biogas plants)/tank (Janatha biogas plants)
- **Digester-**Fed slurry is subjected to anaerobic fermentation with the help of microorganisms inside the digester.
- **Gas holder-**Gas produced is stored in gas holder (Drum in the case of KVIC and in dome in the case of fixed dome biogas plants)
- **Slurry outlet tank/pipe-**Digested slurry is let out from the digester through slurry outlet pipe (KVIC biogas plants)/tank (Janatha biogas plants)
- **Gas outlet pipe-**Stored gas is released and conveyed through the gas outlet pipe present at the top of gas holder.

5. Stages of anaerobic digestion in methane production

- Formerly, methane fermentation designed as a two steps ie., an acid forming stage and methane forming stage.
- Current scheme represents a three stage processes in which three groups of bacteria are involved.

- Hydrolytic and acidogenic bacteria - *Clostridium*, *Pseudomonas, Bacillus, Streptococcus, Desulfobacter.*
- Acetogenic bacteria - *Acetobacterium* spp
- Methanogenic bacteria - *Methanosarcina barkeri, Methanobacterium thermoautotrophicum, M. bryantii.*

Anaerobic digestion

i) Enzymatic hydrolysis

- Facultative microorganisms convert insoluble, complex, high molecular compounds of biomass into simple, soluble, low molecular compounds.
- Polysaccharide, protein and lipid are converted into monosaccharide, peptide, amino acids and fatty acids.
- Further converted into acetate, propionate and butyrate.

ii) Acid formation

- Acid formers (facultative and anerobic microorganisms), hydrolyze and ferment the water soluble substances into volatile acid i.e., acetic acid.
- Then CO_2, acetate or hydrogen, butyric acid and propionic acid are also produced.

iii) Methane formation

- Acetate or hydrogen plus carbon dioxide are converted into mixture of methane gas (CH_4) and CO_2 by the anerobic bacteria methane fermentators.
- Indigestible matter is slurry.
- One kg of dry cattle dung produces approximately 1 m^3 of biogas.
- One kg of fresh cattle dung contains 8% dry biodegradable mass.
- One kg of fresh cattle dung has a volume of about 0.9 litres.
- One kg of fresh cattle dung requires an equal volume of water for preparing slurry.
- Typical retention time of slurry in a biogas plant is 40 days.

6. Factors affecting efficiency of biogas generation

- For anerabic formentation, temperature range of 8-55°C with 35°C as optimum.
- Methane bacteria work best in the temperature range of 35 and 38°C.

- At 8°C, very slow rate of biogas formation.
- For anerabic digestion, temperature variation should not be more than 2 to 3°C.
- pH 6.8 and 7.8.
- C:N ratio should be 25:1 and 30:1 with 30:1 as optimum.
- Water content should be around 90% of the total weight.
- Slurry should contain 8-9% solid organic matter.

13.5.2. Classification of biogas plants

1. Batch type

- Organic waste materials to be digested under anaerobic condition are charged only once into a reactor-digester.
- Feeding is between intervals
- Plant is emptied once the process of digestion is completed.
- Retention time varies from 30 to 50 days.
- Plant needs addition of fermented slurry to start the digestion process.
- Well suited for fibrous material.
- Not economical to maintain.

2. Semi continuous

- A predetermined quantity of feed material mixed with water is charged into the digester from one side at specified interval of time (say once a day).
- Digested material (effluent) equivalent to the volume of the feed, flows out of the digester from the other side (outlet).

3. Continuous type

- Feed material is continuously charged to the digester with simultaneous discharge of the digested material (effluent).
- Main features of this type are continuous gas production, requires small digestion area, lesser period for digestion, less maintenance, etc.

4. The biogas plants used in the villages are of semi continuous type employing animal dung

i) Floating drum type – KVIC model

ii) Fixed dome type model – Deenbandhu model

i) Floating drum biogas plant (Constant pressure)

- Plants digester is made of bricks and circular in shape.
- Partition wall is constructed (dividing the digester into two parts) for higher size capacity plants to avoid short-circuiting of digested slurry with the fresh feed.
- Separate gas holder is fabricated and fixed to store the gas.
- Central guide frame is provided to hold the gas holder and to allow it to move vertically during gas production.
- Drum is made up of mild steel and it constitutes around 60 per cent of overall plant costs.
- Gas outlet pipe is provided at top of the drum to let the gas out of drum.
- KVIC floating drum model is predominantly used in India.

Advantages

- Higher gas production per cubic metre of the digester volume.
- Floating drum has welded braces, which help in breaking the scum by rotation.
- No problem of gas leakage.
- Constant gas pressure.

Disadvantages

- Higher cost, as cost is dependent on steel and cement
- Heat is lost through the metal gas holder.
- Gas holder required painting once or twice a year, depending on the humidity of the location.
- Flexible pipe joining the gas holder to the main gas pipe requires maintenance, as ultraviolet rays in the sun damage it.

ii) Fixed dome biogas plant (Constant volume)

- To reduce the cost fixed dome plants is designed in which dome act as gas holder in place of high cost drum.
- Gas holder and digester constructed as single unit.
- Digesters are completely underground to maintain a perfect environment for anaerobic fermentation
- Also avoids cracking of dome due to difference in temperature and moisture.

iii) Janatha biogas plants

- Developed exclusively in India with complete masonry structure.
- Provision of inlet of raw and outlet of digested slurry is constructed in the form of tank.
- Slurry fed is allowed to undergo anaerobic fermentation in the digester.
- Gas produced rises up and gets collected in the dome.
- As the pressure of gas stored in the dome increases, it pushes up the slurry down and causes the slurry level to increase both in inlet and outlet tanks.

iv) Deenbandhu biogas plant

- Improved version of Janatha biogas plant model.
- Action for Food Production (AFPRO), a voluntary organization in New Delhi, developed this model in 1984.
- Constructed with locally available materials and the plant demands skillful manpower for construction.
- Important considerations for design modification.
- Elimination of the loss of biogas through inlet chamber.
- Maximum utilization of digester volume to make the operational HRT close to the designed HRT.
- Constructed by joining the two spheres of different diameters at their bases, thus reducing the cost of bricks used in construction of digester wall.
- Bottom part of the plant is a designed as a segment of sphere, whereas the top portion as hemisphere.
- Feedstock is fed through concrete pipes and the digested slurry is taken out the digester through tank.
- To avoid the entry of slurry through gas outlet pipe, outlet opening is constructed 150 mm lower than the bottom of gas outlet pipe.
- Gas holding capacity is 33% of total capacity of the plant.
- Cost of deenbandhu is 30 and 45 per cent less than that of Janatha and KVIC biogas plants.

2.5.3. Effluent storage

- Products of the anaerobic digestion of manure are biogas and effluent.
- Effluent is a stabilized organic solution that has value as a fertilizer and other potential uses.
- Waste storage facilities are required to store treated effluent because the nutrients in the effluent cannot be applied to land and crops year round.

2.5.4. Biogas use

- Recovered gas is 60 – 80% methane, with a heating value of approximately 600 - 800 Btu/ft^3 (Btu-British thermal unit).
- Gas of this quality can be used to generate electricity.
- Used as fuel for a boiler, space heater or refrigeration equipment.
- Directly combusted as a cooking and lighting fuel.

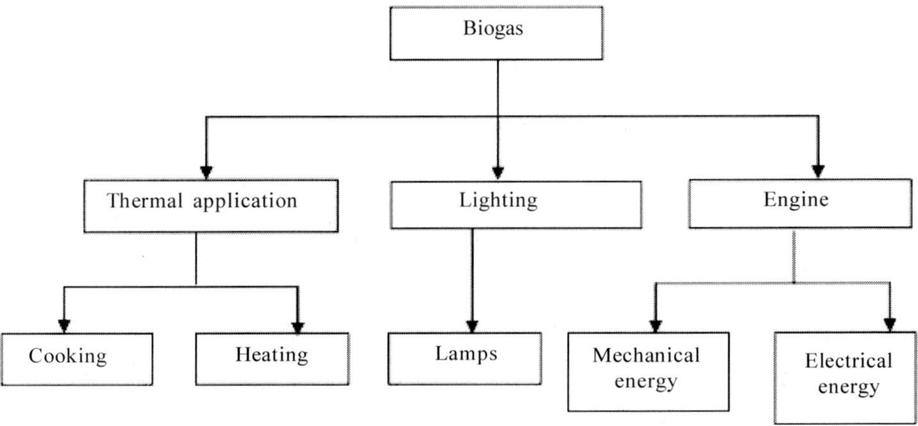

2.5.5. Benefits of biogas technology

- Cost-effective, environment and neighborhood friendly.
- Anaerobically digested manure, resulting in biogas and a liquified, low-odour effluent.
- Reduces BOD (Biochemical Oxygen Demand) and pathogen levels of effluent.
- Noxious odour is removed.
- Organic nitrogen is converted to inorganic N.

Principal reasons for installing biogas system by a farmer

- **On-site farm energy**: Reduces monthly energy purchases from electric and gas suppliers.
- **Reduced odours:** Biogas system reduces offensive odours from overloaded or improperly man aged manure storage facilities.
- **High quality fertilizer:** Organic nitrogen in the manure is converted to ammonium.
- Ammonium is the primary constituent of commercial fertilizer, which is readily available and utilized by plants.

- **Reduced surface and groundwater contamination:** Properly applied, digester effluent reduces the likelihood of surface or groundwater pollution.
- **Pathogen reduction:** Heated digesters reduce pathogen populations dramatically in a few days.

2.5.6. Biogas upgradation

- Biogas is upgraded to Synthetic Natural Gas (SNG) by removing CO_2 and H_2S.
- Family biogas plants usually of 2-3 m^3 capacity.

2.5.7. Advantages

- Initial investment is low for the construction of biogas plant.
- Technology is very suitable for rural areas.
- Use of biogas in village helps in improving the sanitary condition and checks environmental pollution.
- By-products like nitrogen rich manure can be used with advantage.
- Reduces the drudgery of women and lowers incidence of eye and lung diseases.

2.5.8. Biogas is produced mainly from

- Cow dung
- Sewage
- Crop residues
- Vegetable wastes
- Water hyacinth
- Poultry droppings
- Pig manure

2.5.9. Problems related to biogas plant

- No more dangerous than other fuels such as wood, gasoline or bottled gas.
- Explosive when mixed with air in the proportion of one part biogas to 8-20 parts air in an enclosed space.
- A biogas leak can be smelled, if the hydrogen sulfide has not been removed from the biogas.
- Smells like rotten eggs.
- Negative pressure will pull air into the biogas system and the mixture of biogas and air might explode.

3

Mechanization in Agriculture

3.1. Concept of farm mechanization

- To apply the principles of engineering and technology to do the agricultural operations in a better way to increase crop yield.
- Development, application and management of all mechanical aids for field operation, water control, material handling, storage and processing.
- Mechanical aids include hand tools, animal drawn implements, power tillers, tractors, engines, electric motors, grain processing and hauling equipments.

3.2. Scope of farm mechanization

- Improved irrigation facilities, introduction of high yielding varieties, use of higher doses of fertilizers and pesticides have increased the scope for greater farm mechanization.
- Helps for proper utilization of basic inputs like water, seed and fertilizer, optimum placement of the seed and fertilizer, ploughing, removal of weeds, levelling of uneven land and land reclamation.

3.3. Benefits of farm mechanization

- Timeliness of operation
- Precision of operation
- Improvement of work environment.
- Enhancement of safety
- Reduction of drudgery of labour
- Reduction of loss of crops and food products
- Increased productivity of land
- Increased economic return to farmer
- Improved dignity of farmer
- Progress and prosperity in rural areas

3.4. Constraints in farm mechanization

- Small land holdings.
- Less investing capacity of farmers.
- Adequate availability of draft animals.
- Lack of suitable farm machine for different operations.
- Lack of repair and servicing facilities for machines.
- Lack of trained man power.
- Lack of coordination between research organization and manufacturer.
- High cost of machines.
- Inadequate quality control of machine.

4

Tractors and Power Tillers

4.1. Tractor

- Self propelled power unit having wheels or tracks for operating agricultural implements and machines including trailers.
- Tractor engine is used as a prime mover for active tools and stationary farm machinery through power-take off (pto) or belt pulley.

4.1.1. Classification of tractors

- Based on structural design

1. Wheel tractor

- Tractors having three or four pneumatic wheels are called wheel tractors. Four wheel tractors are popular.

2. Crawler tractor

- Track type tractor or chain type tractor.
- In this tractors, there is endless chain or track in place of pneumatic wheels

3. Power tiller

- Walking type tractor.
- Fitted with two wheels only.
- Operations are performed by operator, walking behind the tractor.

4.1.2. Wheel tractor

- On the basis of purpose, wheel tractors are classified into three groups.

1. General purpose tractor

- Used for major farm operations such as ploughing, sowing, harvesting and transporting works.
- Tractors have low ground clearance, increased engine power, good adhesion and wide tyres.

2. Row crop tractors

- Used for row crop cultivation.
- Tractor is provided with replaceable driving wheels of different tread widths.
- Has high ground clearance to save damage of crops.
- Wide wheel track can be adjusted to suit inter row distance

3. Special purpose tractor

- Used for definite jobs like cotton fields, marshy lands, hill sides, garden etc.
- Special designs are there for special purpose tractor.
- Eg. Tractor with winch unit, multi drive tractor and tractor for golf grounds etc.

4.1.3. Selection of tractor

- **Land holding:** 1 hp for every 2 hectares of land i.e., one tractor of 20-25 hp is suitable for 40 hectares farm.
- **Cropping pattern:** Generally 1.5 hectare/hp has been recommended where more than one crop is grown. 30-35 hp tractor is suitable for 40 hectares of land.
- Soil condition
- Climatic condition
- Repair facilities
- Running cost
- Initial cost and resale value

4.1.4. Components

- I.C engine
- Clutch
- Transmission gears
- Differential unit
- Final drive
- Rear wheels
- Front wheels
- Steering mechanism
- Hydraulic control and hitch system
- Brakes

- Power-take-off unit
- Tractor pulley
- Draw bar
- Control panel

4.2. Power tillers

- Also known as **hand tractor** or **walking type tractor**.
- Prime mover in which the direction of travel and its control for field operation is performed by the operator walking behind it.
- Concept of power tiller came in the world in 1920.
- Japan is the first country to use power tiller
- In Japan, the first successful model of power tiller was designed in 1947.
- In India, power tiller was introduced in 1963.
- Used for ploughing, sowing, spraying, harvesting and transporting works.
- Most wanted machine for puddling operation in rice cultivation.
- Makes are Isuki, Sato, Krishi, Kubota, Yanmar and Mitsubishi.

4.2.1. Components of power tiller

- Engine
- Transmission gears
- Clutch
- Brakes
- Rotary unit
- All power tillers are fitted with an I.C. engine.
- Kubota, Mitsubishi, Krishi, Yanmar and Satoh use diesel engine, Isuki make use kerosene engine.

4.2.2. Uses

- For puddling operation in paddy fields using rotary tines.
- For cutting and pulverizing the soil in dry and garden lands.
- For cutting and pulverizing the stubbles of sugarcane, maize and cotton.
- For sowing and intercultivation works.
- For spraying of orchard trees.
- For transporting purposes.

5

Agricultural Implements and Machineries and Their Usage

5.1. Farm machinery implements

5.1.1. Tillage

S.No.	Implements/machineries	Usage/functions/salient features
1	Improved iron plough	Plough is provided with a mould board as an optional attachment for soil inversion. Improved iron plough is suitable for dry ploughing in all types of soil. Operated by pair of bullocks.
2	Low draft chisel plough	Suitable for deep tillage upto a depth of 40 cm for opening hard pan of soil. Operated by any 35-45 hp tractor.
3	Para plough	For mulch tillage and moisture conservation under dry farming condition. Increases the hydraulic conductivity of soil to an extend of 0.72 cm per hour. Loosed the soil by 25% more at 32-45 cm depth than conventional treatment. Conserve more moisture at 45 cm depth. Restricted moisture depletion upto 10 days after rainfall
4	Tractor drawn channel former	To form channels and beds at regular intervals for irrigation Cost of forming irrigation channel at 5 m interval by • Channel former: Rs. 165/ha. • Manual labour : Rs. 385/ha. • Saving in cost : Rs. 220/ha. • Saving in time : 11 man days/ha.
5	Tractor drawn bed furrow former	Forms alternate beds and channels in ploughed fields.

		Suitable for planting crops like sorghum, maize, cotton etc. Efficient irrigation and save water. Facilitates intercropping and intercultural operations.
6	Tractor drawn trencher	To form rectangular trench of 30 x 30 cm Also used for laying drip irrigation pipes by opening trenches. Application of manure in coconut fields. Making drainage around the sugarcane. Cost of operation is Rs. 0.10 per m run of trench as against Rs. 2.00 per m run using manual labour. Cost, time and energy saved by machine trenching in comparison to manual trenching were 95, 99 and 53%, respectively.
7	Power tiller mounted terracer cum leveller	For land levelling, terracing and bund forming. Simple in design, construction and ease of operation and transport. Efficient for land levelling with a transportation efficiency of 86.6%. Cost of moving 1 m^3 of soil to 1 m distance is Rs.3.30. Cost of bund forming is Rs.17.75/m^3 which is comparable with the manual bunding.
8	Power tiller operated slasher cum in situ shredder	For shredding vegetable residues of brinjal, chillies, bhendi, etc. left after harvest and parthenium, etc. Suitable for any make of power tiller with 10-15 hp. Saving in time - 73%. Saving in cost - 75%.
9	Tractor operated pit digger for sugarcane planting	For digging pits. Dig two pits of 90 cm diameter simultaneously at 1.5 m interval to a depth of 30 cm suitable for planting sugarcane setts. Planting of cane in 1.5 x 1.5 m spacing with pit method favours higher cane yield.
10	Tractor operated rotary spading machine	For primary tillage. For deep inter cultivation between rows under hard soils. For use in rice fallow cotton and coconut garden.

5.1.2. Sowing equipments

11	Automated protray sowing equipments	Automating all the steps involved in the sowing of seeds in protrays. 100 per cent saving in cost over the manual sowing.
12	Tractor drawn pulse seeder	Sow pulse seeds by optimizing seed quantity. Uniform hill to hill spacing closer to the recommended level of 10 cm.
13	Manually operated line marker (Pulses)	Adjustable spacing – starting from 15 cm to 150 cm. Reduced seed rate. Reduced cost of cultivation. Optimum plant population. Easy for inter cultural operation.
14	Broadcaster	To broadcast seeds and fertilizers. Suitable for sesame, sorghum, etc. Large area can be sown in less time.
15	Gorru	Line sowing and fertilizer application for crops like groundnut, sorghum, cowpea, bengalgram, greengram, blackgram, etc. in three rows simultaneously. As it is provided with three furrow openers similar to country plough, the coverage is about 3 times more comparing with the country plough.
16	Bullock drawn seed planter	Line sowing of groundnut, sorghum, cowpea, bengalgram, green gram, black gram, etc. in three rows simultaneously. The seed to seed distance in a row can be adjusted. By changing the size of the cup different crops/ varieties of seed can be sown.
17	Power tiller drawn cup feed seeder	To sow seeds of groundnut, maize, sorghum and pulses. Suited for all the makes of 10-12 hp power tiller. Coverage is 0.2 ha per hour.
18	Tractor drawn cultivator seed planter	For sowing groundnut, bengalgram, maize, sorghum, soybean and pulses in larger areas in short time and other seeds. Coverage : 4.0 ha per day Power requirement : 35 hp tractor Cost of operation : Rs. 250/ha (Rs. 350/ha for conventional method) The break even point is 9.65 ha / year.

19	Tractor drawn ridger seeder	For ridge forming and sowing simultaneously.
20	Basin lister cum seeder	To form intermittent basins in dry land for moisture conservation and simultaneous sowing. Increased moisture retention of 10 per cent is achieved Results in 32, 96 and 18% saving in cost, time and energy respectively when compared to conventional method. An area of 3.5 ha can be covered per day
21	Improved broad bedformer cum seeder	Forms raised bed and sows simultaneously Suitable for sowing groundnut, bengalgram, maize, sorghum and pulses in raised beds. Saving in cost : 57% Saving in labour and time : 95%
22	Power tiller operated air assisted seed drill	For sowing small seeds like sesame, cumbu, horsegram and sorghum. Spacing between the rows can be adjusted from 30 to 60 cm. Suitable for all makes of 10 to 12 hp power tiller. Savings in time : 80% Savings in sowing cost : 50%
23	Tractor drawn air assisted seed drill	To drill small seeded crops like, sesame, ragi, maize. Cost of operation : Rs. 150/ha Saving cost : 55% Saving in time : 92 %
24	Improved direct paddy seeder	For uniform seed distribution with respect to time and maintaining uniform plant population per metre square. Reduction in seed rate and the cost of thinning is reduced. Hill dropping of seeds is achieved and continuous drilling is eliminated.
25	Rice cum daincha seeder	Sows paddy in 3 rows and green manure in 3 rows simultaneously. Uniformity in seed sowing and plant population. Reduction in seed rate and the cost of thinning is reduced. Hill dropping of seeds is achieved and continuous drilling is eliminated. Growth of competitive weeds is checked due to green manure crop.

26	Rice transplanter	For transplanting mat type paddy seedlings. Suitable for all transplanted type paddy varieties. Planting can be done in 6 rows at a time.
27	Seed cum fertilizer drill for paddy	For direct sowing of paddy and simultaneous application of fertilizer. The seed rate and fertilizer rate can be adjusted. Can be operated by a 35 hp tractor. By applying the required quantity of fertilizer at root zone, better crop growth and more yields is obtained. Saving in cost : 65% Saving in labour : 84% Cost of operation : Rs.800 / ha
28	Sugarcane sett cutter	For cutting sugarcane sett with single bud. Reduce the cost of seedcane. Saving in cost : 45% Cost of operation : Rs.240 / ha

5.1.3. Weeding and intercultural implements

29	TNAU improved dry land weeder	Useful in dry land and garden land crops. Ideal to remove shallow rooted weeds. The workable moisture content has to be 8 to 10%.
30	Power rotary weeder	For mechanical control of weeds in sugarcane, tapioca, cotton and orchards. Attachments like sweep blades, ridger, trailer can be used with the machine.
31	Tractor drawn weeding cum earthingup equipment	For weeding and intercultural operations in between row crops Weeding and earthing up operations are simultaneously performed in a single pass; row to row distance between the sweep blades and the ridger bottoms are adjustable (45, 60, 75 and 90 cm). Weeding efficiency is 61%.
32	Tractor operated multi row rotary weeder	For weeding and intercultural operations in between row crops like sugarcane, cotton, maize, etc. Weeding and earthingup operations are simultaneously performed in a single pass; row to row distance between the rotor groups are adjustable. Weeding efficiency is 71%.

		The saving in cost and time are 73.12 and 81.5% respectively as compared to conventional method of manual weeding.
33	Cono weeder	For weeding between rows of paddy crop Weeder does not sink in puddled soil. Field capacity 0.18 ha/day.
34	Two row finger type paddy rotary weeder	For weeding in paddy row crops Row spacing can be adjusted for 20 cm and 25 cm One man can easily operate the unit continuously By push pull action, the weeds are buried and soil aerated. Saving in cost :80% Saving in labour : 60% Cost of operation: Rs.250 / ha
35	Battery operated portable wetland weeder	For weeding in SRI field Easy to operate compared to cono weeder Operated without experiencing any drudgery Weeding efficiency: 95% Cost of operation: Rs.625/ha

5.1.4. Plant protection equipments

36	Battery operated low volume sprayer	For spraying chemicals for paddy, groundnut, pulses and vegetables. Requires 50 litres of water per hectare. Repairs and maintenance problems are less. Reduced water requirement
37	Foot wear operated sprayer	For spraying chemicals on garden crops. No separate action is required by the operator except in guiding the nozzle. Ecofriendly Easy to maintain and repair Suitable for ULV and LV application
38	Power tiller operated boom sprayer	For row crop spraying Light in weight and easy handling Suitable for spraying in row crops Results in 55% saving in cost as compared to knapsack power sprayer
39	Sprayer for coconut tree	To spray chemicals in coconut trees and orchards For easy transport, the unit can be kept in a

		horizontal position by folding the frame with telescopic pipes and the same can be erected vertically by winding a cable. Sprayer can be used in orchard trees and for spraying in field crops, by bifurcating the delivery section into spray lines with spray lances and spray guns.
40	Areca sprayer	To spray chemicals in Arecanut trees. It can be used for spray chemicals for arecanut trees Suitable for spraying trees upto a height of 18 metres.
41	Corcyra moth collector	Used for collection of corcyra moth with minimum health hazard Productivity is increased as collection time is reduced. Ensure collection of moth without damage.

5.1.5. Harvesting and threshing machineries

42	Paddy repear harvester	For harvesting and windrowing non-lodging paddy varieties. It is a self propelled unit and width of coverage is 0.75 m. Fuel consumption : 1 litre/ha Height of cut : 50 mm Labour : 1 operator and 2 women labourer (to collect and bundle the cut crop)
43	Mini combine harvester for paddy	For combined operations of harvesting, threshing and winnowing of paddy Suitable for small and marginal farmers Can be easily transported to inaccessible fields Saving in labour: 91%
44	Power tiller operated groundnut harvester	For harvesting groundnut Suitable for harvesting all varieties of groundnut Suitable for all makes of 10 to 12 hp power tiller Savings in time: 90% Savings in cost: 30%
45	Groundnut harvester	For harvesting and windrowing groundnut crop at soil moisture levels of 8 -15 %. Harvesting and soil separation efficiency are 99 and 95%, respectively. Saving in labour cost and time are 32 and 96%, respectively.
46	Power tiller operated turmeric harvester	For harvesting turmeric rhizomes. 65% saving in cost and 90% saving in time.

		Damage caused to the rhizomes is 0.5% as compared to 4.2 % in manual harvesting. Undug rhizomes left in the field is 0.8% as compared to 4.8% in manual harvesting.
47	Tractor drawn turmeric harvester	For harvesting turmeric rhizomes. Results in 70% saving in cost and 90% in time when compared to manual digging. Extent of damage caused to the rhizomes is very much less (2.83%).
48	Fodder sorghum harvester	Harvesting and windrowing the fodder sorghum. Operated by : 5.4 hp light weight diesel engine overage : 0.75 ha/day Saving in labour : 85% Saving in time : 85%
49	Maize husk cum sheller	For removal of husk from maize cob and separation of grains.
50	Castor sheller	For shelling and cleaning castor pods.
51	Coconut tree climber	To harvest the coconuts. Useful for climbing coconut trees for harvesting nuts, cleaning and other operations Any unskilled person including ladies can climb the coconut trees using this unit. Requires 1.5 minutes to climb a tree of 30 to 40 ft height.
52	Oil palm harvesting tool	For harvesting of oil palm. The sickle can be attached to as well as detached from long aluminium poles of various lengths as per the height of the tree. The operator can work for long time continuously without fatigue. Saves cost and time. 50% more bunches harvested than conventional method.
53	Tractor operated fruit-shake harvester	For harvesting fruits by shaking branches. Tractor PTO operated. Suitable for tamarind, lime and other fruits. Harvesting efficiency: 85%
54	Mulberry stem cutter	To cut the mulberry stem for planting. Cut easily without damage.
55	Power tiller operated lawn mower	For mowing lawn grass.

		Simple to operate and easy to handle.
		Results in 50% and 64% saving in cost and time, respectively
56	Lawn / shrub mowing attachment to power weeder	Machine is capable of cutting grasses on the farm roadsides and mowing lawns.
57	Sugarcane detrasher	For detrashing the sugarcane leaves.
		Labour requirement is less.
		Easy for handling.
		Reduced cost of detrashing.
		Also removes the sprouted buds.
		Easy collection of detrashed leaves.

5.1.6. Miscellaneous farm equipments

58	Multi row power weeder for SRI	0.75 to 1.0 ha of weeding per day.
59	Tractor operated pit digger for sugar-cane planting	Pit digger for sugarcane planting.
60	Tractor operated subsoil coir pith applicator	The sub soil coir pith mulch applied at 15-30 cm deep ensured higher moisture retention, crop growth and yield.
		Improved soil structure.
		Enhanced water holding capacity.
		Increased yield of rain fed crops.
61	Hand cum pedal operated chaff cutter	To cut the chaff into bits for easy assimilation by animals.
		Operated by hand and leg simultaneously with minimum effort.
		Uniform sized pits can be optioned.
		Ideal for cutting green fodder, dry fodder and paddy straw.
62	Power tiller operated axial flow pump	To lift water from open water sources.
		Capacity is 2 to 3 times higher than that of centrifugal pumps for lifts between 1.0 and 3.0 metres.
		High efficiency at lower lifts.
63	Power tiller operated heavy duty auger digger	To dig holes for planting tree saplings Results in 16.0 and 91.0% saving in cost and
		time when compared to manual digging of holes.
		It can reach spots where tractor entry and manoeurvability are difficult.
64	Subsoiler attachment for stump removal	To remove roots. Tractor operated equipment.
		Attachment to chisel plough.

65	Banana clump remover	To remove banana clumps. A tractor operated implement. Labour required : 2 (1 driver + 1 helper). Savings in time :85% Savings in labour: 90%
66	Cotton stalk puller	For pulling cotton stalks Counter rotating pulling wheels. Row to row distance can be adjusted from 0.75 to 1.2 m. Savings in time: 97%

5.1.7. Commercial equipments

67	Tractor drawn disc plough	For primary tillage and is especially useful in hard and dry, trashy, stony or stumpy land conditions and in soil where scouring is a major problem.
68	Tractor mounted disc harrow	For primary and secondary tillage.
69	Spring tyne cultivator	For seedbed preparation both in dry and wet soils. Also used for interculture and puddling purposes.
70	Rigid tyne cultivator	For loosening and aerating the soil and preparing seed beds quickly and economically. Useful for subsoil cultivation and also eliminates the use of plough even for hard soils.
71	Tractor drawn ridger	For making furrows and ridges for sugarcane, cotton, potato and other row crops.
72	Post hole digger - tractor operated	To dig holes for planting tree saplings.
73	Rotovator	For use in dry as well as wet land cultivation.
74	Seed cum fertilizer drill	For sowing of groundnut, pulses and other cereal crops in already prepared field.
75	Naveen dibbler	For sowing of bold seeds such as peas and also for gap filling in rows.
76	Manually operated fertilizer broad caster	For broadcasting granular fertilizers like urea, DAP etc., in the field uniformly.
77	Self propelled power weeder	For weeding and intercultural operations in upland row crops like groundnut, maize, soybean, pigeonpea.
78	Wheel hoe	For weeding and interculture of vegetables and other crops sown in rows.
79	Hand operated sprayers/ dusters	Ideal for small nurseries rose plants, kitchen gardens and spraying wettable insecticides and fungicides.

80	Power tiller mounted orchard sprayer	For spraying of foliage, in orchard crops like pomegranate, orange, sweet lime and grapes.
81	Hand rotary duster	For control of pests and diseases by dusting in nursery, vegetable gardens, field crops, tea and coffee plantations, green houses, glasshouses and godowns.
82	Foot sprayer	Suitable for both small and large scale spraying on field crops, orchards, vegetable gardens, tea and coffee plantations, rubber estates, flower crops, nurseries etc
83	Improved sickles of different designs	Harvesting hand tool. Vegetables, cereal crops and cutting of the grass and other vegetative matters.
84	Bhendi plucker	For plucking of bhendi from plant.
85	High capacity multicrop thresher	For threshing of wheat, maize, soybean, sorghum, sunflower, pigeonpea, gram, mustard.
86	Power operated coconut dehusker.	For dehusking of coconut
87	Manual fruit harvester	For harvesting of fruits.
88	Seed treating drum	For thorough mixing of seed treating insecticides and fungicides.

5.2. Agricultural processing equipments

5.2.1. Threshing equipments

S. No.	Implement/Machinaries	Usage/ functions/salient features
1	Areacanut dehusker	• To dehusk the dried arecanut fruits • The unit is portable • Depending upon the size of the fruit, the concave has to be changed.
2	Groundnut decorticator (Power operated)	• To shell groundnut pods and separate kernels • Clearance between the concave and oscillating sector is adjustable to suit the different varieties. • Concave sieves are also replaceable depending upon pod size. • The oscillating sector of the unit decorticates by rubbing action. • The efficiency of the unit is 92%.
3	Groundnut decorticator (Hand operated)	• To shell groundnut pods and separate kernels. • Continuous horizontal type extrusion machine. • Briquettes are formed by compression.

		• Produces hollow briquettes with a calorific value of 3000 k.cal/kg.
4	Sunflower seed sheller	• To shell the sunflower seeds and to separate husk • Power operated centrifugal type sheller. • Quality of oil and cake from shelled seeds is superior. • Efficiency of the unit is 89%.
5	Cotton seed delinting machine	• To delint the fuzzy cotton seed for seed purpose • Batch type operation. • Delinting of the seeds by acid. • Efficiency of the unit is 95%. • Seed borne diseases are minimized. • Acid treated seeds are free to flow.
6	Sorghum pearler	• To remove the hull/seed coat from sorghum and other millets. • Dry milling system to remove the seed coat. • It has cylindrical abrasion disc type stones. • Pearling efficiency is 80 - 85%
7	Hand operated thresher for black pepper	• To separate the pepper berries from the pepper vines. • Manually operated. • Efficiency – 96%. • Suitable for small and marginal holdings.
8	Mechanical thresher for black pepper	• To separate the pepper berries from the pepper vines. • Efficiency – 95%. • Suitable for medium holdings.
9	Paddy winnower	• Cleaning of paddy by winnowing.
10	Maize husker-cum-sheller	• Dehusking and shelling of maize cobs.

5.2.2. Cleaning and grading equipments

11	Grain winnower	• To clean all grains after threshing. • It is a continuous type. • The efficiency of the unit is 97%
12	Rotary sieve multi crop cleaner cum grader	• To clean and grade grains, spices, etc. • It is a continuous type. • The sieves can be changed according to the crop. • Efficiency of the unit is 96%

13	Hand operated rotary type garbling unit for cardamom	• For garbling dried cardamom. • Hand operated unit. • Capacity is 5 kg of cardamom per batch and time taken is 2- 5 minutes per batch. • Percentage broken is less than 5%. • Reduces drudgery to the labourers.
14	Peeler cum washer for production of white pepper	• For the production of white pepper hygienically from ripe pepper berries. • 1 hp power is required for power operated unit. • Water fed inside the peeling champer helps easy peeling and removal of skin after peeling. • Water requirement is less than 50%.
15	Pulper cum washer for coffee	• To pulp and wash the coffee parchments. • Suitable for both pulping and washing. • Requires less water (4 litres per kg of parchments) compared to 14 litres by the conventional pulpers.
16	Seed Cleaner-cum-grader	• Cleaning and grading of all seeds. • Screen suitable for sorghum, pearl millet, paddy and sunflower are available.
17	Groundnut grader	• Grading of groundnut pods
18	Agricultural waste fired mechanical dryer	• To burn agricultural wastes and to produce hot air. • Suitable for drying pods, grains, coconut kernel, etc. • Air temperature and flow rate can be adjusted and controlled.
19	Fluidised bed fryer for mushroom	• To dry the oyster and milky mushroom. • Dries oyster mushroom in 2 hours and milky mushroom in 6 hours.
20	Mini dhal mill	• To split the grain legumes into dhal. • Pre treatment of pulses before milling is essential. • The efficiency of the unit is 90%.
21	Dhal mill cum wet grinder	• Wet grinding, splitting of pulses into dhal and dry grinding. • With a single prime mover any one can be operated using a clutch.
22	Chilli seed extractor	• To extract seeds from dried chilli. • Continuous type.

		• Minimal scorching and pungent smell to labourers. • Separated chilli can be ground and suitable for food purpose.
23	Improved four roller sugarcane crusher	• To extract the juice by crushing sugarcane. • Four roller horizontal crusher. • 60-70% of the available juice can be extracted. • 8-10% additional juice is recovered than the conventional crushers.
24	Hand operated anola seed remover	• To remove seeds from anola. • Mechanical pulping of fresh aonla is feasible without seeds • Saving in cost: 90%

Part-7 Animal Husbandry

1

Livestock Rearing

1.1. Livestock breeds

Denotes and established group of animals / birds having the similar general body shape, colour, structure and characters which produced offspring with same characters.

1.1. Cattle

1. Indigenous breeds

- **Milch:** Sindhi, Sahiwal, Gir and Deoni
- **Dual:** Hariana, Ongole, Tharparkar, Kankrej
- **Draught:** Kangayam, Umblachery, Amritmahal, Hallikar

2. Exotic

- **Milch:** Jersey, Holstein Friesian

1.1.1. Indigenous dairy breeds of cattle

1. Red sindhi

- Red Karachi and Sindhi and Mahi.
- Originated in Karachi and Hyderabad (Pakistan) regions of undivided India.
- Colour is red with shades varying from dark red to light, white strips.
- Milk yield ranges from 1250 to 1800 kg per lactation.
- Age at first calving 39-50 months and inter calving period from 425-540 days.
- Bullocks, despite lethargic and slow, can be used for road and field work.

2. Gir

- Bhadawari, Desan, Gujarati, Kathiawari, Sorthi, and Surati.
- Originated in Gir forests of South Kathiawar in Gujarat, also found in Maharashtra and adjacent Rajasthan.

- Basic colours of skin are white with dark red or chocolate-brown patches or sometimes black or purely red.
- Horns are peculiarly curved, giving a half moon appearance.
- Milk yield ranges from 1200-1800 kgs per lactation.
- Age at first calving 45-54 months and inter calving period from 515 to 600 days.
- This is known for its hardiness and disease resistance.

3. Sahiwal

- Originated in Montgomery region of undivided India.
- Lola (loose skin), Lambi Bar, Montgomery, Multani, Teli.
- Best indigenous dairy breed.
- Colour is reddish dun or pale red, sometimes flashed with white patches.
- Heavy breed with symmetrical body having loose skin.
- Average milk yield is 1400 and 2500 kg per lactation.
- Age at first calving ranges from 37 to 48 months and calving interval is 430 to 580 days.

1.1.2. Indigenous draught breeds of cattle

1. Hallikar

- Originated from the former princely state of Vijayanagarm, presently part of Karnataka.
- Colour is grey or dark grey.
- They are compact, muscular and medium size animal with prominent forehead, long horns and strong legs.
- Breed is best known for its draught capacity and especially for its trotting ability.

2. Amritmahal

- Originated in Hassan, Chikmagalur and Chitradurga district of Karnataka.
- The Maharajahs of Mysore developed this breed.
- Colour is grey, but their shade varies from almost white to near black.
- The muzzle, feat and tail are usually black.
- Horns are long and end in sharp black points.

3. Kangayam

- Also known as kongu and konganad.
- Originated in Kangayam, Dharapuram, Perundurai, Erode, Bhavani and part of Gobichettipalayam taluks of Erode and Coimbatore districts.
- The Kangayam breed was developed by the efforts of the late Pattogar of Palayakottai, Sri N. Nallathambi Sarkari Manradiar.
- Coat (cow's hair) is red at birth, but changes to grey at about 6 months of age.
- Bulls are grey with dark colour in hump, fore and hind quarters.
- Horns are spread apart, nearly straight with a slight curve backwards.
- Cows are grey or white. However, animals with red, black, fawn and broken colours are also observed.
- Eyes are dark and prominent with black rings around them.
- Moderate size with compact bodies.

4. Bargur

- Found around Bargur hills in Bhavani taluk of Erode district.
- Developed for work in uneven hilly terrains.
- They are in brown colour with white markings. Some white or dark brown animal are also seen.
- Animals are well built, compact and medium in size.
- Known for their speed and endurance in trotting.
- Cautious in behaviour and tends to remain away from strangers.

5. Umblachery

- It is otherwise called as Jathi Madu, Mottai Madu, Molai Madu, Therkathi Madu.
- Originated inThanjavur, Thiruvarur and Nagappattinam districts of Tamil Nadu.
- Suitable for wet ploughing and known for their strength and sturdiness.
- Umblachery calves are generally red or brown at birth with all the characteristic white marking on the face, limbs and tail.
- Legs have white markings below the hocks like socks.
- Practice of dehorning of bullocks is peculiar in Umblachery cattle. Unlike in other breeds, the bullocks are dehorned.

1.1.3. Indigenous dual purpose breeds of cattle

1. Tharparkar

- Originated inTharparkar district (Pakistan) of undivided India and also found in Rajasthan.
- Otherwise known as White Sindhi, Gray Sindhi and Thari.
- They are medium sized, compact and have lyre-shaped horn.
- Body colour is white or light grey.
- Bullocks are quite suitable for ploughing and casting and the cows yield 1800 to 2600 kg of milk per lactation.
- Age at first calving ranges from 38 to 42 months and inter calving period from 430 to 460 days.

2. Hariana

- Originated from Rohtak, Hisar, Jind and Gurgaon districts of Haryana and also popular in Punjab, UP and parts of MP.
- Horns are small.
- The bullocks are powerful work animals.
- Hariana cows are fair milkers yielding 600 to 800 kg of milk per lactation.
- Age at first calving is 40 to 60 months and calving interval is 480 to 630 days.

3. Kankrej

- Also known as Wadad or Waged, Wadhiar.
- Originated from Southeast Rann of Kutch of Gujarat and Barmer and Jodhpur district of Rajasthan.
- Horns are lyre-shaped.
- Colour varies from silver grey to iron grey or steel black.
- It is valued for fast, powerful, draught cattle.
- Useful in ploughing and carting.
- They are good milkers, yielding about 1400 kg per lactation.

4. Ongole

- Also known as Nellore.
- Home tract is Ongole taluk in Guntur district of Andhra Pradesh.
- Large muscular breed with a well developed hump.

- Suitable for heavy draught work.
- White or light grey in colour.
- Average milk yield is 1000 kg per lactation.
- Age at first calving is 38 to 45 months and the intercalving period is 470 days.
- Exported to South East Asian and American countries for development of meat cattle.

5. Krishna valley

- Originated from black cotton soil of the watershed of the river Krishna in Karnataka and also found in border districts of Maharashtra.
- Animals are large, having a massive frame with deep, loosely built short body.
- Tail almost reaches the ground.
- Colour is grey white with a darker shade on fore quarters and hind quarters in male. Adult females are more whitish in appearance.
- Bullocks of this breed are powerful animals useful for slow ploughing, and valued for their good working qualities.
- Cows are fair milkers with average yield being about 900 kg per lactation.

6. Deoni

- Also known as Dongerpati, Dongari, Wannera, Waghyd, Balankya, Shevera
- Originated in Western Andhra Pradesh and also found in Marathwada region of Maharashtra and adjoining parts of Karnataka.
- Body colour is usually spotted black and white.
- Age at first calving ranges from 894 to 1540 days.
- Milk yield ranges from 636 to 1230 kg per lactation.
- Caving interval averages 447 days.
- Bullocks are suitable for heavy cultivation.

1.1.4. Exotic dairy breeds of cattle

1. Jersey

- Developed in the Jersey Island, U.K.
- Smallest of the dairy types of cattle.
- In India, this breed has acclimatized well and is widely used in cross breeding with indigenous cows.

- Colour is reddish fawn.
- Dished fore head and compact and angular body.
- Average milk yield is 4500 kg per lactation.
- Age at first calving is 25 to 30 months and calving interval is 13 to 14 months.

2. Holstein Friesian

- Developed in the northern parts of Netherlands, especially in the province of Friesland.
- Ruggedly built and possess large udder.
- Largest dairy breed and mature cows weigh as much as 700 kg.
- Typical marking of black and white that make them easily distinguishable.
- Average production of milk is 6000 to 7000 kg per lactation with fat content of 3.45%.
- Age at first calving is 29 to 30 months and calving interval is 13 to 14 months.

1.1.5. Dairy cross breeds of cattle

1. Jersey cross

- Jersey crosses are produced by cross breeding indigenous breeds with Jersey semen.
- Medium sized have better heat tolerance than other exotic crosses and well adapted to our climate.
- Jersey crosses may show 2 to 3 fold increase in milk yield in the first generation.

2. Holstein Friesian cross

- HF crosses are more suitable for cooler climatic regions like hilly areas as they are less tolerant to heat.
- Have less resistance to tropical diseases than Jersey crosses.
- Although the milk yield is higher in HF crosses, the fat per cent is less.

1.2. Methods of milking

1.2.1. Hand milking

- Milking is done either by stripping or by full hand method.

- **Stripping** is done by firmly holding the teat between the thumb and fore finger, and drawing it down the length of the teat with pressing which cause the milk to flow down in a stream.
- Grasping the teat with all the five fingers and pressing it against the palm for **full hand milking**.
- Full hand method is superior to stripping as it stimulates the natural sucking process by calf.
- Many milkers during milking tend to bend their thumb against the teat that is known as **knuckling** which should be avoided to prevent injuries of the teat tissues.

1.2.2. Machine milking

- Modern milking machines are quick and efficient without injuring the udder.
- It performs two basic functions.
- It opens the **streak canal** through the use of a partial vacuum, allowing the milk to flow out of the teat cistern through a line to a receiving container.
- It massages the teat, which prevents congestion of blood and lymph (water fluid) in the teat.

1. Advantages

- Easy to operate, costs low, saves time as it milks 1.5 to 2 litres per minute.
- Very hygienic and energy conserving.

2. Milking machine

- 352mm Hg in cattle.
- 400mm Hg in buffaloes.

1.3. Feeding schedule

1.3.1. Dairy animals

S. No.	Type of animal	Feeding period	Green fodder (kg)	Dry fodder (kg)	Concentrate (kg)
Cross breed cow					
1	6 to 7 litres milk	Lactation days	20 to 25	5 to 6	3.0 to 3.5
	per day	Dry days	15 to 20	6 to 7	0.5 to 1.0
2	8 to 10 litres milk	Lactation days	25 to 30	4 to 5	4.0 to 4.5
	per day	Dry days	20 to 25	6 to 7	0.5 to 1.0

1.3.2. Different classes of adult cows (approximate body weight-250 kg)

When green grass is plenty			When paddy straw is the major roughage			
Category	**Concentrate**	**Green grass mixture (kg)**	**Concentrate (kg)**	**Green grass mixture (kg)**	**Paddy straw (kg)**	**(kg)**
Dry cows	-	25 – 30	1.25	5.0	5 – 6	
Milking		1 kg for every 2.5 - 3.0 kg of milk	30	1.25 + 1 kg for every 2.5- 3.0 kg of milk	5.0	5 – 6
Pregnant	Production	25 - 30 allowance + 1 to 1.5 kg from 6th months of pregnancy	Maintenance +	5.0 production + 1 to 1.5 kg from 6th months of pregnancy	5 – 6	

1.3.3. Bull

Body weight (kg)	Concentrate mixture (kg)	Green grass (kg)
400-500	2.5-3	20-25

1.4. Feeding shed

1.4.1. Loose housing

- Animals are kept loose in an open paddock (small field with enclosures) throughout the day and night except at the time of milking and treatment.
- Shelter is provided along one side of open paddock under which animals can retire when it is very hot or cold or during rains.

1. Advantages

- Cost of construction is cheaper.
- Future expansion is possible.
- Animals will move freely so that it will get sufficient exercise.
- Animal can be kept clean.
- Common feeding and watering arrangement is possible.
- Clean milk production is possible because the animals are milked in a separate milking barn.
- Oestrus detection is easy.
- At least 10-15% more stock than standard can be accommodated for shorter period.

2. Disadvantages

- Not suitable for temperate Himalayan region and heavy rainfall areas.
- Requires more floor space.
- Competition for feed.
- Attention of individual animal is not possible.
- A separate milking barn is needed for milking of animals.

1.4.2. Conventional barns or stanchion barns

- Housing the animals are confined together on a platform and secured at neck by stanchions (a device that fits loosely around the neck of animal) or neck chain.
- Animals are fed as wells as milked in the same barn.
- These barns are completely covered with roofs and the sidewalls are closed with windows or ventilator located.
- At suitable places to get more ventilation and lighting.
- Applicable for temperate and heavy rainfall region.
- Same type of housing can be utilized for tropical region with slight modification.

1. Advantages

- Animals and men caring for animals are less exposed to harsh environment.
- Animals can be kept clean.
- Diseases are better controlled.
- Individual care can be given.
- Separate milking barn is not required.

2. Disadvantages

- Cost of construction is more.
- Future expansion is difficult.
- Not suitable for hot and humid climatic conditions.

1.5. Livestock buildings/shed

1.5.1. Dairy cow building

Milch animal shed should have the following parts

- Feeding passage

- Manger (open container in which animal feed is kept)
- Standing space
- Gutter or drainage channel
- Milking passage

1. Single row system

- 12-16 numbers of animals can be kept.

2. Double row system

- 50 animals can be maintained in a single shed.
- Distance between two sheds should be greater than 30 feet or it should be twice the height of the building.
- In this system there are two methods

Tail to tail system

- Cleaning and milking of animals easy.
- Supervision of milking also easy.
- Less chance for transmission of diseases from animal to animal.
- Animals can get more fresh air from outside.

Head to head system

- Getting animals into the shed is easy.
- Feeding of animals also easy.
- Disinfection of gutter will be more due to the direct fall of sun rays over the gutter.
- Animals are better exhibited to visitors
- Milking supervision is difficult.
- Possibilities of transmission of disease are more.

1.5.2. Milking barn or parlour

- This is a barn (farm building) where milch animals are milked and is fully covered.
- Should be located at the centre of the farm with all other farm buildings arranged around it.

- There shall be an individual standing in the milking barn and the number of standings required should be 25% of total number of milch animals.
- Milking operation should be carried out in batches.

1. Dimensions of milking barn

- Length of standing space : 1.5 – 1.7 m
- Width of standing space : 1.05 – 1.2 m (80% of length of standing space)
- Width of central passage : 1.5 – 1.8 m
- Width of feed alley : 0.75 m
- Width of gutter : 0.30 m
- Overhang : 0.75 m

1.5.3. Down calver shed/ calving pen

- Pregnant animals are transferred to a calving pen 2 to 3 weeks before the expected date of calving.
- Calving pen of 3 x 4 m is essential to keep the animals in advanced stage of pregnancy.
- Should be located nearer to the farmer's quarters for better supervision.
- Number of calving pens required is 10% of total breedable female stock in the farm.

1.5.4. Calf pen

- For housing young calves separately.
- Located either at the end or on the side of the milking barn.
- If more calves, separate calf shed should be arranged and located nearer to milking barn.

1.5.5. Young stock/heifer (young females) shed

- Meant for housing young heifers separately.
- Older heifers of six months of age to breeding age are housed separately from the suckling calves.
- When there are large number of young stocks, they should be divided into different age groups and each group housed separately.

1.5.6. Dry animal shed

- In large farms, milch and dry cows are housed separately.
- Floor in the covered area should be made of cement concrete.
- Under Indian conditions, in smaller farms, milch and dry animals can be housed together.

1.5.7. Bull shed

- Should be located at one end of the farm.
- There shall be one shed for each bull.
- One bull for every 50 breedable females.
- When artificial insemination service facilities are available, not necessary to keep the bulls on the farm.
- Bull shed shall have covered 3x4 metre dimensions, leading into a paddock of 120 square metres.

1.5.8. Isolation shed

- Separation of sick animals from healthy animals to avoid transmission of diseases.
- Should be located at the corner of the shed.

1.5.9. Quarantine shed

- Should be located at the entrance of the farm.
- Newly purchased animals entering into the farm should be kept in quarantine shed for of 30 to 40 days to watch out for any disease occurrence.

1.5.10. Milk room

- It is essential to keep the milk with chilling facility in larger dairies having 400 to 700 litres production capacity.
- Requires 3.7 x 5 m size room and an additional 0.37 m^2 for every 40 litres of milk production.
- For a dairy unit below 100 litres, a small room with a dimension of 3.75 x 3 m can be sufficient for storing milk and concentrate feed.

1.5.11. Hay or straw shed

- An adult animal consumes about 5 to 10 kg of hay or straw per day, while young stock consumes about 2 to 5 kg of hay or straw per day.

1.6. Diseases of cattle

Infectious or contagious	Non infectious or non contagious
Bacterial	**Metabolic**
Anthrax, black quarters, hemorrhagic septicemia, T.B., brucellosis	Milk fever or hypocalcaemia acetonemia or hypoglycemia, ketosis.
Viral	**Dietary**
Rinderpest, foot and mouth	Tympanites or bloat impaction
	Non specific enteritis
Parasitic	
Ectoparasite: Tick, lice and mite	
Endoparasite: Tapeworm, roundworm	
Fungal	
Aflatoxicosis	

1.6.1. FMD (Foot and mouth disease)

- Highly contagious disease.
- Susceptible to cattle, goats, sheeps, water buffalo and pig.
- Caused by Apthovirus.

Smallest of the animal virus: 7 types virus: O, A, C Asia I, SAT 1, 2, 3

Transmission: Direct contact, through water, manure, pasture and cattle attendant.

Symptoms: Incubation period 2 – 5 days, temperature 40°C.

- Drooping of saliva and excess salivation.
- Formation of vesicles (fluid filled blisters) in tongue, gum, interdigital space, dental pad, nostrils, muzzle, udder and teat.
- Rough coat with long hair, panting.
- Animal looses appetite and body weight.
- Reduced milk production.
- Lamness on account of painful foot lesions.
- Abortion, if pregnant cows.
- Young calves may die.

Treatment: External application of antiseptics contributes to the healing of ulcers and wards off attacks by flies.

- Antibiotics may be administered to counter bacterial infections.

Prevention: Thorough disinfection of shed, utensils and clothes of attendants.

Control: Vaccination of polyvalent once in 4 months or varies with type of vaccine.

1.6.2. Anthrax

- Peracute (very severe, very short duration and proving quick fatal) disease affecting cattle, sheep, horses and pigs.
- Disease is always fatal in animals.

Causative organism: *Bacillus anthracis* bacterial disease, sporulation occurs outside the body. Spores highly resistant and not killed by heat, light and disinfectant.

Symptoms: Peracute – death occurs within minutes and animal collapses with convulsions.

- Hemorrhagic septicemia (HS) is an acute and highly fatal which occurs within 48-96 hrs.
- Pregnant animals may abort.

Acute: High rise of temperature, shivering abdominal pain, before death blood oozes out of rectum and nostrils.

- All the orifices usually exudate dark tarry blood which does not clot.
- Spleen is enlarged 10-15 times its normal size.

Diagnosis: Sudden death, acute bloat, exudation of blood through orifices (an opening, mouth or anus).

- Blood smear, presence of large blue rods with pink capsule.

Control: Annual vaccination.

- A live spore vaccine prepared from a virulent uncapsulated strain of *Bacillus anthracis* @ 1ml.
- Organism is susceptible to penicillin G, tetracyclines, erythromycin and chloramphenicol.

Prevention: Hygiene and sanitation, carcass of the animals suspected of dying due to anthrax should never be opened.

- Carcass is burnt or buried in a deep pit and lime is applied.

1.6.3. Milk fever: (Parturient paresis, metabolic disease in cows soon after calving)

Cause: Serum calcium levels fall in cows after calving as a result of failure to mobilize calcium reserves and development of negative calcium balance in late pregnancy.

Symptoms: Disease flares up with in 72 hours of calving.

- Initially the cow's show excitement, in coordination of movement, muscular tremors in limbs and head, lying in recumbent position with her head directed towards flank (Flank: Animals body between the ribs and the hip).
- In final stages subnormal temperature, dilatation of the pupil, impalpable pulse, coma and death.

Treatment and control: Dramatic recovery by intravenous administration of 300-400 ml calcium borogluconate with Vitamin D_3 injected intramuscularly.

- Continued mixing of ½ litre of supernatant lime water for cow may reduce the incidence.

1.6.4. Bloat: (Tympany)

- Rumen and reticulum is over distended with the gases of fermentation.

Cause: Excess intake of fresh legumes and feeding of high grain ration lead to frothy bloat.

- Obstruction to normal expulsion of gases from rumen by choking the oestophageal passage by corn cob, turnip and sugar beet which cause free gas bloat.

Symptoms: Sudden death before rendering any aid to the affected animal.

- Distension of the rumen occurs quickly, the flank and the whole abdomen is enlarged.
- On percussion (striking with hand) the left flank, produces a drum like sound.
- Initially the animal frequently gets up and lies down, kicks at belly and even rolls.
- Breath becomes difficult and is evidenced by oral breathing, protrusion of tongue and salivation.
- When the distension of abdomen becomes extreme, the animal exhibits uncoordinated movement, inability to stand, falls all of a sudden.
- Collapse and death occur quickly.
- In chronic tympany, the distension of abdomen and intra abdominal pressure are not serious.

Diagnosis: Characteristic symptoms of distension of abdomen and distress by the affected animal.

Control and treatment: In acute cases, puncture the rumen with a sharp knife or with a trocar and canula to expel the gases (Trocar: A sharp point surgical instrument filled with a cannal).

- Administer orally oil of turpentine 60 ml well mixed with one litre of groundnut oil or gingelly oil or coconut oil.
- After six to eight hours, administer powdered ginger 30grams, asafoetida 30grams, well mixed to jaggery.
- Fresh legumes should be wilted and then fed to animals.
- Feed dry roughages to avoid bloating.

1.6.5. Black quarter

- Acute infectious disease, but not contagious, inflammation of muscle, severe toxaemia.

Causative organism: Bacteria – *Clostridium chauvoei*, gram positive anaerobic spore.

Symptom: Animals may die without showing symptom obvious sign.

- Crepitant swelling in hind and fore quarters which crackles when rubbed due to gas in the muscle.
- Lameness, fever, twitching of muscle, affected region is hot and painful, but becomes cold and painless.
- Skin over affected area is dry, hard and dark.

Diagnosis: Affected part is black or dark red, characteristic rancid smell.

Control: Hygiene and prophylaxis (prevention of disease) control.

Prevention: Vaccination before onset of rainy season with 5ml of polyvalent (*Clostridium sp.*).

- Antibiotics like penicillin and tetracycline may be given.

1.6.6. Vaccination schedule

Disease	Age	Interval	Month
FMD	3rd month	Every six month	Jan-Feb, June-July
BQ	6th Month	Every Year	Aug-Sep
HS	6th Month	Every Year	Sep-Oct
Anthrax	6th Month	Every Year (Affected area only)	April - May
Brucellosis	4-8th month of heifer	—	Mar - April

2

Poultry Rearing

2.1. Breeds of poultry

Asiatic	American	English	Mediterranean
Aseel	Plymouth rock	Sussex	Leghorn
Kadaknath	Wyandotte	Orpington	Minorca
Ghagus	Rhode island red	Australorp	Ancona
Chittagong	New Hampshire	Cornish	Spanish
Mini			Andalusian
Brown desi			
Denki			
Naked neck			
Brahma			
Cochin			
Langshan			

2.1.1. Based on utility

1. Layer – Leghorn, Minorca
2. Broiler – Orpington, Cornish
3. Dual – Plymouth, Rhode island red

2.1.2. Based on the utility and performance

1. Layer – Babcock 300, Hyline-WS 36, Bovans.
2. Broiler – Ross, Vencobb, Hybro.

2.2. Rearing house

2.2.1. Cage system

- Super intensive system providing floor area of 450-525 sq.cm. (0.6-0.75 sq.feet) per bird.
- In cage, the birds are kept in one, two or three per cage, arranged in single or double or triple rows.

2.2.2. Deep litter system

- Deep litter system is commonly used in all over the world.

2.2.3. Confinement rearing

- **Size of flock:** Larger size units are more economical than smaller ones under commercial conditions.
- A unit of 2000 layers is usually considered as economical for commercial egg production. (Flock: A number of birds of one kind feeding, resting or travelling together).

2.2.4. Floor space, feeding space and watering space for chicks

Age in weeks	Floor space (sq.ft./chick)	Feeding space (inches/chick)	Watering space (inches/chick)
1	0.2	1.5	0.5
2	0.2	2.0	0.7
3	0.3	2.0	0.7
4	0.4	2.5	0.8
5	0.6	2.5	0.8
6	0.8	3.0	1.0
7	0.8	3.0	1.0

Age in weeks	Temperature under hover, at 5 cm above floor (°C)
0-1	35
1-2	32
2-3	29
3-4	26
4-5	23

2.2.5. Litter management

- Litter materials such as wood shavings, saw dust, paddy husk, peanut shell, paddy chaff, chopped straw and other materials that absorb moisture well.
- Spread the litter to a depth of 5 cm on the floor
- Build it up to a depth of 15 cm by adding litter material 2 cm per week.
- Require approximately 10 kg of litter material/sq.m.
- Use dry lime at of 10 kg/per 10 m^3 and rake the litter.

2.2.6. Light

- Dim light of a 40-watt bulb for every 250 chicks can be provided during the night for broiler chicks.

2.3. Feeding of poultry

- Major expenditure (60-70%) in poultry rearing is feed cost.
- More than 40 nutrients are required by the poultry.

2.3.1. Feed ingredients

- **Cereals:** Maize, rice, wheat, oat, barley,
- **Cereal byproducts:** Wheat bran, or rice polish.
- **Animal and vegetable protein:** Fish meal, meat meal, soybean oil meal, groundnut cake.

2.3.2. Average feed consumption of egg type birds during growing period

Age in weeks	Feed consumed (g/bird/day)
10	53.0
11	58.0
12	60.0
13	60.0
14	60.0
15	62.0
16	62.0
17	65.0
18	70.0
19	75.0
20	75.0

2.4. Vaccination for chicken

Name of vaccine	Route	Age of birds
La sota or F vaccine ranikhet	Intranasal drop	3 to 7 days
Marek's vaccine (in hatchery)	Intramuscular	1 day
Infectious bronchitis (1st dose)	Eye drops	2-3 weeks
La sota ranikhet	Drinking water	5-6 weeks
Fowl pox (1st dose)	Wing web	7-8 weeks
R2B ranikhet	Sub cut or intramuscular	9-10 weeks
Infectious bronchitis	Eye drop or drinking water	16 weeks
Fowl pox (2nd dose)	Skin scarification	18 weeks
La sota (if necessary) ranikhet	Drinking water	20 weeks
La sota (if necessary) ranikhet	Drinking water	40 weeks
IBD (Infectious bursal disease)		
Mildly invasive vaccine	Drinking water	0-3 day
Intermediately invasive vaccine	Drinking water	15th day
Intermediately invasive vaccine	Drinking water	28-30th day

2.5. Poultry management

2.5.1. Brooder management

1. Brooder house

- Brooder house should be draft-free, rain-proof and protected against predators. Brooding pens should have windows with wire mesh for adequate ventilation.

  (Brooder: Heated house for chicks); (Draft: Air current)

2. Sanitation and hygiene

- Malathion spray, blow lamping or both can be used to control ticks and mites.

3. Litter

- Saw dust and paddy husk should be spread up to 5 cm. Should be stirred at frequent intervals to prevent caking.
- Wet litters, if any, should be removed immediately and replaced by dry new litter.
- To prevent ammoniacal odour.

4. Brooding temperature

- During the first week, the temperature should be 95°F (35°C) which may be reduced by 5°F per week during each successive week till 70°F (21.1°C).
- Brooder should be switched on for 24 hours before the chicks arrive.
- As a rule of thumb, the temperature inside the brooder house should be approximately 20°F (-6·7°C) below the brooder temperature.
- Infra-red lamps are also very good for brooding.

5. Brooder space

- Brooder space of 7 to 10 sq inch (45-65 cm^2) is recommended per chick.
- Thus, a 1.80 m hover can hold 500 chicks.

6. Brooder guard

- To prevent baby chicks from the source of heat, hover guards are placed at 1.05 to 1.50 m from the edge of hover.
- It is not necessary after 1 week.

7. Floor space

- Floor space of 0.05 m^2 should be provided per chick to start with, which should be increased by 0.05 m^2 after every 4 weeks until the pullets are about 20 weeks of age.
- For broilers, 0.1 m^2 of floor space and for female chicks 0.15 m^2 should be provided till 8 weeks of age.

  (Pullet: Young chicks less than a year)

2.6. Care and management of layers

Laying period: From the point of lay to one year is called laying period.When first egg laid by Pullet, it is known as pullet egg.

- Floor space: 2 sq.ft.
- Feeder space: 4 sq.ft.
- Water space: 2 sq.ft.
- Nest space : 1 box for 5 birds
- Litter depth: 6 box for 5 birds.

Feeding: Layer mash (More of key nutrients, rather than grain based) is fed during this period - 18% protein.

- Calcium is supplied to the birds in feed.

Ration of layer mash for chicken

Ingredients	Percentage
Yellow maize	47
Soyabean meal	12
Gingelly oil cake	4
Groundnut oil cake (expeller)	6
Rice polish	13
Wheat bran	4
Fish meal/dried unsalted fish	6
Dicalcium phosphate	1
Salt	0.25
Mineral mixture	1.75
Shell meal	5
Total	100.00

Lighting: Layer birds have to be kept with a light period of 16 hours a day (12 hrs day + 4 hrs night).

2.7. Broiler management

- Broiler is defined as the tender meated chicken of either sex which grow from 35 to 40 g of initial weight to 2kg or more in 6 weeks of age by consuming around 4 kg of feed.

	0-4 weeks	4-8 weeks
Floor space	½ sq.ft	1 sq.ft
Feeder space	3 sq.ft	6 sq.ft
Water space	2 sq.ft	4 sq.ft

2.8. Diseases of poultry

2.8.1. Ranikhet disease (New castle disease, virus- Paramyxoviridae)

Symptoms: Very important disease affecting poultry during rainy season in India.

- Peracute without symptoms and sudden death.
- Outbreak depression is observed, characterised by prostration, closed eyes, drooping wings and loss of appetite.
- Greenish or yellowish diarrhoea.

Prevention and control: Chicks should be vaccinated with F strain or lasota strain on the first day or within 5 days after hatch with a booster dose at 8-10 weeks.

- RDVK [Ranikhet disease vaccine K (Komorov)] strain is usually administered at 8 weeks of age.
- In layer flocks, booster dose of ranikhet vaccine is given every 2 months.

2.8.2. Bacterial diseases (*Coli bacillosis*)

***Escherchia coli* infection:** Aggravated by other stress factors.

Symptoms: Diarrhoea, swelling of joints, comb and wattle.

Moratility: Very high

Prevention: Proper sanitation and management, avoiding addition of antibacterials and antibiotics in feed and water.

2.8.3. Infectious coryza (*Haemophilus gallinarum*)

Symptoms: All ages are affected, acute respiratory infection, high morbidity (diseased) and low mortality, odema of face, wattle and comb, discharge from nostrils, recovered birds carriers.

Prevention and control: Better hygiene, addition of antibacterials and antibiotics – Sulpha in feed, tylosin, tetracycline.

2.8.4. *Salmonellosis pullorum/S.gallinarum*

- *S.pullorum*- pullorum disease
- *S. gallinarum*-typhoid/bacillary white diarrhoea.
- *S .typhimurium*- paratyphoid

Symptoms: Chalk like diarrhoea, huddling, weight loss, pasted vent, pullorum visceral organs affected.

Treatment: Supha drugs, hygienic management, hatchery hygiene is important.

2.8.5. Protozoan

Coccidiosis: *Eimeria tenella, E.necatrix*

Symptoms: Severe upto 10 weeks of age, due to poor litter management, bloody droppings, high mortality, production performance is hampered.

Prevention and control: Anticoccidials, litter management and hygiene, amprolium (sulpha drugs), Coccidiostats may be mixed with feed.

2.8.6. Nutritional deficiencies and control

1. **Vit A:** Xeropthalmia (dry eyes), gout (Joint pain), retarded growth, discharge from eyes and nose.
- **Control measures:** Cod fish liver oil, fish liver oil and vit A supplementation.
2. **Vit D3:** Rickettsia, leg weakness, swollen hock joints, rubbery beak and thin shelled eggs.
- **Control measures:** Cod fish liver oil, fish liver oil and vit D3 supplementation.
3. **Vit E:** Encephalomalacia (Loss of brain tissue), crazy chick disease, paralysis of leg, retraction of head, convulsions and death.
- **Control measures:** Vegetable oils and synthetic vit E.
4. **Vit B1** (Thiamine): Polyneuritis (Pain due to several nerves), paralysis of wing and neck.
- **Control measures:** Yeast products and synthetic vit B1.
5. **Vit B2** (Riboflavin): Curled toe paralysis, tendency to walk on hocks (Hocks: Joint of hind legs), dermatitis on corners of mouth, vent and foot pads.
- **Control measures:** Fish products, vit B2 and rice bran.

6. **Vit B12** (Cyanacobalamine): Retarded growth, increased mortality, drop in production and hatchability.

- **Control measures**: Fish meal, meat meal and synthetic B12.

7. **Choline**: Fatty liver syndrome, poor feed utilization, ruffled feathers and increase in liver fat.

- **Control measures**: Fish, meat and groundnut meal.

2.8.7. Mineral deficiency

1. **Manganese:** Slipped tendon (deforming leg weakness) and deformity of hock joints.

- **Control measures:** Fish and meat meal.

2. **Zinc and magnesium deficiency (Goose stepping)**: Bone formation is affected.
3. **Calcium and phosporous deficiency:** Leads to deficiency of vit D, imbalance in calcium causes deficiency during laying.
4. **Phosporous:** Poor shell formation, curved beak and bone deformities.

- **Control measure:** Supplementation with Ca and P.

2.8.8. External and internal parasites (Lice, ticks and mites)

- **Control measure:** Deticking, delicing round worm, tape worm infestation and deworming regularly.

UNIT-V
CROP IMPROVEMENT
Part-1 Plant Breeding

1

Principles of Plant Breeding

1.1. Plant breeding

- An art and science, deals with ways and means to change the genetic architecture of plants so as to attain a particular objective.
- For ensuring food security, it is important to breed and develop new varieties that are high-yielding, pests diseases and drought resistant or regionally adapted to different environments and growing conditions.

1.2. Objectives

- Crop improvement
- Improved agronomic characters
- Resistance against biotic and abiotic stress

1. Increased yield

- Developing more efficient genotypes. Eg. Dee Gee Woo Gen in rice and Norin10 in wheat.
- Identification and utilization of male sterility

2. Improving the quality

- Rice-milling, cooking quality, aroma and grain colour
- Wheat-milling and baking quality and gluten content.
- Pulses-protein content and improving sulphur containing amino acids
- Oilseeds-PUFA (Polyunsaturated fatty acid) content

3. Elimination of toxic substance

- HCN content in jowar plants
- Lathyrogen content in *Lathyrus sativus* (Grass pea)
- Erucic acid in brassicas
- Cucurbitacin in cucurbits

4. Resistance against biotic and abiotic stresses

- **Biotic stress**: Evolving pests and diseases resistant varieties thereby reducing cost of cultivation, environmental pollution and saving beneficial insects.
- **Abiotic stress**: Breeding resistant varieties is the easy way to combat abiotic stress.

5. **Change in maturity duration** - Evolution of early maturing varieties
6. **Improved agronomic characters** - Production of more tillers –Rice, Bajra,
7. **Reducing the plant height to prevent lodging** - Rice
8. **Photo insensitivity** - Redgram, sorghum
9. **Non-shattering nature** - Greengram, brassicas
10. **Synchronized maturity** - Pulses
11. **Determinate growth habit** - Pulses
12. **Elimination or introduction of dormancy** - Groundnut

1.3. Scope of plant breeding

- As cultivable land is shrinking, only solution to meet out food requirement is by increasing yield through genetic improvement.
- Two ways by which yield improvement is possible.

1.3.1. Enhancing the productivity of crops

- Adopt high yielding varieties bred by appropriate genetic manipulation.
- Proper management of soil and crops following good agronomic.

1.3.2. Stabilizing the productivity achieved

- Use varieties have wide adaptation or specific to unfavorable environmental zones.

1.4. Past, present and future scopes

- **Rice**: Indica x japonica cross derivative, ADT 27, is the first high yielding rice of Tamil Nadu.
- Identification of Dee Gee Woo Gen and release of wonder rice IR 8 (Peta x DGWG) changed poverty
- **Wheat**: Identification of dwarfing gene in Japanese wheat variety Norin-10 by Norman Borlaug and breeding of Mexican dwarf wheat varieties led to the release of Kalyan sona in India.

- **Pearl millet**: Introduction of male sterile line Tift 23A (Developed by Burton and his coworker at Tifton, Georgia) to India led release of hybrids HB1 to HB4.
- **Sorghum**: Introduction of male sterile line combined kafir 60A to India led to the release of first hybrid CSH 1 (CK 60A x IS 84) during 1970s.
- Search of alternate source of cytoplasm to develop resistance to new pest and diseases; future of plant breeding is a challenging task.
- Deployment of innovative breeding techniques will be a new tool to assist conventional breeding.

1.5. Undesirable effects of plant breeding

1.5.1. Genetic erosion

- Disappearance of land races due to introduction of high yielding varieties.
- Introduction of IR 20 rice led to disappearance of land races of samba rice.

1.5.2. Narrow genetic base

- Genetic vulnerability to pest and diseases.
- Pearl millet: Tift 23A - Susceptible downy mildew.
- Maize: T cytoplasm - Susceptible to Helminthosporium

1.4.3. Minor disease and pest become major due to intensive resistance breeding

- RTV (Rice Tungro Virus)
- Grey mold in bengalgram

1.5.4. Attainment of yield plateau

- No more further increase in yield.

1.6. History of plant breeding

- 700 BC - Babylonians and Assyrians artificially pollinated the date palm.
- 1717 - Thomas Fairchild produced the first artificial hybrid.
- 1760 - 1866 - Joseph Kolreuter made extensive crosses in tobacco and solanum and studied the progenies in detail.
- 1759-1835 - Thomas Andrew Knight was first to produce several new fruit varieties using artificial hybridization.

- 1843 - A farmer, Le Coutier published results on selection in wheat and concluded that progenies from single plants were more uniform.
- 1857 - Vilmorin proposed individual plant selection based on progeny testing in sugar beets (*Beta vulgaris*); known as Vilmorins principle of progenys testing. Ineffective in wheat. Demonstrated difference between cross and self pollinated crops in selection.
- 1890 - Sval of Sweeden refined the single plant selection.
- 1903 - Johansen proposed the famous pure line theory in *Phaseolus vulgaris*, which states that a pure line is progeny of a single self fertilized homozygous plant.
- 1908 - G.H. Shull work in maize is the forerunner for the present day hybrid maize programme; detailed the effect of inbreeding.
- 1960's - Norman Borlaug, the Nobel laureate developed Mexican semi dwarf wheat varieties, which paved the way for green revolution. Dwarfing gene isolated from Mexican dwarf Norin-10 introduced to India by Dr. M.S.Swaminathan and developed Kalyan Sona, Sharbathi Sonara.
- 1965 - In rice at IRRI, using dwarf Dee Gee Woo Gen from a tall rice variety (Identified by a Taiwan farmer), the wonder rice IR 8 was released.
- Nobilisation in sugarcane by C.A. Barber and T.S.Venkataraman is another monumental work in plant breeding.

2

Plant Domestication and Introduction

2. 1. Plant domestication

- Process of bringing wild species under human management.

2.1.1. Changes in plant species under domestication

- Elimination or reduction in shattering of pods, spikes, etc.
- Elimination of dormancy
- Decrease in toxins of other undesirable substances. Eg. Bitter principle of cucurbits.
- Cultivated plants show altered tillering, branching, leaf characters.
- Decrease in plant height. Eg. Cereals and millets.
- Increase in plant height. Eg. Jute and sugarcane.
- Shortening of life cycle. Eg. Cotton, redgram
- Increase in size of grains or fruits.
- Increase in economic yield.
- Asexual reproduction has been promoted. Eg. Sugarcane, potato.
- Preference for polyploidy under domestication.

2.2. Plant introduction

- Introduction of new species (or) a genotype into new environment
- **Foreign introduction or exotic collection**: Introduced from foreign countries.
- **Indigenous collection or introduction**: Introduction from one state to another.

2.2.1. Types

- **Primary:** Introduction of variety to new environment
- Eg. Semi dwarf wheat varieties like Sonara – 64 and Lerma rojo
- Semi dwarf varieties rice like IR 8, IR 36 and TN 1
- **Secondary**: Introduction of variety after selection as superior variety and released to new environment.
- Eg. Kalyan sona and Sonalika of wheat selected from material introduced from CIMMYT

2.2.2. Procedure

- **Procurement**: Introduced germplasm must be routed through the NBPGR, New Delhi
- **Quarantine**: Isolate materials to prevent the spread of diseases, pests etc.
- **Cataloguing**: Given with information like entry number, name of species, variety, place of origin and adaption.
- Exotic collections denoted as EC.
- Indigenous collections denoted as IC.
- Indigenous wild collections denoted as IW.
- **Evaluation**: To assess the potential of new introductions.
- **Acclimatisation**: Adaptation of a variety to a new environment.
- Determined by
- Mode of pollination
- Range of genetic viability present in the original population
- Duration of crop life cycle.
- **Multiplication and distribution**: Promising introductions or selections multiplied and released as varieties after necessary trials.

3

Breeding Methods for Self Pollinated Crops

3.1. Selection

- To isolate desirable plant types from a population.

3.1.1 Progeny test

- Evaluation of growth of plants on the basis of performance of their progenies
- Real value of plant could be known by studying the progeny.

3.1.2. Functions

- To determine the breeding behaviour of a plant
- To find out whether the character for which the plant was selected is heritable.

3.2. Pure line selection

- Progeny of a single, homozygous, self-pollinated plant.
- Large number of plants were selected from a self pollinated crop and are harvested individually.
- Individual plant progenies are evaluated and the best progeny is a pureline variety.

3.2.1. Procedure

- Selection of individual plants from a local variety
- Visual evaluation of individual plant progenies
- Yield trials

3.2.2. Merits

- Achieves the maximum possible improvement over the original variety.
- Extremely uniform; easily identified in seed certification programmes.

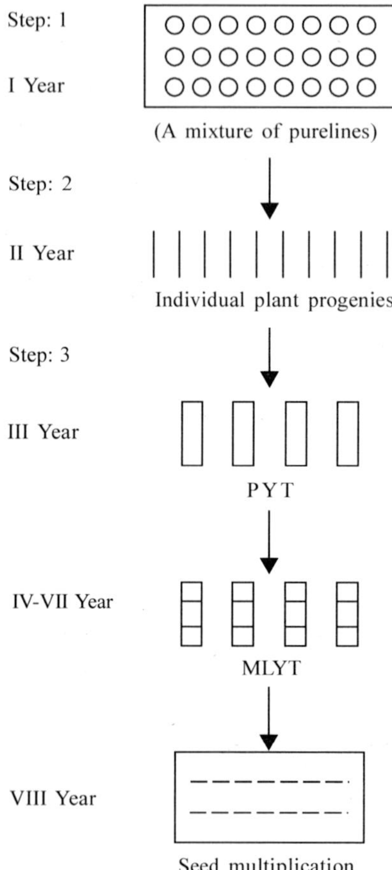

i) 200-3000 plants are selected on the basis of their phenotype
ii) Seeds are harvested separately

i) Individual plant progenies are grown
ii) Undesirable progenies are rejected
ii) Superior progenies are harvested separately

i) Remaining progenies are planted in a preliminary yield trial (PYT)
ii) Inferior progenies rejected

i) Replicated yield trials (RYT) conducated at several location
ii) Inferior progenies rejected
iii) Disease resistance and quality tests are done

i) Best progeny is selected as a variety
ii) Seed multiplied for distribution

3.2.3. Demerits

- Do not have wide adaptation.
- Upper limit on improvement is set by the genetic variation present in the original population.
- Breeder has to devote more time.

3.3. Mass selection

- Large number of plants of similar phenotype are selected and their seeds are mixed together to constitute a new variety.

3.3.1. Procedure

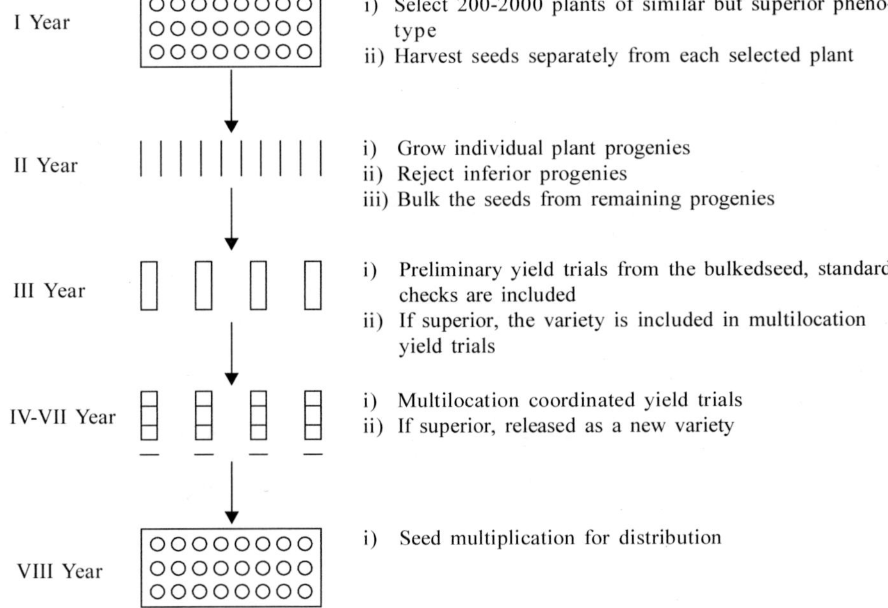

3.3.2. Merits

- Widely adapted than pure lines.
- Extensive and prolonged yield trials are not necessary.
- Retains considerable genetic variability.

3.3.3. Demerits

- Not as uniform as pure lines
- Improvement is generally less than pure line selection.
- Utilizes variability already present in a variety or cannot generate variability.

3.3.4. Difference between pure line selection and mass selection

Pureline selection	**Mass selection**
New variety is a pure line.	Mixture of pureline
Highly uniform	Moderately uniform
Narrower adaptation and lower stability	Wide adaptation
Selection within a pure line variety is ineffective	Selection within a variety is effective due to genetic variation
Easily identified	Difficult to identify
More demanding because careful progeny tests and yield trials have to be conducted.	Less demanded due to selection of larger number of plants.

3.4. Hybridization

Matting or crossing of two plants or lines of dissimilar genotype.

3.4.1. Objectives

- To create genetic variations.
- To transfer of one or more characters into a single variety from other varieties; governed by oligogenes or polygenes.
- Improving yield or its contributing characters through transverse segregation.
- F_1 is more vigorous and high yielding than parents.

3.4.2. Types

1. Inter varietal hybridization (Intraspecific hybridization)

- Parents belong to same species; they may be two strains, varieties or races of the same species.
- In crop improvement, this is most commonly used.
- **Simple cross**
- Two parents are crossed to produce F_1
- Eg. A x B ® F_1
- **Complex cross**
- More than two parents are crossed to produce the hybrid.
- Also known as convergent crosses; bringing together genes from several parents into a single hybrid.
- 3 parents (A, B, C)
- A x B
- (A x B) x C ® F_1

2. Distant hybridization

- Between different species of the same genera or of different genera.

3. Inter specific hybridization

- Two species of same genus are crossed.
- CO 31 rice variety was developed from *Oryza sativa* var. *indica* x *Oryza perennis* cross.

4. Inter generic hybridization

- Two different genus are crossed.
- Eg. Triticale is a cross between *Triticum* sp. and *Secale cereale.*

3.4.3. Procedure

- **Choice of parents**: Depends upon the objective of the breeding program.
- **Evaluation of parents**: If a performance of a parent is known, evaluation is not necessary
- If not known, evaluation is necessary.
- **Emasculation**: Removal of stamens or anthers of a flower without affecting female reproductive organs.

Hand emasculation	Removal of anther with the help of forceps.
Suction method	Removal of anther using thin rubber or glass tube attached to a suction hose
Hot water emasculation	Hot water treatment is given before anther dehiscence and prior to the opening of flowers. Sorghum: 2–48°C Rice: 40–44°C
Alcohol treatment	Immersing the inflorescence of flowers in alcohol followed by rinsing with cold water.
Cold water treatment	Kills pollen without damaging gynoecium. Rice: 0-6°C
Genetic emasculation	Used to eliminate the necessity of emasculation.

- **Bagging**: Immediately after emasculation, cover with suitable bags to prevent cross pollination.
- **Tagging**: Tags should contain date of emasculation, date of pollination and parentage.
- **Pollination**: Mature fertile and viable pollen should be placed on a receptive stigma.

3.5. Pedigree method

- Description of ancestors of an individual.
- To develop pure line varieties.

3.4.1. Procedure

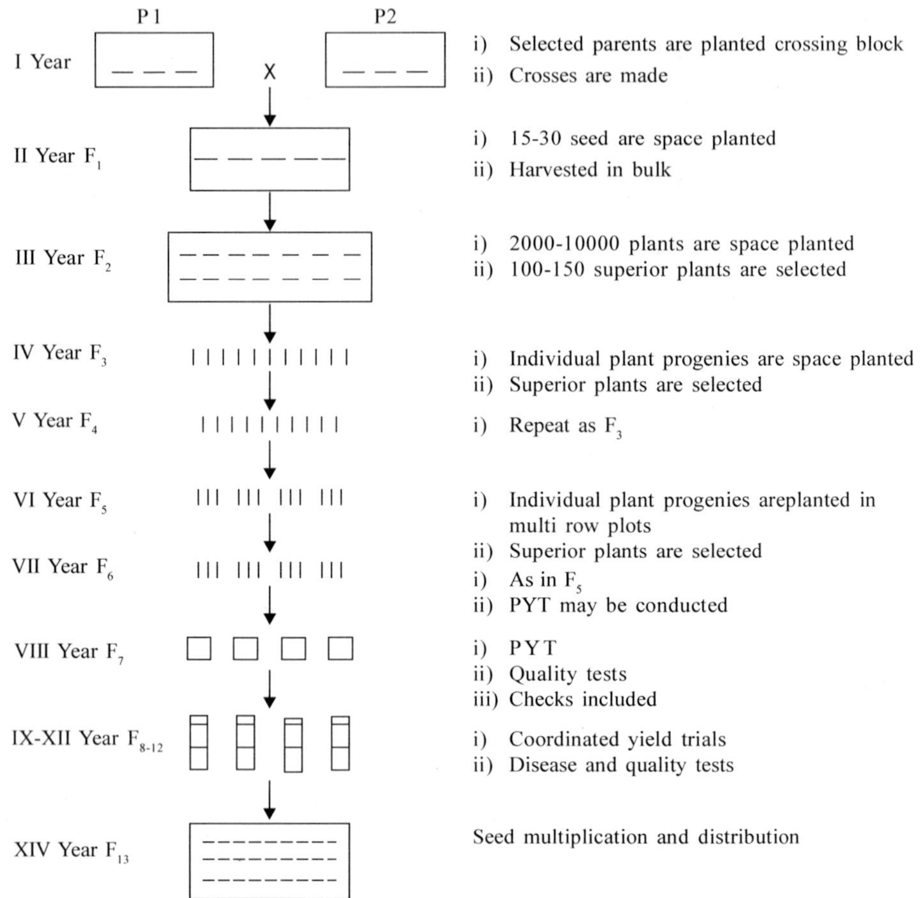

3.5.2. Merits

- Maximum opportunity for the breeder to use his skill and judgement for the selction of plants.
- Well suited for improvement of characters.
- Takes less time than the bulk method to develop a new variety.

3.5.3. Demerits

- Maintenance of pedigree records takes more time
- Laborious and time consuming.
- Success of the methodology depends upon the skill of the breeder.

3.6. Bulk method

- Also known as mass method or the population method.
- F_2 and subsequent generations are harvested in mass or as bulks to raise the next generation.
- Duration may vary from 6-30 or more generations

3.6.1. Procedure

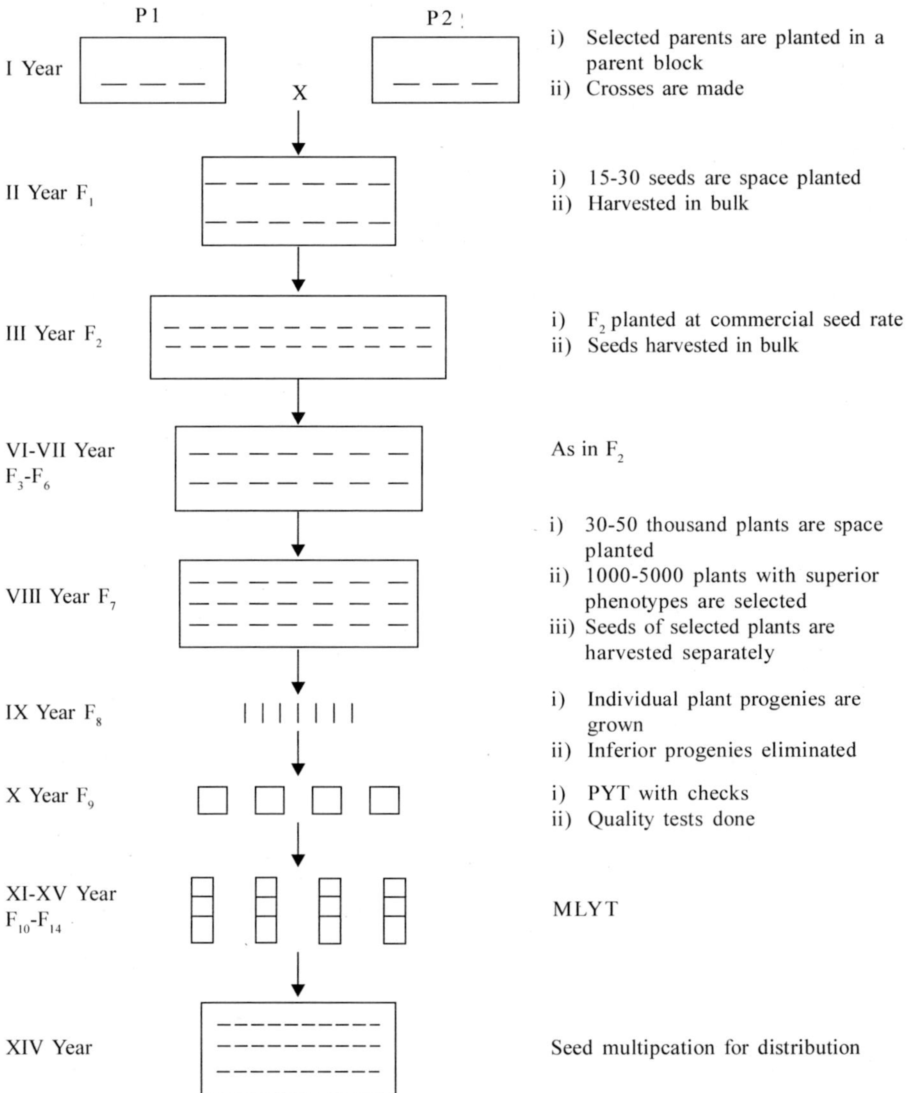

3.6.2. Merits

- Simple, convenient and inexpensive method
- Artificial or natural disease epiphytotics, winter killing etc. eliminate undesirable type.
- Artificial selection may be practiced to increase the frequency of desirable types.

3.6.3. Demerits

- Takes much longer time to develop a new variety.
- Provides little opportunity for the breeder to exercise his skill or judgement in selection.

3.7. Modified bulk method

Modification of bulk method based on artificial selection in F_2 and the subsequent generations.

3.7.1. Procedure

F_2 and F_3 generations

- Select 1000-5000 desirable plants from F_2 and F_3 plant population.
- Bulk seeds from the selected plants.

F_4 generation

- Select 1000- 5000 desirable plants from F_4.
- Harvest seeds separately.

F_5 generation

- Grow individual plant progenies.
- Harvest in bulk seeds from each selected progenies.

F_6 generation

- A PYT is conducted.
- Quality tests may be done on superior progenies.

F_7 generation

- Superior progenies are space planted.
- Individual plants are selected from the superior progenies and their seeds are harvested separately.

F_8 generation

- Individual plant progenies are grown
- Inferior progenies are rejected.
- Selected progeny is harvested in bulk.

F_9 generation

- PYT is conducted at several locations with standard varieties as checks.
- Quality test is done to eliminate undesirable progenies.

F_{10} - F_{13} generation

- RYT are conducted at several locations with standard varieties as checks.
- The lines superior to the standards are released as new varieties.

F_{15} generation

- Seeds are multiplied for distribution

3.7.2. Merits

- Provides an opportunity for the breeder to use his skill and judgement in selection of superior types in earlier generations (F_2 and F_4).
- Does not involve record keeping.

3.8. Single seed descent method

- Modification of bulk method.
- To rapidly advance the generations of crosses.

3.8.1. Procedure

- Single seed from each of 1000-2000 F_2 plants are bulked.
- In F_3 and F_4, one random seed selected from every plant is planted in bulk in next generation.
- In F_5 or F_6 g, plants will nearly be homozygous.
- In F_5 or F_6, 100-150 individuals plants selected are grown in next generation.
- PYT & quality tests begin in F_7 and F_8 and coordinated yield trials in F_8 or F_9.

3.8.2. Merits

- Raising F_3 and later generations from a bulk of one seed to ensure that each F_2 plant is represented.

- Advances the generation with the maximum possible speed.
- Requires very little space, effort and labour.

3.8.3. Demerits

- No form of selection is possible during the segregating population.
- In each successive generation, population size becomes smaller due to poor germination and plant death.
- Some superior genotypes may lose.

3.9. Back cross method

- A cross between a hybrid (F_1 or a segregating generation) and one of its parents.

3.9.1. Objective

- To improve one or two specific defects of a high yielding variety and has other desirable characteristics.

3.9.2. Procedure

1. Transfer of a dominant gene

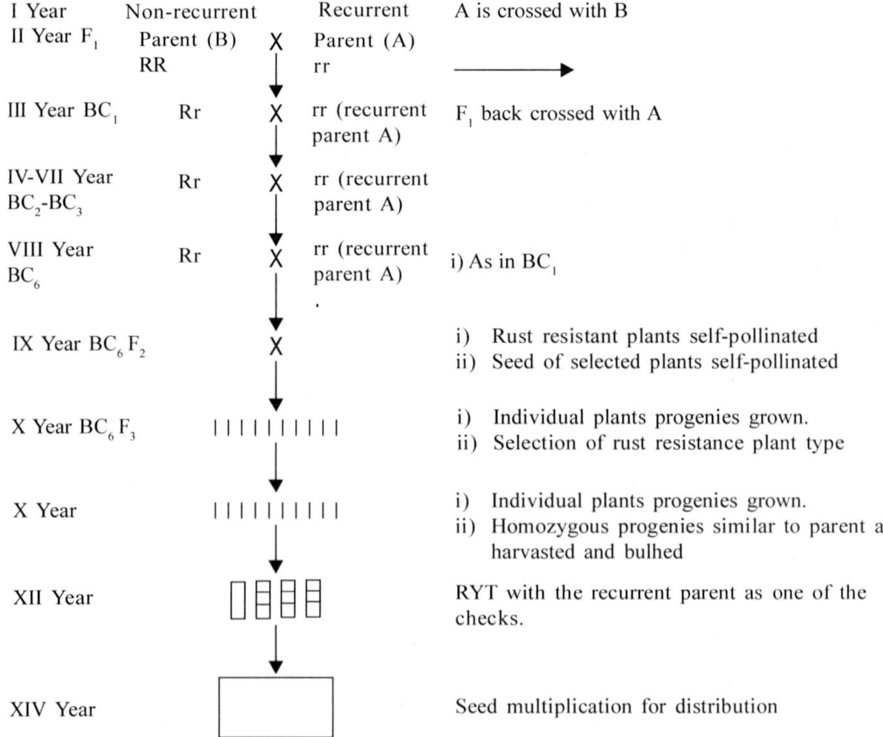

2. Transfer of recessive gene

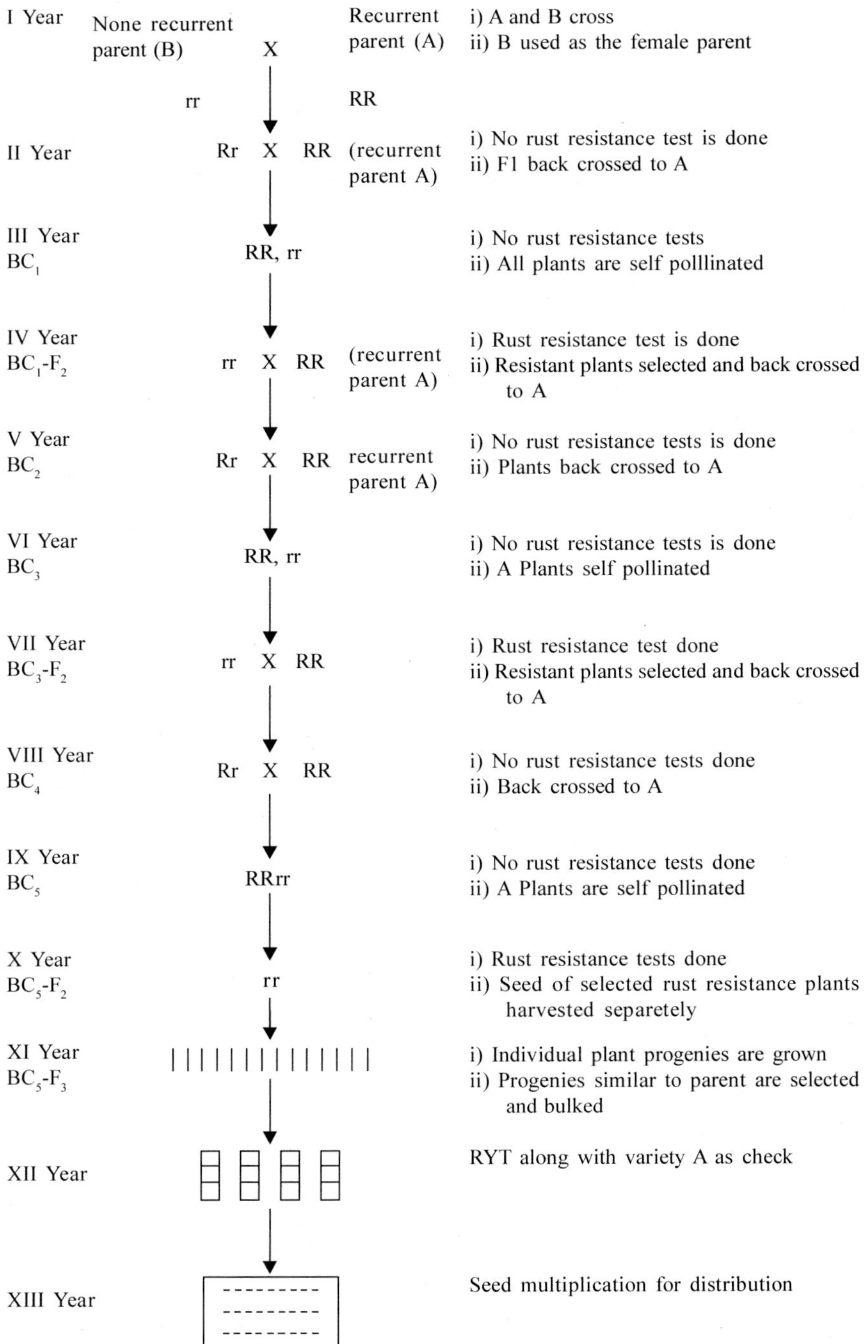

3.10. Multiline varieties

- Mixture of several pure lines of similar height, flowering and maturity dates, seed color and agronomic characteristics, but having genes for disease resistance.

3.10.1. Procedure

1. Back cross method

- Different resistance genes may be transferred in different lines.
- After 6 back crosses, selection may be practiced for uniform lines to make multilines.

2. Restricted back cross method

- Only 3 back crosses may be made for selecting segregates with uniform agronomic features.

3.10.2. Merits

- All the lines are most identical to the recurrent parent.
- Only one of a few lines become susceptible to the pathogen in any one season.

3.10.3. Demerits

- A new race may attack all the lines of a multiline variety.

3.11. Population approach

- Two obvious limitations of breeding methods based on self pollination of the hybrid (Pedigree method and bulk method).
- Recombination is limited to two or three generations.
- No possibility for further changing the genotype of segregants

3.11.1. Procedure

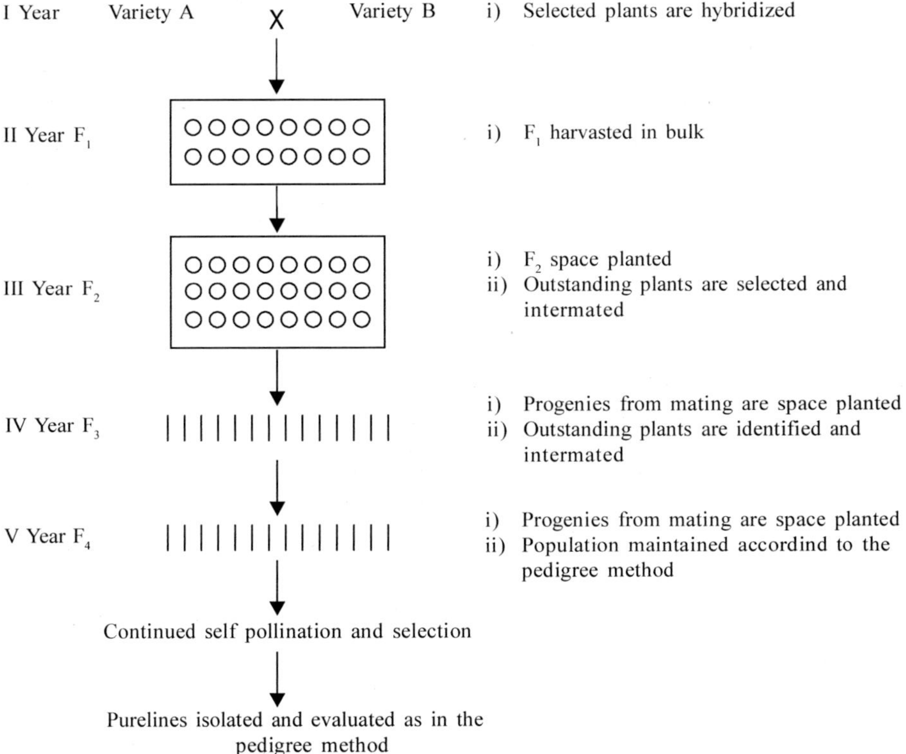

3.11.2. Merits

- Provides greater opportunities for recombination.
- Helps in the accumulation of desirable genes in the population.

3.11.3. Demerits

- Success depends upon the identification of desirable plants in F_2 and subsequent generations.
- Time consuming is higher compared to the pedigree method.

3.12. Hybrid

- Self and often cross pollinated crops show little inbreeding.
- Hybrids are more common in cross pollinated than self pollinated crops for two reasons
- Cross pollinated crops are well suited for hybrid seed production in larger quantities.

- In self pollinated crops, superior pure line varieties give high yield and easily multiplied and maintained.

3.12.1. Use of CGMS

- CMS is used commercially in sorghum to produce hybrid seed.

1. Development of male sterile lines

- A lines or new male sterile lines developed by back crossing
- A line is identical with the male fertile line, known as B line or maintainer line.

2. Maintenance of MS lines

- A line are maintained by crossing them with B lines (male fertile)

3. Production of hybrid seed

- Male sterile line is pollinated with the pollen from a R (Restorer line).

4

Breeding Methods for Cross Pollinated Crops

4.1. Synthetic varieties

- Produced by crossing in all combinations, a number of lines that combine well with each other.

4.1.1. Production method

1. I Method

- Equal amounts of seeds from the parental lines (Syn_0) are mixed and planted in isolation.
- Open pollination is allowed in all combinations.
- Seed from this population is harvested in bulk.
- Population raised from this is Syn_1 generation.

2. II Method

- All possible crosses among the selected lines are made in isolation.
- Equal amounts of seeds from each cross are composited to produce synthetic variety.

4.1.2. Merits

- Offers a feasible means of utilizing heterosis in crop species where pollination control is difficult.
- Good reservoirs of genetic variability.
- Cost of seed is lower than that of hybrid varieties.

4.1.3. Demerits

- Performance of synthetic varieties is usually lower than that of the single or double cross hybrids.

- Eg. Bajra ICMS 7702, coconut BCA – SIN – VAR BCA – SIN-VAR-00 (syn)

4.2. Composite varieties

- Advanced generation seed mixture of an inter varietal cross.
- Same seed is used for 3-4 years compared to hybrids.

4.2.1. Procedure

- Select diverse varieties
- Make all possible crosses among them.
- Select the heterotic crosses and advance them to F_2, F_3 and F_4.
- Evaluate F_1, F_2, F_3 and F_4 and compare with the best hybrids or other varieties.
- Select high yielding crosses, where inbreeding depression is least.
- Mix seeds of selected lines and open pollinate in isolation and release.
- Eg. Amber, Vijay, Vikram, Jawahar, Tarun, Kisan, Shakti, Vikas, Sona, Protina.

4.3. Heterosis and inbreeding depression

4.3.1. Inbreeding

- Average reduction in fitness or of a character due to inbreeding.

1. Effects of inbreeding

- **Appearance of lethal and sub lethal alleles**: Includes chlorophyll deficiencies, rootless seedlings, defects in flower structure.
- **Reduction in vigour**: Plants become shorter ad weaker.
- **Reduction in reproductive ability**: Many lines reproduce so poorly that they cannot be maintained.
- **Increase in homozygosity**: Lines which are almost homozygous due to continued inbreeding
- **Reduction in yield**: Inbred lines yield less than that of the open pollinated varieties.

4.3.2. Heterosis

- Superiority of an F_1 hybrid over both of its parents in terms of yield or some other character.

4.3.2.1. Manifestation of heterosis

- Increased yield
- Increased reproductive ability
- Increased size and general vigour
- Better quality
- Earlier flowering and maturity
- Greater resistance to pest and diseases
- Greater adaptability

4.3.3. Hypothesis of heterosis and inbreeding depression

- **Dominance hypothesis:** At each locus, a dominant allele has a favourable effect, while the recessive allele has an unfavourable effect.
- In heterozygous state, the deleterious effects of recessive alleles are masked by their dominant alleles.
- Thus, heterosis results from the masking of harmful effects of recessive alleles by their dominant alleles.
- **Over dominance:** Single gene hypothesis, super dominance, and cumulative action of divergent alleles, stimulation of divergent alleles.

5

Breeding Methods for Asexually Propagated Crops

5.1. Characteristics of asexually propagated crops

- Majority of them are perennials
- Many of them show reduced flowering and seed set.
- Invariably cross pollinated.
- Mostly polyploids.
- Many species are inter specific hybrids.

5.2. Methods of improvement of asexually propagated species

5.2.1. Clonal selection

- Phenotypic value of a clone is due to effect of genotype (G), environment (E) and G x E interaction

5.2.2. Clone

- Group of plants produced from a single plant through asexual reproduction.

5.2.3. Characteristics

- All the individuals belonging to a single clone are identical in genotype.
- Phenotypic variation within a clone is due to the environment only.

5.2.4. Procedure

First year

- From a mixed variable population, few hundred to few thousand desirable plants are selected.
- Plants with obvious weakness are eliminated.

Second year

- Clones from the selected plants are grown separately.
- Desirable clones are separated.

Third year

- A replicated PYT is conducted, with standard checks.
- Selection for quality, disease resistance etc., are done
- Few outstanding clones are selected.

Fourth year-seventh year

- MLYT with standard checks are done
- Best clones is identified for release as a new variety.

Eighth year

- Best clone is released as a new variety
- Seed multiplication for distribution

5.2.5. Merits

- Only method of selection applicable to clonal crops.
- Clonal selection can be combined with hybridization

5.2.6. Hybridization

- Clonal crops are generally improved by crossing two or more desirable clones, followed by selection in the F_1 progeny and in the subsequent clonal generations.

1. Inter specific hybridization in the improvement of clonal crops

- Potato variety Kufri Kuber was developed from a complex cross (*Solanum curtilobum* x *Solanum tuberosum*) x *Solanum andigena.*

6

Resistance Breeding

- Done for both biotic and abiotic stresses.
- Biotic stresses are disease and insect pests.
- Abiotic stresses are drought, salt, flood, heat, cold, etc.

6.1. Breeding for disease resistance

- **Selection**-Selection for resistant plants from commercial varieties.
- Eg. Potato-Kufri red selection from Darjeeling red round
- Bhendi-Pusa Sawani selection from a Bihar variety resistant for YVMV
- **Introduction**-Introducing the resistant varieties in a new area.
- Eg. Groundnut-Early varieties resistant to tikka leaf spot introduced from USA.
- **Hybridization**-Transferring disease resistance from one variety or species to the other.
- Eg. Cotton-Laxmi resistant to red leaf blight was developed by pedigree method using susceptible parent Gadag 1 and resistant parent Coimbatore compodia 2 (Gadag1 x CC2).
- **Mutation breeding**-Induce resistance to a susceptible variety, which is widely adapted and agronomically desirable.

6.2. Breeding for insect resistance

- **Introduction**-Introduction of *Phylloxera vertifolia* resistant grape rootstocks from USA.
- **Selection**-Selection for resistance to potato leafhopper
- Spotted alfalfa aphid in two broad based germplasm of alfalfa.
- **Hybridization**-Insect resistance variety is crossed with an agromonically superior, but insect susceptible variety to develop a superior and insect pest resistant variety.

6.3. Breeding for abiotic stress resistance

6.3.1. Drought resistance

- Ability of a crop variety to perform better over other varieties under specific drought conditions.

1. Procedure

- Selection of drought escaping varieties
- Selection of genotypes under defined drought conditions by applying appropriate selection pressure.
- Hybridization: Locate and transfer drought resistance to adapted high yielding variety.
- Eg. Sahbhagithan (rice variety) developed in India

6.3.2. Salt tolerance

- Ability to reduce or overcome possible injurious caused directly or indirectly by presence of soluble salts in root zone.

1. Procedure

- Selection of genotypes under defined stress conditions, field and/or under stimulated field conditions.
- Transferof salt tolerant traits into agronomically superior varieties.
- Eg. Salt tolerant varieties - CSR -10, 13, 23, 27, 30, 36.
- Salt resistant varieties - Sumathi vartha 1, 2, 3; USAR DHAN 1, 2, 3; Vytulla 1, 2, 3, 4; Panvel 1, 2 –

6.3.3. Flood resistance

- Screening and/or development of varieties under defined stress conditions.
- Identification of suitable donor parents and incorporation of submergence tolerance.
- Better performing germplasm may be isolated and released.
- Eg. Rice - Swarna, sub-1 (India), MTU-10, MTU-1001, MTU-1140 developed at Kattak (CRRI)
- Maruteru – MTU in Andhra Pradesh.

6.3.4. Cold tolerance

1. Procedure

- Conducting field survey tests to screen plants with winter hardiness.
- Conducting trials over a large number of plants that have survived the rigours of winter.
- Select source materials for the various traits separately and to recombine the selected sources.

7

Mutation Breeding

- Refers to entire operation of the induction and isolation of mutants.
- Term mutation was introduced by Hugo de vries in 1900 in connection with sudden heritable changes observed by him in *Oenothera*.

7.1. Causes of mutation

- **Spontaneous mutation**-Occurs naturally without any apparent cause
- **Induced mutations**-Artificial induction with mutagens.

7.2. Procedure

7.2.1. Selection of variety

- Select variety (i.e) best suited for mutagenesis

7.2.2. Treatment of plant parts

- Both sexually and asexually propagated materials.
- Seed, pollen grains and vegetative propagules
- **Chemical mutagens**: EMS (Ethyl Methane Sulphonate), MMS (Methyl methyl sulphonate), NH (Nitrous hydrazide), diethyl nitrozomide, diethyl sulphonate, nitrosol methyl urea; seeds are presoaked for a few hours to initiate metabolic activities, exposed to desired mutagen and then, washed in running tap water to remove mutagen present.
- **Physical mutagens:** Electro magnetic radiations - x rays, UV light, gamma rays
- **Principle**: Changes structure of DNA
- **Dosage of mutagen**: LD-50 (Lethal dead – point 50) (Dose of a mutagen, which would kill 50% of the treated individuals)

Handling of the mutagen treated population

I year (M_1)	Treated seeds are space planted
	Seeds from individual plants are harvested
II year (M_2)	Individual plant progenies are grown
	Plant from rows containing the mutant allele are harvested separately
III year (M_3)	Individual plant progenies are grown.Superior mutant lines are harvested in bulk, if they are homogenous; harvested separately, if they are heterogeneous progenies.
IV year (M_4)	PYT with a suitable checkSuperior lines are selected
V- VIII year (M_5-M_8)	RYT in several locationsOutstanding lines are released as a new variety
IX year (M_9)	Seed multiplication for distribution among farmers

7.3. Merits

- Mutation creates inexhaustible variations when no improvement is possible, this method has to be adopted.

7.4. Demerits

- Frequency of desirable mutations is very low (0.1%)
- Desirable mutations are commonly associated with undesirable side effects.
- Most of the mutants are recessive.

8

Polyploidy Breeding

- Refers to induced chromosome manipulation
- **Aneuploidy**-One or few chromosomes extra or missing from 2n.
- **Autoploidy**-Genomes identical with each other.

8.1. Procedure for aneuploidy production

- **Spontaneous**: Originate spontaneous at a low frequency.
- Meiotic irregulation lead to the formation of 2n + 1 and 2n-1 gametes.
- **Triploid plants**: Best source of the aneuploids.
- Unreduced 2n gametes unit with normal haploid gamete to form triploid zygote.
- 2n + 1n = 3n - triploid
- Eg. Banana, watermelon (does not produce functional seed)
- **Tetrasomic plants**: Four homologous chromosomes present instead of 2 chromosomes.
- 4n
- There are about 4 alleles in loci or locus.
- Eg. Potato
- **Tetrasomic plants**

8.2. Procedure for autoploidy production

- **Spontaneous**-Chromosome doubling occurs occasionally in somatic tissues.
- **Physical agents**-Heat or cold treatments, centrifugation and x- ray or g ray irradiation may produce polyploids in low frequencies.
- **Regeneration of *in vitro***-Polyploids plants are regenerated from callus and suspension cultures.
- **Colchicine treatment**-Most effective treatment for chromosome doubling.

8.2.1. Method of colchicine treatment

- Concentrations are from 0.001 to 1%.
- Soaked in a shallow container to facilitate aeration.
- Growing shoots apices are treated with 0.1 to 1% colchicine by brush.

9

Modern Tools in Crop Improvement

9.1. Molecular breeding and marker assisted selection

- Using conventional breeding, developing new varieties takes atleast 15-20 years.
- Molecular breeding shortened the time to 7-10 years.
- One of the tools for faster variety development is Marker Assisted Selection (MAS).
- MAS-refers to use of DNA markers to assist phenotypic screening.
- Plants that possess particular genes or quantitative trait loci (QTL) may be identified based on their genotype rather than their phenotype.

9.1.1. Advantage of MAS

- Greatly increase the efficiency and effectiveness of breeding compared to conventional breeding.
- It is time and labour saving, when compared to conventional breeding, which involves time consuming field trial.
- DNA marker selection is more reliable, while in field trials the influence of environmental factors are more.
- DNA markers may be more cost effective and in some cases too expensive.
- Selection may be carried out at seedling stage itself.
- Single plants may be selected with high reliability.
- Reduction in total number of lines that needs to be tested.

9.1.2. Molecular makers

- RFLP - Restriction Fragment Length Polymorphism
- RAPD - Random - Amplified Polymorphic DNA

- STS – Sequence Tagged Sites
- AFLPS – Amplified Fragment Length Polymorphism
- SSR – Single Sequence Repeats or Microsatellites
- SNP – Single Nucleotide Polymorphism

Comparison of different marker systems

Feature	RFLPs	RAPDs	AFLPs	SSRs	SNPs
DNA required	10	0.02	0.5-1.0	0.05	0.05
DNA quality	High	High	Moderate	Moderate	High
PCR-based	No	Yes	Yes	Yes	Yes
Number of polymorph loci analyzed	1.0-3.0	1.5-50	20-100	1.0-3.0	1
Ease of use	Not easy	Easy	Easy	Easy	Easy
Amenable to automation	Low	Moderate	Moderate	High	High
Reproducibility	High	Unreliable	High	High	High
Development cost	Low	Low	Moderate	High	High
Cost per analysis	High	Low	Moderate	Low	Low

9.1.3. Importance of QTL mapping for MAS

- Identification of genes and quantitative trait loci and DNA markers that are linked to the genes is accomplished via QTL mapping.
- QTL is the foundation for the development of markers (MAS)
- Widely accepted that QTL confirmation, validation and/or additional marker steps may be required after QTL mapping and prior to MAS.

1. Steps in QTL mapping

- **Marker conversion**: by using marker genotypic method, the reliability of MAS is improved.
- **QTL formation**: Testing the accuracy of the results from the primary QTL mapping study.
- **QTL validation**: Refers to the verification that QTL is effective in different genetic back ground.
- **Marker validation**: Testing the level of polymorphism of most tightly linked markers and also the reliability of the markers to predict genotype.

  (Polymorphism – ability of an object to take on many forms).

9.2. MAS in plant breeding

9.2.1. Marker assisted back crossing

- Three levels of selections in which markers may be applied in back cross breeding.
- In the first level, markers are used to screen for the target trait which have laborious phenotypic screening procedure or recessive alleles
- In second level, selecting back cross progeny with target gene and tightly linked flanking markers to minimize linkage drag. This is known as recombinant selection.
- Third level involves selecting back cross progeny with back ground markers.
- In conventional breeding, it takes 5-6 generations to recover the recumbent parent.
- Using markers, three to four back cross generations can be saved.

9.2.2. Marker assisted pyramiding

- Process of simultaneously combining multiple genes / QTLs together in a single genotype.
- It is also possible in conventional breeding. But, it is a time consuming because; it involves phenotypic screening which is impossible in early generations.
- DNA markers facilitate early selection because of non-destructive and markers for multiple specific genes/QTL can be tested using a single DNA sample without phenotyping.
- Pyramiding has wide spread application for combining multiple disease resistant genes to develop durable disease resistant varieties. So, it has wide spread application.

9.2.3. Early generation maker assisted selection

- Simply discard, many plant with unwanted gene combinations especially those that lack of essential disease resistance traits and plant height.
- Helps in selection of plants at an early generation.

9.2.4. Combined approaches

- Combinations of phenotypic screening and mass approach may be useful
- To maximize genetic gain
- Level of recombination between marker and QTL.

- To reduce populations size for traits where marker genotype cheaper and easier than phenotype screening.

9.2.5. Obstacles in adoption of MAS

- MAS in rice are too expensive (prohibiting) cost.
- Equipment and consumbles needed to establish and maintain a marker labs is more expensive.
- Large initial cost in the development of markers.
- Cost involved in marker assisted back crossing is more expensive.
- Low reliability of marker to determine phenotype.
- Level of integration between molecular geneticists and plant breeders may not be adequate to ensure that markers are effectively applied for line development.

9.2.6. Commercially cultivated MAS varieties developed in India

Plant	Cultivar /breeding line	Year of release	Institute	Trait
Maize	Vivek QPM9	High quality protein	2008	ICAR
Pearl millet	HHB67 improved	Disease resistance	2005	ICAR /ICRISAT
Rice	MAS 946-1	Drought tolerant	2007	UAS
Rice	Pusa 1460	Disease resistance	2007	IARI
Rice	RP Bio 226 (Improved sambha mahsuri)m	Disease resistance	2007	DRRI/CCMB

9.3. Genetic transformation

- Refers to heritable change in a cell (or) organism, due to introduced DNA.
- Its nothing but genetic engineering – introduction of novel gene this GT includes.
- Access to novel molecules.
- Ability to change the level of gene expression.
- Then capability to change their expression pattern of genes
- Develop transgenic plants with novel genes.

9.3.1. Basic requirements for genetic transformation

- Target genome
- A candidate gene

- Vector to carry the gene
- Tissue culture and regeneration system.
- Modification of the foreign DNA to increase the level of gene expressions.
- Method to deliver the plasmid DNA
- Protocols to identify the transformed cells.
- Characterization of the putative transgenic plants at molecular and genetic levels.

9.3.2. GMO or transgenic plant

- Plant that has novel combination of genetic material obtained through the use of modern biotechnology.
- Plant contains genes, which have been artificially introduced, instead of the plant acquiring them through pollination.
- **Transgenic plant-**Plants that have been genetically engineered (Breeding approach that uses recombinant DNA techniques) to create plants with new characters.

9.3.3. Steps involved in development of transgenic plants

1. DNA delivery
2. Selection of transgenic tissues
3. Recovery of transgenic whole plants

9.3.4. Methods of gene delivery

- Indirect gene transfer or vector mediated gene transfer

1. Agro bacterium mediated delivery

- Eg. *Agrobacterium tumifaciens*
- Gram positive bacterium that uses horizontal gene transfer to cause tumour in plants.
- It has strong ability to transfer DNA between itself and plants.
- Plays an important tool for improvement of genetic engineering
- When bacterial DNA integrated into a plant genome, it becomes the part of the plant chromosome and integrated DNA sequence stably expressed and inherited.

2. Vectorless or direct gene transfer

- Particle bombardment method
- Chemical mediated gene transfer
- Micro injection
- Electroporation
- Pollen transformation
- Laser induced transformation
- DNA delivery via growing pollen tubes
- Fibre mediated gene transformation
- Liposome mediated gene transfer or lypofection
- **Particle bombardment method (gene gun)**
- Alternative to *Agrobacterium* mediated gene transfer method
- Aim: Penetrate cell or tissues with accelerated metal microspheres (spherical particles of solid or hollow having diameter of 1-200 micron, made up of glass, ceramic, carbon or plastics which are coated with DNA.
- Microspores of tungsten or gold (high density metals) having diameter of 1-4 mm is found to penetrate cells at high velocities.
- DNA is released and stably integrated into chromosome.
- This technique is widely used in tobacco, soyabean, rice, maize and barley.

9.3.5. Benefits of GM plants

- Enhanced resistance to insects results in significant reduction of chemical pesticides. Eg. Bt cotton, Bt corn.
- Tolerance to herbicides eg. Roundup ready soybean
- Increased resistance to viral, bacterial and fungal disease with improved safety for human consumption. Eg. Cucumber and papaya
- Increased tolerance to heat, cold, drought, salinity, water logging etc (Environmental stresses).
- Increased ability of plants to remove toxic materials from soil (bioremediation)
- Production of more bio degradable industrial products.

9.3.6. Consumer benefits

- Development of novel oils starches and industrial products. Eg. Soybean with high level of oleic acid.

- Enhancement of vitamins and minerals in food grains. Eg. Vitamin A – golden rice
- Elimination of certain allergens and antinutritional compounds from foods.
- Production of pharmaceutical products including edible vaccines, recombinant antibodies and anti coagulant compounds. Eg. Banana
- Improved transport and shelf life of products. Eg. Controlled ripening in melons, peas, peppers and tomatoes.
- First GM crop – Flavr Savr tomato for human consumption by Calgene Monsanto Company in California.

Traits inserted in transgenic plant

Crops	Trail	Gene product
Onion, Tobacco	Insect tolerance	Bt toxins
Cowpea		Protease inhibitors
Tobacco, Tomato	Virus tolerance	Coat proteins Antisense ribozyme
Rice	Better nutritional quality (Golden rice)	Vitamin A
Potato	Tolerance to fungi Tolerance to bacteria	Chitinases Phytoalexins Lysozymes
Mustard	Male sterility Drought tolerance	Ribonuclcase Gluconase Proline Bcuine
Wheat, Tobacco	Herbicide tolerance Oxidative stress tolerance	Superoxide dismutasc Glutathione reductase
Tobacco	Cold tolerance Salinity tolerance	Glycerol-1 -phosphate acyltransferase Mannitol-phosphate dehydrogenase

10

Heterosis Breeding and Hybrid Seed Production Technologies

10.1. Heterosis

- Occurrence of a genetically superior offspring from mixing the genes of its parents.
- Superiority of F_1 hybrids over its both parents in terms of yield and vigour.

10.1.1. Manifestation of heterosis

- Increased yield
- Increased reproductive ability
- Increased vigour
- Better quality
- Early flowering and maturity
- More resistance to pest and
- Wider adoptability disease

10.2. Inbreeding depression

- Average reduction of a character due to selfing
- Self and often cross pollinated crops show little or no loss in vigour and yield due to in breeding.
- Often cross and self pollinated crops: Inbreeding depression is low
- Cross pollinated crops: Inbreeding depression is high and high hybrid vigour.
- Cross pollination are best suited for hybrid seed production.

10.2.1. Effects of inbreeding

- Appearance of lethal and sub lethal and alleles
- Chlorophyll deficiency

- Rootless seedling
- Defect in flower structure
- Reduction in vigour
- Plant become shorter and weaker
- Reduction in reproduction ability.
- Increase in homozygosity
- Reduction in yield.
- Yield of inbreds is always less than varieties

10.3. Types of heterosis

10.3.1. On the basis of origin and nature

- Euheterosis or true heterosis
- Mutational heterosis
- Balanced heterosis
- Pseudoheterosis or luxuriance

1. Mutational heterosis

1. Lethal (mostly), recessive, adaptively unfavorable mutants are either eliminated or sheltered by their non-lethal, dominant and adaptively superior alleles in cross pollinated crops.
2. Naturally occurring mutants are generally of recessive and less adaptive to environmental conditions.

2. Balanced heterosis

- Well balanced gene combinations which are more adaptive to environmental conditions
- Has application in hybrid production.

3. Pseudoheterosis

- Progeny possess superiority over parents in vegetative growth, but not in yield and adaptation, usually sterile or poorly fertile.
- Cannot be utilized in hybrid production.

10.3.2. On the basis of types of estimation

1. Average or relative heterosis

- Estimated over mid parental value *i.e.* average of two parents.

- Average heterosis = [(F_1 - MP) / MP] x 100
- Where, F_1 = Value of F_1
- MP = Mean value of two parents

2. Heterobeltiosis

- Estimated over better parent.
- Heterobeltiosis =[F_1 – BP] / BP x 100
- Where, F_1 = Value of F_1,
- BP = Value of better parent

3. Standard or economic heterosis

- Estimated over standard commercial hybrid.
- Standard heterosis = [(F_1 - SH)/ SH] x 100.
- Where, F_1 = Value of F_1
- SH = Value of standard hybrid

10.3.3. Diagrammatic representation of heterosis

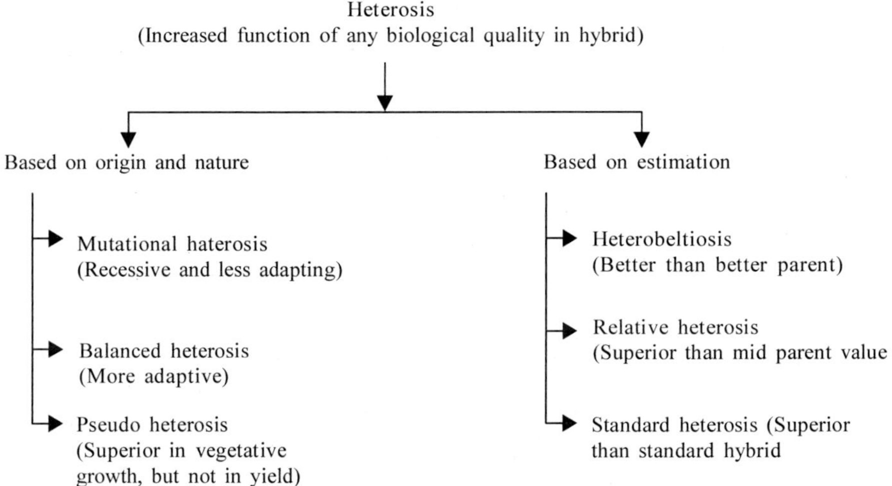

10.4. Genetic basis of heterosis

10.4.1. Dominance hypothesis

- Superiority of hybrids to suppression of undesirable (deleterious) recessive alleles from one parent by dominant alleles from other. AAaay

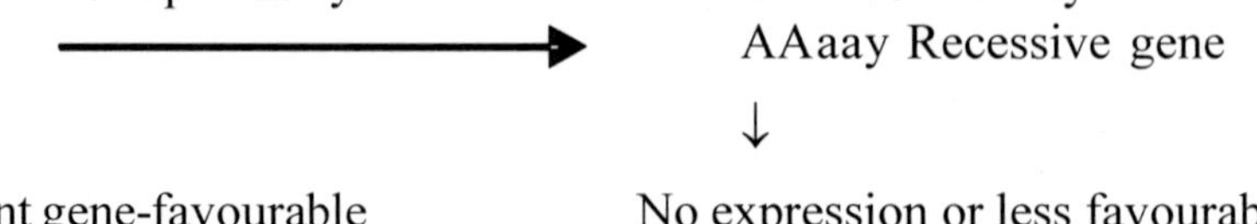

AAaay Recessive gene

↓ ↓

Dominant gene-favourable effect

No expression or less favourable effect

10.4.2. Overdominance hypothesis

- Some combinations of alleles (Crossing two inbred strains) are advantageous when paired in a heterozygous individual.
- In heterozygous, deleterious of effect of recessive allele are masked by dominated alleles.

Inbreds (homozygous) AAAA x aaaa

Female ↓ male

AAaa

Heterozygous Deleterious effect of recessive gene marked by dominance

10.5. Hybrids

- Any offspring resulting from the mating of two distinctly homozygous individuals
- Very common in cross pollinated crops than self pollinated crops - Vegetables, maize, sorghum, pearl millet, sunflower.
- Often cross pollinated crops - Cotton, redgram, chilli, brinjal, bhendi
- Self pollinated crops - Rice

10.5.1. Attributes and benefits of F_1 hybrids

- Maximum performance under optimal condition
- Stability of performance under stress
- Proprietary control of parents
- Often, reduced time to cultivar development
- Joint improvement of traits

- Good vigour
- Heavy yields
- Uniform growth and maturity
- Early maturing
- Disease resistance
- Good yield holding ability

10.5.2. Steps in breeding

- Develop inbred homozygous lines
- Find good F_1 combination between inbreds
- Produce F_1 seed in large scale for growers

10.5.3. Types of hybrids

1. Inter varietal hybridization

- Crossing of two parents from the same species (Two varieties, strains or races of same species)
- Eg. CSH 5,9, K Tall, JKHY-1, Suguna.
- **Single cross hybrid**: Cross between two different homozygous lines produces an F_1 hybrid (F_1 = Filial 1; meaning "first offspring").
- F_1 is heterozygous having two alleles, contributed by each parent and one is dominant and other recessive.
- Phenotypically homogeneous.
- A x B = F_1
- Eg. Maize-COH1 (UMI 29 x UMI 51).
- **Double cross hybrid**: Cross between two different F_1 hybrids.

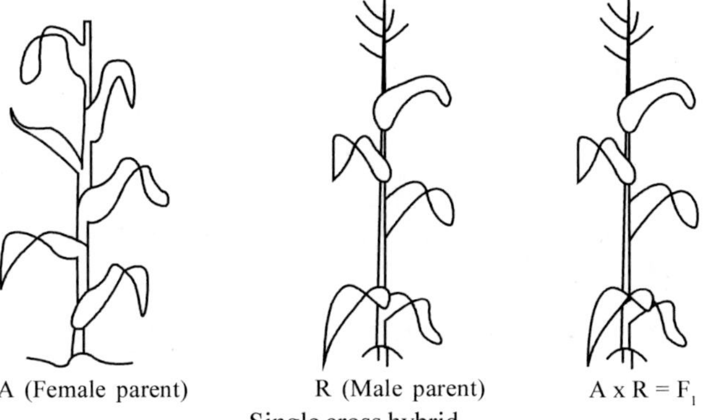

Single cross hybrid

- (A x B) x (C x D) = Double cross
- Eg. Deccan maize (CM 104 x CM 105) x (CM 202 x CM 201)
- **Three-way cross hybrid**-Cross between F_1 hybrid and an inbred.

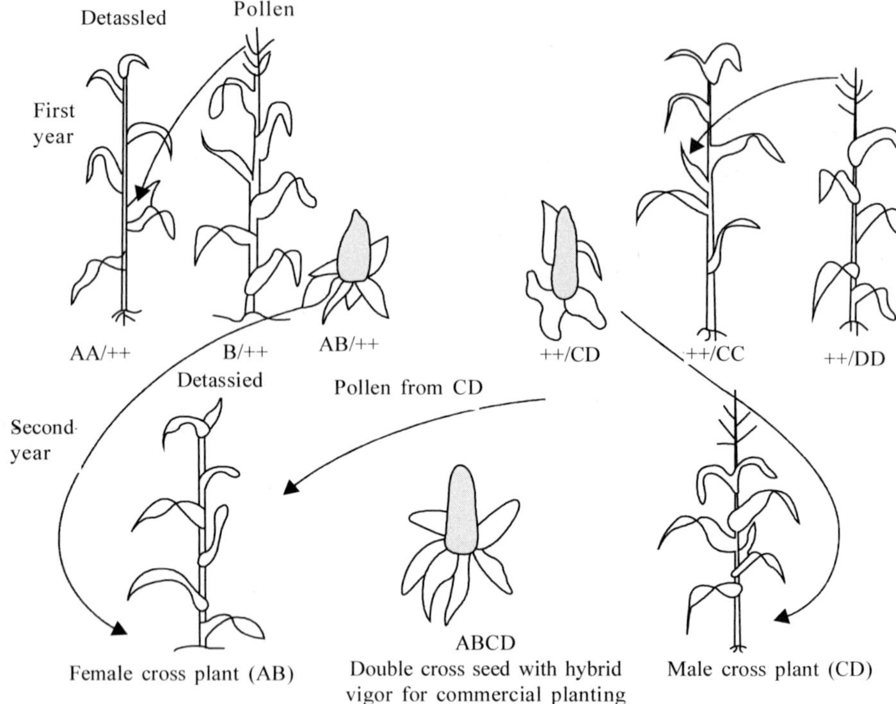

Double cross hybrid

- (A x B) x C = Three way hybrid
- Eg. Maize – Ganga (CM 202 x CM 111) x CM 500
- **Triple cross hybrid**: Crossing of two different three-way cross hybrids.
- **Top cross hybrid**: Cross between inbred line and OPV.

2. Distant hybridization (Population hybrids)

- **Inter generic hybridization**: Crossing between parents from the two different genera
- **Inter specific hybridization**: Crossing between parents from entirely two different species
- Eg. Cotton - Varalakshmi

- Laxmi x SB 289 E
- (G. hirsutum) x (G. barbadense)
- DCH 32 or Jayalakshmi
- D.S. 28 x SB (YF) 425
- (G. hirsutum) x (G. barbadense)
- Tomato (pusa red pulp)
- *L. esculentum* x *L. pimpinellifolium*

10.6. Techniques of hybrid seed production

10.6.1. Single line breeding

- One parent is sufficient to maintain or multiply the particular hybrid.
- Hybrid produced is multiplied by asexual methods using stolons, rhizomes, stem cuttings.
- Eg. NB 21 grass (Cross between napier grass and bajra (Pearl millet)).
- Sugarcane
- Through tissue culture in many plantation crops and spices like turmeric, pepper etc.

10.6.2. Two line breeding

- Two parental lines are involved for production of hybrids.

1. Physical / manual method

- **Hand emasculation and pollination**: Male reproductive part (androecium) of female parent (seed parent) is removed before opening of flower.
- Requires a lot of labour and is costlier.
- Two different methods.
- Hand emasculation and hand pollination
- Eg. Cotton, tomato, brinjal
- Hand emasculation and natural pollination
- Eg. Rice, maize, sorghum, bajra, sunflower
- **Detaseling**: Removal of male inflorescence (Tassel in maize) from female parent before anthesis.
- Eg. Maize

2. Use of genetic male sterility

- Fertility is controlled by dominant gene (MSMS) and sterility is controlled by recessive gene (ms ms).
- Eg. Redgram, castor, groundnut, etc.
- 100 per cent male sterility cannot be maintained; fertile plants are removed during flowering.
- More labour required
- More fertilizer required

3. Use of chemical hybridizing agents (CHA)

- Physiological male sterility is induced by chemical hybridizing agent or gametocide or chemical pollen suppressant.
- Eg. Maleic hydrazide (MH), ethrel, sodium methyl arsenate, zinc methyl arsenate (ZMA), etc.

4. Use of environmental sensitive genic male sterility (EGMS)

- Production of male sterility in different environmental conditions.
- **Temperature sensitive genic male sterility (TGMS)**
- Sterile plants: Day and night temperature exceeds 32°C and 24°C, respectively
- Fertile plants: When temperature falls below 24°C and 18°C at day and night temperature, respectively.
- Utilized in tropical countries like India.
- **Photoperiod sensitive genic male sterility (PGMS)**
- Sterile plants: When day length exceeds 14 h.
- Fertile plants: When the day length is less than 13 h.
- **Photo thermo sensitive genic male sterility**
- Male sterility is controlled by the interaction of photo period and temperature.
- Used both in tropical and temperate regions.

10.6.3. Three line breeding

Use of cytoplasmic genic male sterility

- Sterility is controlled by cytoplasm.
- Most widely used system for hybrid seed production.

- 100 % male sterility can be maintained in this three line breeding system.
- Eg. Rice, sorghum, pearl millet, sunflower.
- Threc parents involved
- A line is a sterile (Female) line
- B line is a fertile (Maintainer) line
- A (no pollen) and B (Pollen) are isogeneic lines except pollen fertility.
- R line is restorer line

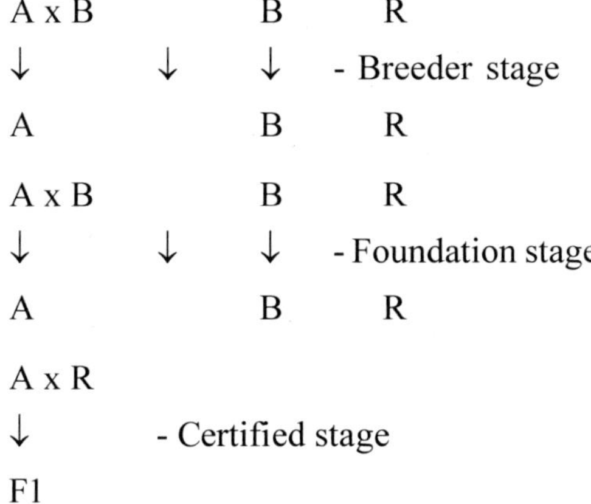

10.6.4. Diagrammatic representation of development of hybrids

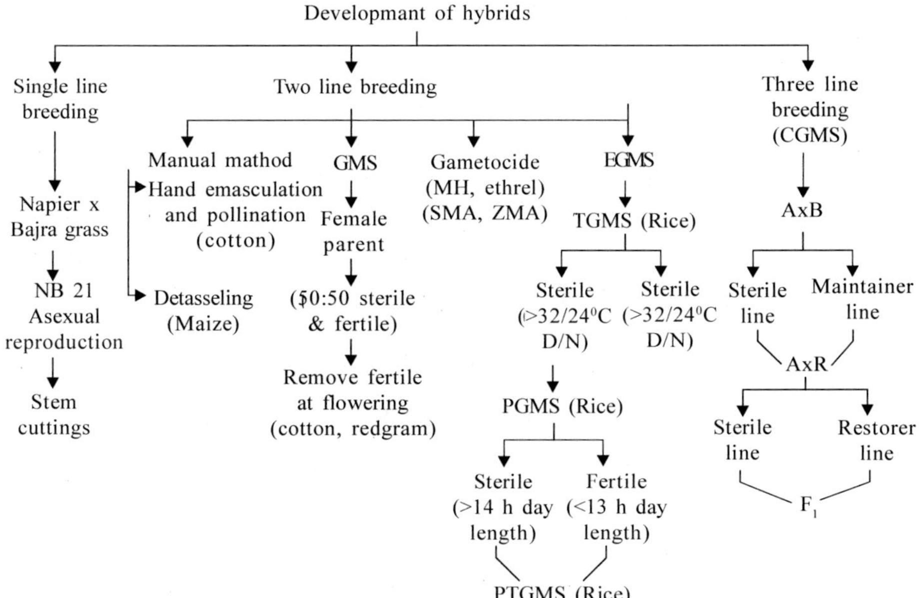

10.7. Constraints for hybrid technology

- Lack of appropriate policy on hybrid at National level including processing and marketing
- Insufficient R&D support
- Poor coverage under hybrids due to non-availability of quality seed.
- Lack of information and material exchange
- Lack of participation of private sectors
- Non-availability of trained man power in some countries
- Absence of required regional co-operation

Three steps

1. Development of inbreed
2. Evaluation of inbreeds and production of hybrid seed
3. Inbreed line: homozygous genotype produced by repeated selfing with selection over several generations.
4. In breed lines (or) individual with same generations mostly pedigree method. Breeding is used to develop inbreed lines.
5. It needs 6-7 generations to attain homozygous line.
6. In addition, bulk method (or) single seed descent method (or) back cross method are also used.

Procedure

- I year-select many plants with desirable phenotypes from a population. Self pollinate: selected individuals
- II year-about 30-40 plants are space planted from the selfed seed from each of the selected plant
- III – IV year-repeat the process of II year
- VII year-discontinue the selfing, individual plant progenies which are homozygous that form the inbreed lines.
- Selfing can be done by may be full in mating (or) half sib mating

Evaluation of inbreeds

1. Phenotype evaluation
2. Phenotypic performance of the inbreeds is very effective and have high heritability. Reject poor performing inbreeds.

Top cross test

- The superior phenotypic individuals are crossed to a tester having wider genetic base.
- Evaluate the performance of top cross progeny by replicated yield trial (RYT)
- Then about 50% of the inbreeds are eliminated (having interior performance)

Single cross evaluation

- Inbreeds are crossed in a di-allele to produce all possible single crosses. di-allele means-each inbreed is crossed with every other inbreed lines.
- Evaluate the performance of single cross is replicated yield trial.
- Identify the outstanding inbreed lines.

Production of hybrid varieties (or) seed

- Crossing of two unrelated homozygous inbreed lines.
- Inbreeds are mostly evaluated for GCA-General combining ability and SCA (specific combining ability.

GCA-is evaluated by top cross and single cross

- SCA-is evaluated by single cross and crossing of tester (open pollinated variety)-inbreed line and with another specific inbreeds to get additional genetic value.

11

Breeding for Climate Resilience Varieties

- Crop production is challenged by abiotic and biotic stresses results in low yield and high production costs.
- Magnitude and intensity of stresses are increasing dramatically due to climate change causing recurrent droughts, increased temperature (heat) and emergence of new diseases and pests.
- Potential of crop breeding is must to develop new varieties with resistance/ tolerance to drought, heat, and major diseases and insect pests to combat climate change.
- Crop breeding program at ICARDA (International Centre for Agriculture Research in Dry Areas) applies both conventional and molecular approaches including focused identification of germplasm strategy, targeting mega environments, shuttle breeding, doubled haploids, marker assisted selection, and key location phenotyping.

11.1. Breeding for climate resilience varieties

- Genetic diversity of crops is foundation for sustainable development of new varieties for present and future challenges.
- Resource-poor farmers have been using genetic diversity intelligently over centuries to develop varieties adapted to their own stress environments.
- Common bean biodiversity has been used by plant breeding to develop both heat and cold tolerant varieties grown from hot Durango region in Mexico to cold high altitudes of Colombia and Peru.
- Corn genetic resources have been used in breeding varieties adapted to cultivation from sea level to over 3,000 masl as in Nepal.
- The Sub1 rice tolerant to flood developed by Bangladesh breeders can survive total submersion for more than two weeks, with great benefits to farmers.

- Plant breeding has been developing varieties for heat, drought and flood stresses, but with more severe and frequent challenges from aggravated climate change it needs extra immediate support to overcome the challenges.

12

Variety Release Procedure

12.1. Variety release

- Superior cultivars are to be purified by breeder and tested for genuiness in 7 generations.
- Then, they are forwarded to MLT after RYT, RRYT, PYT and CYT in respective research station.
- Based on results of the MLT, strain is forwarded to Adaptive Research Trials (ART).

12.1.1. Row Yield Trial (RYT)

- For every 10th row, there is a check and non replicated.

12.1.2. Replicated Row Yield Trial (RRYT)

- From RYT, best cultures will be tested in RRYT along with appropriate check.
- Best centres from RRYT will be forwarded to PYT.

12.1.3. Preliminary Yield Trial (PYT)

- Replicated trial with appropriate checks.
- Conducted normally for two seasons.
- Best entries will be nominated to All India Trials also.
- Screening for biotic and abiotic stresses will be done during PYT stage.
- Best entry will be carried to comparative yield trial.
- Eg. Sorghum - SPV (Sorghum Project Variety)
- Rice - IET (Initial Evaluation Trial.

12.1.4. Comparative Yield Trial (CYT)

- Replicated one conducted with more than one check.
- Repeated for 3 seasons.

- Entry proved to be superior in all the 3 seasons will be proposed for Multi Location Trial (MLT).

12.1.5. Multi Location Trial (MLT)

- Centres for MLT will be decided at crop scientist meet held once in a year.
- Each station will propose its own entry.
- Based on discussion of merits and demerits of each culture, the centres will be nominated.
- MLT will be conducted at Research Stations of State Agricultural Universities.
- Best entries will be proposed for Adaptive Research Trial (ART).

12.2. Varietal description

- Varietal description is required for.
- Carrying out the effective seed certification program.
- For the implementation of plant breeders right (PBR).
- Also necessary for consumer protection.
- Varieties are tested for Distinctness, Uniformity and Stability (DUS).
- National variety release committee often advises on release based on performance (MLT) and DUS.

12.3. Dus test

- To establish whether or not a variety is sufficiently distinct from all other varieties and sufficiently uniform and stable.

12.3.1. Distinctness

- New variety must be different from all existing one.
- Recognizable not only in field inspection, but also for seed growers and farmers.
- For granting property rights (PBR), the variety must be clearly recognizable.
- Distinctness may be morphological, physiological, cytological or chemical

12.3.2. Uniformity

- Must be uniform as possible to guarantee constant quality and field inspection purposes.
- Degree of uniformity depends on the mode of reproduction.

- Varieties of self pollinated crops are more uniform than varieties of cross pollinated crops
- In a highly mechanized agricultural system in a region that is agriculturally and climatologically homogenous, a high degree of uniformity may be desirable, but under other conditions a certain degree of variability may be advantageous.

12.3.3. Stability

- During breeder to certified seed multiplication, variety should not loss its distinctive characters.
- Genetic makeup should remain as near as possible the same.
- Varieties of self pollinated crops are more stable than varieties of cross pollinated crops.
- Hybrids are not stable and new hybrid seed must be produced each year for farmers.

13

Plant Genetic Resoures

13.1. Germplasm

- Also known as genetic resources or gene pool or genetic stock.
- Sum total of hereditary material i.e. all the allele of various genes, present in a crop species and its wild relatives

13.2. Important features of germplasm

- Represents entire genetic variability or diversity available in a crop species.
- Consists of land races, modern cultivars, obsolete cultivars, breeding stocks, wild forms and wild species of cultivated crops.
- Includes both cultivated and wild species and relatives of crop plants.
- Collected from centres of diversity, gene banks, gene sanctuaries, farmer's fields, markers and seed companies.
- Basic material for launching a crop improvement program.
- May be indigenous (Collected within country) or exotic (Collected from foreign countries).

13.3. Germplasm conservation

- Protection of genetic diversity of crop plants from genetic erosion.

13.3.1. *In - situ* conservation

- Conservation under natural conditions
- Achieved by protecting the area from human interference - called natural park, biosphere reserve or gene sanctuary.
- NBPGR, New Delhi, established gene sanctuaries in Meghalaya for citrus, north Eastern regions for *Musa*, Citrus, *Oryza* and *Saccharum*.
- **Merits:** Wild species compete natural or semi natural ecosystems are preserved together.

- **Demerits**-Each protected area will cover only very small portion of total diversity of a crop species, hence several areas will have to be conserved for a single species.
- Management of such areas also poses several problems.
- Costly method of germplasm conservation.

13.3.2. *Ex - situ* conservation

- Preservation of germplasm in gene banks.
- Most practical method of germplasm conservation.
- **Merits**-Possible to preserve entire genetic diversity of a crop species at one place.
- Handling of germplasm is also easy.
- Cheap method of germplams conservation.

13.3.3. Types of conservation

1. Seed banks

- Quite easy, relatively safe and needs minimum space.
- **Orthodox seeds:** Dried to low moisture content and stored at low temperature without losing their viability for long periods of time
- Eg. Corn, wheat, rice, carrot, papaya, pepper, chickpea, cotton, sunflower.
- **Recalcitrant:** Drastic loss in viability if seed moisture content decrease below 30 to 35%
- Eg. Citrus, cocoa, coffee, rubber, oilpalm, mango, jack fruit etc.
- **Seed storage**
- Based on duration of storage, seed bank collects are classified into three groups.
- **Base collections:** Conserved for long term (50 to 100 years) at about -20°C with 5% moisture content.
- Disturbed only for regeneration.
- **Active collection:** Stored medium duration, i.e., 10-15 years at 0°C temperature and the seed moisture is between 5 and 8%.
- Used for evaluation, multiplication, and distribution of the accessions.
- **Working collections:** Stored for 3-5 years at 5-10°C and the usually contain about 10% moisture.
- Regularly used in crop improvement programmes.

2. Plant Bank (Field or plant bank)

- An orchard or a field in which accessions of fruit trees or vegetatively propagated crops are maintained.
- **Limitations**
- Require large areas
- Expensive to establish and maintain
- Prone to damage from disease and insect attacks
- Man - made
- Natural disasters
- Human errors in handling

3. Shoot tip banks

- Conserved as slow growth cultures of shoot tips and node segments (Meristem culture).
- **Merits:** Conserved indefinitely free from virus or other pathogens.
- Eg. Potato, sweet potato, cassava etc., because seed production in these crops is poor
- Saved from natural disasters or pathogen attack.
- Long regeneration cycle can be envisaged.
- Extremely easy.
- Recalcitrant speciss can be easily conserved.

4. Cell and organ banks

- A germplasm collection based on cryopreserved (at -196°C in liquid nitrogen) embryogenic cell cultures, somatic/ zygotic embryos.

5. DNA banks

- DNA segments from the genomes of germplasm accessions are conserved.

13.3.4. Germplam evaluation

- Screening of gemplasms in respect of morphological, genetical, economic, biochemical, physiological, pathological and entomological attributes.
- Evaluation of germplasm is essential
- To identify gene sources for resistance to biotic and abiotic stresses, earliness, dwarfness, productivity and quality characters.

- To classify the germplasm into various groups
- To get a clear pictures about the significance of individual germplasm line.

1. Biodiversity International (IPGRI)

- Biodiversity International (IPGRI) Rome has developed model list of descriptors (= characters) for which germplasm accessions of various crops should be evaluated.
- Evaluation of germplasm is done in three different places viz., in the field, in green house and in the laboratory.
- IPGRI is now renamed as biodiversity international in 2006.
- **History**
- 1974 – IBPGR (International Board of Plant Genetic Resources)
- 1991 – IPGRI (International Plant Genetic Resources Institute)
- 1994 – Merged with INIBAP (International Network for the Improvement of Banana and plantain)
- 2006 – IPGRI and INIBAP merged to form Biodiversity International

13.3.5. Germplasm cataloguing, data storage and retrieval

- Each germplasm accession is given an accession number.
- Number is pre fixed, with IC (Indigenous collection), EC (exotic collection) or IW (Indigenous wild).
- Information on the species and variety names, place of origin, adaptation and on its various feature or descriptors is also recorded in the germplasm maintenance records.
- Catalogues of the germplasm collection for various crops are published by the gene banks.
- Its compilation, storage and retrieval is now done using special computer programmes.

13.3.6. National Bureau of Plant Genetic Resources (NBPGR)

- NBPGR established in 1976 is the nodal organisation in India for planning, conducting, promoting, coordinating and lending all activities concerning plant.
- Collection
- Introduction
- Exchange

- Evaluation
- Documentation
- Safe conservation
- Sustainable management of germplasm

Vegetable crop responsibilities and germplasm activities at NBPGR

- Vegetable crops are subjected for evaluation, documentation and maintenance of active collections besides their long term storage

Solanaceous	Brinjal, tomato, chillies
Cucurbitaceous vegetables	Pumpkin, melons, gourds and cucumber
Leguminous vegetables	Cowpea, pea, lablab bean, winged bean, faba bean, French bean
Bulb crops	Garlic, onion
Root vegetables	Radish, carrot, turnip
Okra	-
Miscellaneous vegetables	Cole crops, Chinese cabbage, spinach beet, spinach

- India is one of the important centres/regions of variability of vegetable crops.
- Number of vegetable crops of economic importance and their wild relatives originated in this region.
- Genetic resources possess genes for wide adaptability, high yield potential including resistance/tolerance to biotic and abiotic stresses.
- Indian sub-continent thus holds prominence as one of the twelve regions of variability in crop plants in global perspective.

13.3.7. Gene banks for various crops in India

Institutes	Crops
Central Institute for Cotton Research, Nagpur	Cotton
Central Plantation Crops Research Institute, Kasargod	Plantation crop
Central Potato Research Institute, Simla	Potato
Central Tobacco Research Institute, Rajahmundry	Tobacco
Central Tuber Crops Research Institute, Thiruvananthapuram	Tuber crops other than potato
Central Rice Research Institute, Cuttack	Rice
Directorate of Oilseeds research, Hyderabad	Oilseeds
Directorate of Wheat Research, Karnal	Wheat
Indian Agricultural Research Institute, New Delhi	Maize
Indian Grassland and Fodder Research Institute, Jhansi	Forge and fodder crops
Indian Institute of Millets Research, Hyderabad	Sorghum
International Crops Research Institute for Semi-Arid Tropics	Groundnut, Pearl millet, sorghum, pigeon pea and bengal gram

17.3.8. List of important International Institutes conserving germplasm

Name	Institute	Activity
IRRI	International Rice Research Institute, Los Banos, Philippines	Tropical Rice collection: 42,000
CIMMYT	Centre International de-Mejoramientsed maize Trigo, El Baton, Mexico	Maize and wheat (Triticale, barely, sorghum) Maize collection – 8000
CIAT	Center International de-agricultural Tropical Palmira, Columbia	Cassava and beans, (also maize and rice) in collobaration with CIMMYT and IRRI
IITA	International Institute of Tropical Agriculture, Ibadan, Nigeria.	Grain legumes, roots, and tubers, farming systems.
CIP	Centre International de-papa-Lima. Peru	Potatoes
ICRISAT	International Crops Research Institute, for Semi-Arid Tropics, Hyderabad, India	Sorghum, Groundnut, Cumbu,Bengalgram, Redgram.
WARDA	West African Rice Development Association, Monrovia, Liberia	Regional Cooperative Rice Research in Collaboration with IITA and IRRI
IPGRI	International Plant Genetic Research Institute, Rome Italy	Genetic conservation.
AVRDC	The Asian Vegetable Research and Development Centre, Taiwan	Tomato, Onion, Peppers Chinese cabbage.

13.3.9. Exchange of plant genetic resources

1. Import/ Introduction of germplasm

- Every year over 20,000 accessions of germplasm and 50,000 samples of trial material are being received.
- Introduced crops like apple (USA), buffalo gourd (USA), bull oke (Australia), french bean, kiwi fruit (New Zealand), guayle (USA), peach, pepper mint (Russia), prickly pear (Mexico and USA), sea buckthorn (Russia and China), sugarbeet (European countries), tomatillo (USA) became major crops

2. Germplasm introduced for abiotic stress resistance/ tolerance traits traits

- Submergence, salinity , zinc deficiency, phosphorus deficiency, iron toxicity, aluminium toxicity , aerobic, drought (rice), frost, lodging, low temperature, diverse climatic conditions, cold, good pre harvest sprouting, blue aleurone- a strong xenia effect (wheat), acid soil tolerant (maize), drought tolerant (barley), lodging drought heat tolerant lines (Soybean), siff stem, lodging (Pea) ,lodging (Linseed), heat, cold, humidity (Tomato), heat, cold , drought, humidity (Chilli), cold tolerant lines (Cauliflower), lodging resistance (Sunflower)

3. Germplasm introduced for biotic stress resistance/ tolerance traits

- Brown plant hopper, bacterial blight, tungro virus, blast, kernel smut, rice water weevil, rice stalk borer, multiple disease, stem borer, grassy stunt virus, necrosis virus, root rot, sheath blight, bacterial panicle blight (rice), sawfly, powdery mildew, stripe rust, leaf rust, yellow rust, brown rust, black rust , fusarium race-1, barley yellow dwarf virus, net blotch, scald, covered smut, False loose smut, Common root rot, New version lines carrying genes resistant to Ug 99, Hessian fly (Wheat), thrip, stripe rust, barley yellow dwarf virus, powdery mildew, net blotch scald, leaf rust, surface borne smuts (Barley), white fly, downy mildew, stem canker, soybean mosaic virus, southern stem canker, soybean cyst nematode, reniform nematode, sudden death syndrome, frog eye leaf spot, white fly transmitted gemini virus, bacterial wilt, fusarium wilt, tomato mosaic virus, gray leaf spot pathogen (Soybean), ascochyta blight, fusarium wilt, leaf minor, bruchids, cyst nematode (Chickpea), downey mildew, loose smut, aphids (Sunflower), anthracnose, bacterial wilt, bacterial black spot, potato virus Y, pepper venial mottle virus, chilli veinal mottle virus (chilli), powdery mildew, downy mildew (Muskmelon), root knot nematode, bacterial wilt, tomato mosaic virus, fusarium wilt gray leaf spot, tomato leaf curl virus (Tomato), leaf spot, leaf scorch, powdery mildew, red stele, botrytis, verticillium wilt, spotted spider mite, fruit rot (Strawberry), fire blight, apple scab (Apple)

4. Germplasm introduced for Agronomic traits

- Early flowering, early maturing, weed competitive (Rice), high grain yielding (Wheat), early maturing (Maize), early maturing, high yielding, high grain yielding, low shattering (Barley), early flowering (Chickpea), high yielding, low shattering (Soybean), high yielding (French bean), early maturing (Tomato).

5. Value Added

- Low phytic acid, gold hull, new plant types (Reduced number of litters, all tillers are fertile, increased numbe of grains per panicle and larger parcels size) good aroma, cooking quality of basmati 370, long grained, superior parboiling, canning quality, high iron content (Rice), exceptionally high grain protein content, high grain weight, superior bread baking quality, excellent end use qualities for bread and noodle production, strong gluten, waxy, lower sucrose retention capacity (Wheat), higher plump seed percentage (Barley), high amylose content, high lysine content, waxy germplasm, higher seed protein (Maize), low lipoxygenase, less than 5.5% linolenic acid (Soybean), cayenne type large thick red, non pungent, longer shelf life, ornamental types, excellent culinary applications (Chilli), high TSS content, orange fleshed, netted exterior (Musk melon), high beta carotene rich, high lycopene content (Tomato), large fruited, good flavour, outstanding fruit quality (Strawberry).

6. Lines for heterosis exploitation,

- Monogenic lines, introgression lines, isogenic lines, pyramided lines, mutant lines, CMS, maintainer, TGMS and restorer lines (Rice), maintainers of the A1, cytoplasmic genetic male sterility system, restorers of the A1, cytoplasmic genetic male sterility system (Wheat), male sterile, restorers (Chilli), CMS lines; male fertile line (Cauliflower)

7. National Supply

- Catering to germplasm needs of different researchers/scientists working in various research organizations in country.

8. Facilitating Access to germplasm and information to other countries

- During 1976-2017, introduction of PGR was facilitated from more than 147 countries and several International Agricultural Research Centres (IARCs).
- During this period 34,37,720 samples of seed/planting were imported and 8,27,874 samples were exported to different countries; while 4,82,770 samples of various crops were supplied to various users in different institutes/ organizations across the country.

14

Protection of Plant Varieties and Farmers Rights Act

14.1. PPVFR Act

- TRIPS (Trade Related Aspects of Intellectual Property Rights) agreement of WTO (World Trade Organization) was signed.
- Enacted PPVFR act in 2001.
- Regulation formulated in 2003.
- PPV & FRA (Protection of Plant Varieties and Farmers Variety Authority) established in November 2005 adopting ***sui generis*** system.
- Conformity with International Union for the Protection of New Varieties of Plants (UPOV), 1978
- Nodal agency for PPVFR act.
- Chairperson is the Chief Executive of the Authority.
- Besides,Authority has 15 members, as notified by the Government of India (GOI).
- Eight of them are ex-officio members representing various Departments/ Ministries,
- Three from SAUs and the State Governments, one representative each for farmers, tribal organization, seed industry and women organization associated with agricultural activities; nominated by the Central Government.
- Registrar General is the ex-officio Member Secretary of the Authority.
- Contains 11 chapters, 4 schedules and 97 section.

14.2. Objectives

- Protection of varieties
- Protection of farmers
- Right forfarmers variety

- Protection of plant breeders right
- Protect public interest
- Encourage development of new variety
- Promotes food security

14.3. Registration of varieties

- Act created national variety register and covers all plant varieties except microbes.
- Registration started in May 2007 in 12 plant species.
- Rice, bread wheat, maize, sorghum, pearl millet, pigeon pea, greengram, blackgram, chick pea, lentil, field peas, kidney bean.

14.4. Farmers right

- Section 39 - Farmers can save, use sow, re-sow, exchange, share. Sell his farm produce including seeds of variety protected under act.
- Registration of varieties - Exempted from fees.
- Recognization and rewarding for conservation of farmers variety, provided with DUS (Distinctness, Uniformity, Stability).

14.5. Breeders right

- Section 27 and 28
- Protects breeders and his registered variety on payment of fees and having rights to produce, sell, market, distribute, import and export of variety.
- Breeders variety also confirms novelty distinctness uniformity.

14.6. Extant varieties

- Released by breeders, original variety of breeder.
- Notified by central and state government

14.7. EDV (Essentially derived variety)

- Variety derived from initial variety that differs from initial variety for atleast one character and meets out DUS.
- Through single seed gene transfer.
- Recurrent back cross

14.8. Durations of protection under this PPVFR

- Trees and vines - 18 years.
- Initially certificate is given 9 years later on it is renewed on payment fees.
- For other crops -15 years
- Initial certificate for 6 years and then, renewed
- Extend variety - 15 years

14.9. National gene fund

- **Contributions:**
- Benefit sharing received in the prescribed manner from the breeder of a variety or an essentially derived variety registered under the Act, or the propagating materials
- Annual fee payable to the Authority by way of royalty
- Compensation deposited by breeders
- Contribution from any National and International organizations and other sources.
- **Utilized for**
- Any amount to be paid by way of benefit sharing,
- Compensation payable to the farmers/community of farmers.
- Expenditure for supporting the conservation and sustainable use of genetic resources including *in-situ* and *ex-situ* collections and for strengthening the capability of the panchayat in carrying out such conservation and sustainable use,
- Expenditure of the schemes relating to benefit sharing.

14.10. Benefit sharing

- One of the most important ingredients of the farmers' rights.
- Section 26 provides benefits sharing and the claims can be submitted by the citizen of India or firms or non-governmental organization (NGOs) formed or established in India.
- Depending upon the extent and nature of the use of genetic material of the claimant in the development of the variety along with commercial utility and demand in the market of the variety breeder will deposit the amount in the Gene Fund.
- Amount deposited will be paid to the claimant from National gene fund.

- Authority also publishes the contents of the certificate in the Plant Variety Journal of India (PVJI) for the purpose of inviting claims for benefits sharing.

14.11. Rights of community

- Compensation to village or local communities for their significant contribution in the evolution of variety which has been registered under the Act.
- Any person/group of persons/governmental or non-governmental organization, on behalf of any village/local community in India, can file in any notified centre, claim for contribution in the evolution of any variety.

14.12. General functions of the Authority

- Registration of new plant varieties, essentially derived varieties (EDV), extant varieties;
- Developing DUS (Distinctiveness, Uniformity and Stability) test guidelines for new plant species;
- Developing characterization and documentation of varieties registered
- Compulsory cataloging facilities for all variety of plants
- Documentation, indexing and cataloguing of farmers varieties
- Recognizing and rewarding farmers, community of farmers, particularly tribal and rural community engaged in conservation, improvement, preservation of plant genetic resources of economic plants and their wild relatives
- Maintenance of the National register of plant varieties
- Maintenance of National gene bank

14.13. Publications of Authority

- Plant variety Journal of India
- General and crop specific DUS test guidelines
- Technical bulletin
- Gene Bank Manual
- Agro-biodiversity (Hotspots book (Two volumes)
- A video CD entitled Seed of Sustenance highlighting various provisions of the PPV & FR Act, 2001
- Annual Reports
- A video CD on Krishak Adhikar

- A compendium of varieties registered under PPV & FR Act 2001 (from 2009 to 2012)
- A book entitled Cultivated plants & their wild relatives in India:An Inventory

14.14. Fees for registration

Type of variety	Fees of registration
Extant variety notified under section 5 of the seeds Act, 1996	Rs.1000/-
New variety /Essentially Derived Variety (EDV)	Individual Rs.5000/- Educational Rs.7000/- Commercial Rs.10000/-
Extant variety about which there is common knowledge (VCK)	Individual Rs.2000/- Educational Rs.3000/- Commercial Rs.5000/-

- Registration of a variety is renewable subject to payment of annual fee

14.15. DUS Test Centers

- Authority has 122 DUS test Centers for different crops with a mandate for maintaining and multiplication of reference collection, example varieties and generation of database for DUS descriptors as per DUS guidelines of respective crops.

14.16. Plant Variety Journal of India

- Monthly bilingual (Hindi and English) on the first working day of each month on its official website.
- Contents includes official and public notices, passport data of plant varieties, DUS test guidelines of crop species, details of certificate of registration and other related matters.

Part-2 Plant Biotechnology

1

Applications of Biotechnology in Agriculture

- Term Biotechnology was first coined in 1919 by Karl Ereky which means products are produced from raw materials with the aid of living organisms.
- Bio means life.
- Technology means application of knowledge for practical use.
- Biotechnology was first coined by Karl Ereky in 1919.
- Product is developed from raw material with aid of living organism.
- Use of living organism to improve a product.
- Brings true-to-type plant.

1.1. Stages of biotechnology development

- **Ancient biotechnology** - 800 - 4000 BC - early history related to food and shelter includes domestication
- **Classical biotechnology** - 2000 BC; 1800 - 1900 AD - Built on ancient biotechnology; fermentation promoted food production and medicine.
- Genetics - Darwin and Mendal - 1900 - 1953
- DNA research and science explodes - 1953 - 1976
- **Modern biotechnology** - 1977
- Manipulation of genetic information in organism is called as genetic engineering.

1.2. Agricultural biotechnology

- To produce genetically modified plant by removing genetic information from an organism.
- Manipulating it in the lab.
- Transferring it into the plant to change certain characteristics.

- Two objectives of agricultural biotechnology: Crop improvement and nutritional value of the crop.

1.2.1. Crop improvement

- Herbicide tolerance, pest resistance, drought tolerance, nitrogen fixing ability, acidity and salinity tolerance, improved colour and quality.

1.2.2. Nutritional value of crops

- Improved nutrition and taste.
- Healthier cooking oils by decreasing the concentration of saturated fatty acids in vegetable oils.
- Improving healthier foods. Eg. Golden rice
- Improved handling qualities. Eg. Delayed ripening in tomato
- Pharmaceuticals: Developing plants for edible vaccine.
- Industrial: Plants that produce plastics, fuels and other products.
- Plants for environmental cleanup.
- Developing pesticides from naturally occurring micro organisms and insects.

1.3. Application of biotechnology in animals

- Increased milk production
- Leaner meat in pork.
- Market ready fish: Growth hormones in farm raised fish.
- Pharmaceuticals: Animals engineered to produce human proteins for drugs including insulins and vaccines.
- Breeding purpose: Disease resistance, exact copies of desired stock and increased yield.
- Health: Microorganism introduced into feed for growth purpose.
- Diagnostics tool for disease and pregnancy detection.
- Animals engineered to produce organ, which is suitable for transplantation into humans.

1.4. Technologies in plant biotechnology

- Genetic engineering or recombinant DNA technology.
- Tissue culture
- Molecular breeding: Marker Assisted Selection (MAS)

1.4.1. Genetic engineering

- Manipulation of gene is called genetic engineering or Recombinant DNA technology.
- Removal of genes from one organism and either
- Transfers them to another organism.
- Puts them back in the original with a different combination.

1. Techniques used in genetic engineering

i) Agrobacterium mediated gene transfer

- Isolate desired trait from DNA of original organism.
- Insert into Agrobacterium.
- Target plant is infected
- Cells that accept DNA are grown into plants with the new trait.

ii) Gene gun

- Coat the DNA of desired trait on to tiny particles of tungsten.
- Fire into the plant cell.
- Cells that accept DNA are grown into plants with the desired trait.

1.4.2. Tissue culture method

- Manipulates a cell, anther, pollen grains or other tissues under laboratory condition to become whole living growing organism.

1.4.3. Marker Assisted Selection (MAS)

- To study DNA sequences to identify genes, QTLs (quantitative trait loci), and other molecular markers and to associate them with organism functions, i.e., gene identification.
- Identification and inheritance tracing of previously identified DNA fragments through a series of generations.

1.5. Achievements in Agricultural Biotechnology

1.5.1. Insect resistance

- Produces Bt. Gene (*Bacillus thuringiensis*) that contains toxic protein and not harmful to human.
- Eg. Bt. cotton and Bt. maize, cowpeas, sunflower, soybean, tomato, tobacco, sugarcane, walnut, rice.

- Toxic protein – Cry gene (crystal protein)
- Cry gene is located to plasmid of *B.thuringiensis*.

1.5.2. Herbicide resistance

Tobacco, cotton

1.5.3. Disease resistance

Virus resistance in cassava, maize and sweet potato.

1.5.4. Temperature tolerance

Papaya tolerant to hot and cold conditions.

1.5.5. Other traits

Genetically modified crops for water use efficiency, nitrogen use efficiency, salt tolerance, nutritional or dietary value, improved food processing and storage and elimination of toxins and allergins.

1.6. GMO crops status in the world and India

- US have more of GMO crops in soybean, maize, canola (rape seed), sugar beet, papaya, alfalfa, cotton, apple and potato.
- GMO apple: Artic apples are non browning apple that reduces food waste and brings flavour.
- India planted Bt. cotton in 10 million ha during 2011which results in 50% reduction in insecticide application.
- In 2014, India and China planted the Bt. cotton in 15 million hectares.

1. Herbicide tolerant GMO's

GMO	Use	Countries approved in	First approved	Notes
Alfalfa	Animal feed	USA	2005	Approval withdrawn in 2007 and then re-approved in 2011
Canola	Cooking oil	Australia	2003	
	Margarine	Canada	1995	
	Emulsifiers in packaged foods	USA	1995	
Cotton	Fibre	Argentina	2001	
	Cottonseed oil	Australia	2002	
	Animal feed	Brazil	2008	
	Except in India, where cotton-seed oil used for human consumption	Columbia	2004	
		Costa Rica	2008	
		India	2002	
		Mexico	2000	
		Paraguay	2013	
		South Africa	2000	
		USA	1994	
		Argentina	1998	
		Brazil	2007	
		Canada	1996	
		Colombia	2007	
		Cuba	2011	
Maize	Animal feed	European Union	1998	Grown in Portugal, Spain, Czech Republic, Slovakia and Romania
	High-fructose corn syrup			
	Corn starch	Honduras	2001	
		Paraguay	2012	
		Philippines	2002	
		South Africa	2002	
		USA	1995	
		Uruguay	2003	
		Argentina	1996	
		Bolivia	2005	
		Brazil	1998	
Soybean	Animal feed	Canada	1995	
	Soybean oil	Chile	2007	
		Costa Rica	2001	
		Mexico	1996	
		Paraguay	2004	
		South Africa	2001	
		USA	1993	
		Uruguay	1996	
Sugar Beet	Food	Canada	2001	Commercialized 2007, production blocked 2010, resumed 2011.
		USA	1998	

2

Tissue Culture and Its Significance

2.1. Biotechnology institutes

- Regional office of ICGEB (International Centre for Genetic Engineering and Biotechnology) at New Delhi (Headquarters at Trieste, Italy established in 1983).
- Lal Bahadhur Shastri Centre for Advanced Research in Biotechnology at IARI, New Delhi
- National Institute of Animal Biotechnology, Hyderabad
- Rajiv Gandhi Centre for Biotechnology, Trivandrum
- Indian Institute of Agricultural biotechnology, Ranchi
- Indian Institute of Chemical Biology, Kolkata
- CCMB – Centre for Cellular and Molecular Biology, Hyderabad
- National Dairy Research Institute, Karnal
- Indian Veterinary Research Institute, Izatnagar
- The Biotechnology centre of IARI is called National Research Centre for Plant Biotechnology

2.2. Plant biotechnology

- Aims at improving genetic makeup, phenotypic performance or multiplication rates of economic plants.
- Two basic techniques used in plant biotechnology are
- Plant tissue culture
- Genetic engineering

2.3. Plant tissue culture

- Collection of techniques used to maintain or grow plant cells, tissues or organs under sterile conditions on a nutrient culture medium of known composition.

2.3.1. History of plant tissue culture

Scientists	Contribution
Schleiden and Schwann (1938 and 1939)	Cell theory - Cell is the functional unit of living organisms. Worked in Tradeschantia with knop's solution.
Habertland (1902)	First attempt of plant tissue culture
	Father of tissue culture
White (1934)	First reported for the successful continuous cultures of tomato root tips in liquid culture and obtained indefinite growth.
Morel and Martin (1952)	Use of meristem tip culture to obtain virus free Dahlia's
Gautheret (1959)	First hand book on plant tissue culture
Murashige and Skoog (1962)	Development of MS medium
Guha and Maheshwari (1962)	Development of haploids through anther and pollen culture for the first time.
Power et al. (1970)	First protoplast fusion
Larkin and Scowcroft (1981)	Introduction of the term somaclonal variation.

2.3.2. Terms and terminology

1. **Totipotency**-Regeneration capacity of plant cell to develop into the whole plant.
2. **Callus-**A mass of unorganized, regenerated cells in the culture medium
3. **Dedifferentiation-**Conversion of mature cells in to meristamatic state leading to callus formation.
4. **Re-differentiation-**Unorganized cells become specialized in form (root and shoot) function as a whole plant.
5. **Protoplasm (naked cells**)-Cell without cell wall

2.3.3. Techniques of plant tissue culture

1. Explants

- Plant tissue or organ (anther) exercised and used for *in vitro* culture

2. Surface sterilization

- Explant surface should be sterilized to eliminate bacteria , fungi
- 1-2% solution of sodium or calcium hypochlorite or 0.1% solution mercury chloride.

- Rinse the explant in sterilized distilled water to remove disinfectant.
- Under aseptic conditions- laminar flow chamber.

3. Sterilization

- Microbes present in culture media, culture vessels, and instrument are inactivated by suitable treatment.
- Flame-forceps, scalpel, needles- 95% alcohol and flamed.
- Dry heat-test tube, culture flasks- heated on a burner
- Ethanol (70%)- Laminar air flow chamber, culture vessels, hands of worker
- Autoclaving: Culture media, culture vessel at 121°C and 15 psi for 15-20 min.
- Airfilter-air blowing the laminar tools is sterilized by HEPA filter.
- Filter-thermobiable constituents like ABA, GA_3, enzymes 0.45μm pore size.

4. Nutrient medium/culture medium

- Plant cell and organs are cultured
- Contain inorganic salts, vitamins, carbon sources (sucrose)
- Growth regulators viz, auxin (Root formation) and cytokinin (Shoot formation)
- Auxin: 2,4 D (0.5-2.0 mg/l), NAA and IAA (Natural), kinetin and benzyl amino purine (BAP)
- Commonly used kinetin is zeatin.
- Organic supplements: coconut water, casein hydrolysate and yeast extract
- pH of medium- 5.5 using1N KOH or HCL.
- Solidify agent - Agar (6 g/l): Commonly used media (Murashige and Skoog).
- Callus culture: The cells on the agar medium develop into an unorganized mass- callus.
- Suspension culture: In liquid medium, a suspension of free cells and small cell masses. The medium is autoclaved at 15psi for 15-20 min.

5. Environmental conditions

- Organ and cell culture evaluated under controlled temperature, light, RH.
- Temperature-18-250C
- Light is not essential but beneficial for plantlet regeneration
- Culture room/ incubator.

6. Subculturing

- After a period of time, it is necessary to transfer organs and tissue to fresh media.
- In tissue and cell culture, portion of tissue is used to inoculated in new culture tubes
- Subculture of callus culture every 4-6 weeks
- Suspension culture every3-14 days.

7. Plant regeneration and transfer to soil

- Production of roots and shoots from cell and tissue culture- organ regeneration or organogenesis.
- Rice, maize, barley, oats, sugarcane, colocassia, potato, pea, soyabean, chickpea, alfalfa.
- Transfer of whole plants from test tube to soil is easy.
- Plants may be pretreated prior to transfer to soil to make them hardy.
- In laboratory, they have transferred in small pots and cover with vessel eg. Inverted breakers to prevent excess transpiration.
- After 3-4 days, covers are removed and kept in diffuse light for 5-10 days.
- Hardening on a large scale mist chamber.
- Plants then, transferred into greenhouse and after 1-2weeks planted in soil and kept in sunlight

2.3.4. Stages of micropropagation

- Murashige proposed three (I to III) stages
- Debergh and Maene added stage '0'.
- Currently, there are accepted five stages (0 to IV).

1.Stage 0

- **Selection and maintenance of stock plants for culture initiatio-**Grown under more hygienic conditions to reduce the risk of contamination.

2. Stage I

- Initiation and establishment of aseptic culture
- **Explant isolation-**Vegetative parts (Shoot tip, meristem, leaves, stems, roots) or reproductive parts (Anthers, pollen, ovules, embryo, seed, spores).
- Shoot tip and auxiliary buds are most often used.

- Size, physiological age and developmental age of explant, and age of stock plant,decide success rate of stage I.
- **Surface sterilization:** Ethyl alcohol, bromine water, mercuric chloride, silver nitrate, sodium hypochlorite, calcium hypochlorite etc. can be used as disinfectant.
- **Washing:** Washed with water.
- **Establishment of explant on appropriate medium:** Modifications of Murashige and Skoog basal medium (Murashige and Skoog, 1962) are most frequently used.

3. Stage II

- **Multiplication of shoots or somatic embryo formation (rapid) using a defined culture medium**-Rapid multiplication of the regenerative system is carried out for obtaining large number of shoots.
- About 4.3 x 10^7 shoots can be produced from a single explant in a year.
- Cultures obtained from stage I are placed on same medium used in stage I and II, but cytokinin proportion is increased for stage II to produce numerous shoots.
- Repeated for few cycles until an desired number of shoots are developed to carry out for rooting.
- **Medium**-Inorganics: Macronutrients (N, P, K, Ca, Mg) and micronutrients (B, Co, Cu, Mn, I, Fe, Zn)
- Organics: carbon source - needed since plants do not seem to photosynthesize well in culture
- Vitamins-Thiamine (essential), myoinisitol, B vitamins, folic acid and biotin
- Growth regulators-Cytokinins, auxins, GA, ABA rarely used in general: Cytokinins induce shoot bud formation and auxins induce root formation.
- Complex organics: natural orange juice, coconut milk, bananas
- Inert supports: Agar, foam rubber, filter paper bridge, liquid

4. Stage III

- **Rooting of regenerated shoots or germination of somatic embryos *in vitro*:** Shoots or shoot clusters from stage II are prepared to transfer to soil.
- Shoots are separated manually from clusters and transferred on a rooting medium containing an auxin.
- Elongation of shoots prior to rooting, rooting of shoots (individual or clumps), and prehardening cultures to improve survival are some of the activities.

5. Stage IV

- Plantlets are removed from plant media and transferred to soil or potting compost
- **Hardening:** When taken out of culture, plantlets need time to adjust to more natural environmental conditions.
- Slowly weaning plantlets from a high humidity, low light, warm environment to what would be considered a normal growth environment for the species in question.

Diagrammatic representation of stages of micropropagation

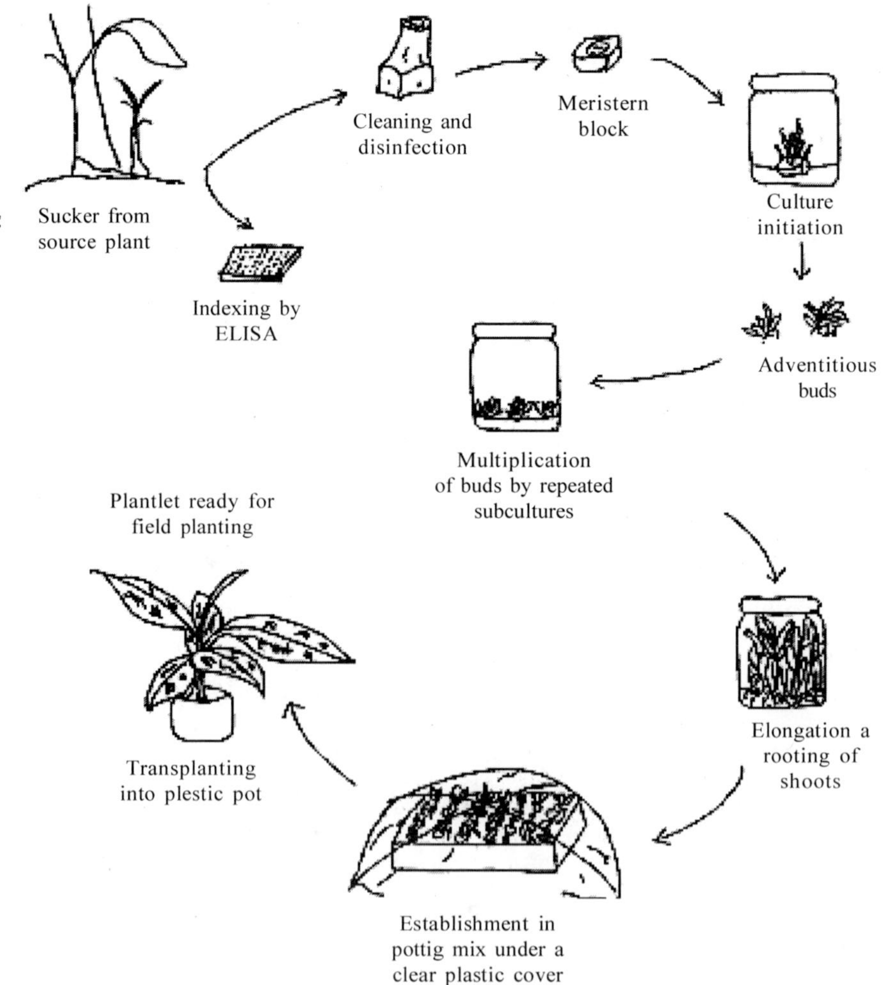

2.3.5. Classification of plant tissue culture techniques

- Based on the plant part used as explants and type of development *in vitro*
- Embryo culture
- Meristem culture
- Anther or pollen culture
- Tissue and cell culture

1. Embryo culture (Hanning, 1904 - Cruciferae/Brassicaceae)

- Young embryos are excised from developing seeds and placed on a suitable nutrient medium to obtain seedlings
- Two types of embryo culture
- Culture of immature embryos (hybrid seed): Eg. Soybean, barley.
- Culture of mature embryos: Eg. Orchids, iris.
- **Applications**
- **Recovery of distant hybrids**
- Prevention of embryo abortion in wide crosses by using embryos from incompatible crosses.
- In inter generic and inter specific hybridization, post zygotic barriers prevent normal seed development and cause embryo abortion.
- Eg. *Hordeum vulgare* x *Secale cereale*; *H. vulgare* x *Triticum* spp.
- Tetraploid and hexaploid wheat carry two dominant genes kr1 and kr2, prevent seed development with secale.
- **Recovery of haploid plants from interspecific crosses**
- *H. vulgare* x *H. bulbosum* - zygote formed but during embryo development
- *H. bulbosum* is eliminated and haploid embryos of *H. vulgare* formed – Bulbosum technique.
- **Propagation of orchids**
- Orchid seeds are naked - lack stored food.
- Embryo development is incomplete at the time the seeds mature
- Young/mature embryos removed and develop into seedling either directly or through callus formation.
- **Shortening the breeding cycle**
- Embryos develop directly into seedlings without undergoing formation of seeds.

- Thus, next generation may be grown one or two weeks early than seeds.
- **Overcoming dormancy**
- Eg. Iris, *Prunus*, *Tarus*.

2. Meristem culture (5-10 mm shoot apical with leaf primordia)

- Regeneration shoot apical meristem for adventitious roots in a suitable culture medium.
- Widely used for quick vegetative propagation of large number of plants.
- **Applications**
- Micro-propagation (mass production of clonal progeny through tissue culture) in banana (commercially successful), strawberries, citrus and *Dalbergia sissoo* (Sisam).
- Production of virus free plants. Eg: Potato, sugarcane, and tapioca.
- Meristems are suitable for cryopreservation (-196 °C for long term storage).

3. Anther culture/pollen /microspore culture (Guha and Maheswari, 1964 in *Datura*)

- Haploid plants may be obtained from pollen grains by placing anthers or isolated pollen grains on a suitable culture medium.
- In tobacco, exposure of excised flower buds at 5°C for 72 hrs enhances recovery of haploid plants.
- Brief exposure of anthers to 35°C for 24 hrs in *B. campestris*.
- Media requirements vary with species. Eg. Pollen grains of *Datura* and tobacco produce embryo on agar medium 2-4 % sucrose, 3% for barley, 6 % wheat and potato.
- Sucrose is essential for anther culture.
- Optimum stage of pollen: Eg. *Datura*, tobacco - just before or after first pollen mitosis.
- Cereals - Early or mid uninucleate stage
- Tomato and *A .thaliana* - PMC's are meiosis-I.
- Brassica -Trinucleate pollen grain.
- Maintained by altering periods of light (12-18 hrs; 5000-10000 lux) at 28°C and darkness (12-6 hrs) at 22°C .
- Embryo or callus with 3-5 cm long is transferred to medium for good root development.

- **Applications**
- Obtaining haploid plants
- Useful in cytological studies
- Doubling the chromosome number of haploids
- Eg. Rice- Xin Xiu, Shin shu
- Wheat- Hua pei 1, Lung Hua 1

4. Tissue and cell cultures

- Shoot regeneration
- Shoot buds differentiate from cell culture.
- Shoot buds are unipolar i.e., have only plumule and have vascular connections with callus tissue
- High cytokinin to auxin ratio supports shoot regeneration, while a high auxin to cytokinin ratio promotes root regeneration
- In alfalfa, two step process of shoot regeneration occurs; first callus is produced on a medium having high 2, 4-D to kinetin ratio, second shoot regeneration occurs in growth regulator-free medium.
- Gibberlliins suppress shoot regeneration
- Eg. Tobacco, rice, wheat, barley, coffee
- **Somatic embryogenesis**
- Somatic embryos originate from single peripheral or deep-seated cells of callus.
- Forms group of cells and divides and progresses through globular, heart shaped, torpedo shaped and cotyledonary stages similar to zygotic embryos.
- Eg. Carrot, wheat, soybean, coffee
- **Application of tissue and cell cultures**
- **Clonal propagation**
- Quick vegetative propagation of plant species.
- Virus and disease free plants produced
- Used for asexual propagating crops
- In oil palm and date palm, somatic embryogenesis offers only route for micropropagation.
- Individual SEs/shoot buds may be enclosed in beads of a suitable matrix, calcium alginate - Artificial seeds
- **Difficulties**

- Occurrence of genetic variation among the regenerated plants
- Reduced by using young tissue cultures
- **Somoclonal variation**
- Plants regenerated from cell cultures show heritable variation for both qualitative and quantitative trait
- Eg. Sugarcane, potato, tomato - Jointless pedicel mutant useful for mechanical picking
- **Mutant isolation**
- Fiji disease in sugarcane line isolated from cell culture and released as a new variety (Ono)
- Bacterial wilt resistant tomato line was also released
- Wild fire disease of tobacco - *Pseudomonas tabaci*; cells resistant to methionine sulfoximine - similar to pathogenic toxin were isolated; clones from this resistant to wild fire disease
- **Biochemical production**
- Callus and suspension cultures accumulate certain biochemicals known as secondary metabolites.
- They are organic compounds that are not directly involved in normal growth, development or reproduction of an organism.
- Useed as secondary metabolites as medicines, flavourings and recreational drugs.
- Shikonin, drug and precious natural dye for silk and cosmetics, produced from cell cultures of *Lithospermum erythrozhizon*.
- Berberine (*Coptis japonica*) and taxol (*Taxus* sp.) are also produced from cell cultures.

2.3.6. Somatic hybridisation

- An alternative to sexual hybridization is somatic cell hybridization or parasexual hybridization.
- Fusion of cells takes place through protoplasts
- Protoplasts are naked cells or cells without cell wall.
- Somatic cells are manipulated by removing the cell wall and protoplasts are fused.

1. Techniques of somatic hybridisation

- Four steps
- **Isolation of protoplasts**
- Treating cells/tissues with a suitable cell wall degrading enzymes (Mixture of 0.1- 1.0% pectinase or macroenzyme and 1-2% cellulase.
- **Protoplast fusion**
- In chemical method, PEG, high PH (10.5) and high calcium ions (50mmol^{-1})
- In electrical method, electric current is passed to induce fusion.
- **Selection of hybrid cells**
- After fusion, there is a mixture of parental types viz., homokaryons and heterokaryons
- Union of proptoplasts of same species - Homokaryons
- Union between protoplasts of two different species - heterokaryons
- Hybrid cells can be identified by
- Hybrid vigour at the callus stage
- Morphology which is intermediate between two parents
- Biochemical and molecular techniques (isozyme pattern and RFLP and PCR based markers).
- **Culture and regeneration of hybrid cells**
- Hybrid cells are cultured in a Murashige and Skoog (MS) medium with suitable modification
- In culture medium, protoplasts regenerate a new cell wall within 2-4 days and cell division starts within 2-7 days.
- Somatic incompatability also occur in some somatic combinations.
- Some somatic hybrid plants retain full or nearly full somatic complements of the two parental species - symmetrical hybrids
- *Solanum tuberosum + Lycopersican esculentum*
- *Arabidopsis thaliana + Brassica campestris*
- Many somatic hybrids exhibit full somatic complement of one species, while the other species lost during mitotic divisions - asymmetric hybrids
- *Nicotiana tabacum + Daucus carota.*

2.3.7. Cybrids

- Also known as cytoplasmic hybrids
- Cells containing nucleus of one species, but cytoplasm from both the parental species.

- **Applications**
- Transfer of plasmagenes of one species into the nuclear background of another species in a single generation
- Sexually incompatible combinations
- Recovery of recombination between parental mitochondrial or chloroplast DNAs
- Production of wide variety of combinations
- Used for transfer of cytoplasmic male sterility from *Nicotiana tabacum* to *N. sylvestris*.

2.3.8. Advantages of micro-propagation

- An alternative approach to conventional methods of vegetative propagation.
- A million of shoot tips can be obtained from a small, microscopic piece of plant tissue within a short period of time and space.
- Shoot multiplication usually has a short cycle (2-6 weeks) and each cycle results in logarithmic increase in the number of shoots.
- More advantageous in case of bulb or corm pro-ducing plants, because mini-tubers or mini corms for plant multiplication are available throughout the year irrespective of season.
- Smaller size of propagules is advantageous for storing and transporting as it takes lesser space.
- Propagules can be maintained in soil free environment which facilitates their storage on a large scale.
- *In vitro* technique helps to raise pathogen free plant and to maintain them.
- Very useful for dioecious plants, because there the seed progeny yield is 50% male and 50% female.
- Thousands to millions of plantlets can be maintained within the culture vials.
- Through seed production genetically uniform progeny is not possible always; but the micro propagation method will help to maintain the genetic uniformity in the propagules.
- Newer tissue material obtained through rDNA technology or haploid culture or somatic hybridization can be the source of tissue material for micro-propagation, as it is the easiest method for obtaining the multiple propagules.

2.3.9. Demerits of micropropagation

- Cost involved in setting up and laboratory maintenance is very high.
- Tissue culture techniques require skill and manpower.
- Slight infection may damage entire lot of plants.
- Some genetic modification (mutation) of the plant may develop with some varieties.
- Seedling grown under artificial condition may not survive when placed under environmental condition.

2.3.10. Commercially propagated plants through micropropagation in India

Category	Cropsm
Fruits	Banana, pineapple, strawberry, cash crops sugarcane, potato spices, turmeric, ginger, vanilla, large cardamom, small cardamom
Medicinal plants	*Aloe vera*, geranium, stevia, patchouli, neem
Ornamentals	Gerbera, carnation, anthurium, lily, syngonium, cymbidium
Woody plants	Teak, bamboo, Eucalyptus, populus
Biofuel	Jatropha, Pongamia

2.3.11. Explants and medium used

Species	Explant	Medium
Gerbera	Shoot-tip	MS
Rose	Shoot tip	MS
Carnation	Meristem tip	MS
Rhododendron	Shoot tip	Anderson formula
Anthurium	Vegetative buds	MS
Rosemary	Shoot tip	MS
Banana	Suckers	MS
Eucalyptus	Shoot tip	WPM
Melaluca (Bottle brush)	Shoot-tip	WPM

3

Transgenic Plants and Their Applications

3.1. GM or transgenic crop

- Plant that has a novel combination of genetic material obtained through the use of modern biotechnology.
- GM crop contains a gene(s) that has been artificially inserted instead of plant acquiring through pollination.
- In 1994, Calgene's delayed ripening tomato (Flavr-Savr) is first GM food crop.

3.2. Some success stories

3.2.1. Transfer of genes for herbicide tolerance

- In 1986, Shah and his co-workers isolated a cDNA clone encoding an enzyme 5-enolpyruvyl-shikimate phosphate (EPSP) synthase from a glyphosate tolerant *Petunia hybrida* cell line.
- In 1987, De Block and his co-workers transferred a gene conferring resistance to bialaphos and phosphinothricin called bar gene isolated from *Streptomyces hygroscopicus* into tobacco, tomato and potato.
- This gene encodes for an enzyme, phosphinothricin acetyltransferase which prevents toxicity due to phosphinotricin.
- This gene has been transferred to tobacco, tomato and potato with the help of 35 S promoter of cauliflower mosaic virus (CaMV).

3.2.2. Expression of insect tolerance in transgenic plants

- *Bacillus thuringiensis* is a bacterium that produces proteinaceous crystals during sporulation.
- These crystal proteins have insecticidal properties especially to lepidopteran insects.

- Use of *Bacillus thuringiensis* as a microbial insecticide offers advantages over chemical control agents in that the species-specific action of its insecticidal crystal proteins (ICPs) makes it harmless to non-target insects, to vertebrates, to the environment and the user.

3.2.3. Expression of coat protein genes for virus protection

- In agriculture, cross protection is a common practice to protect the plants from viruses
- Coat protein of viruses have an important role in systemic cross protection.
- Abel and his coworkers introduced a chimeric gene containing a cloned cDNA of the coat protein (CP) gene of TMV in to tobacco cells on a Ti plasmid of *Agrobacterium tumefaciens* from which tumour inducing genes have been removed.
- Plants regenerated from transformed cells expressed TMV mRNA and CP as a nuclear trait.

3.2.4. Expression of antisense RNA in transgenic plants

- In 1987 Rottstein and his co-workers demonstrated the inhibition of the expression of the nopaline synthase (NOS) gene in tobacco.
- Transgenic plant having the NOS gene was transferred with a NOS antisense gene construct with CaMV 35 S promoter.
- Transformed plants were analysed for NOS - 3 - activity and the enzyme activity varied depending on the tissue used.

3.2.5. Expression of reporter genes in plants

- Luciferase gene from firefly (*Photinus pyralis*) is a novel tool for this purpose.
- This gene encodes an enzyme that catalyses the light producing ATP-dependent oxidation of luciferin.
- In 1986, Ow and his co-workers introduced the luciferase gene into tobacco.
- Transgenic plants showed the expression of the luciferase gene by producing light when watered with the substrate luciferin.
- This reporter gene system provides a simple tool for rapid screening of large numbers of transgenic plants.

3.3. Development of transgenic crops

- Made through a process known as genetic engineering.
- Introducing trans genes into plant genomes either by a gene gun or using a bacterium.

3.3.1. Methods of gene delivery

1. Particle bombardment method

- Metal microspheres (Tungsten or gold - high densities with 1-4mm diametre) coated with DNA penetrate
- DNA-coated micro projectiles are placed on a macro projectile and accelerated towards the tissue following an explosive discharge to deposit DNA.
- Successfully used in tobacco, soybean, rice, maize, barley and wheat.

2. *Agrobacterium* mediated delivery

- *Agrobacterium tumafciens* is a gram-negative bacterium that uses horizontal gene transfer to cause tumors in plants.
- When bacterial DNA is integrated into a plant, it becomes part of plant chromosomes and integrated DNA sequence stably expressed and inherited.

3.3.2. Method

1. First step

- Extraction of DNA from the organism known to have the trait of interest.

2. Second step

- Gene cloning, which will isolate the gene of interest from the entire extracted DNA, followed by mass-production of the cloned gene in a host cell.
- Once it is cloned, the gene of interest is designed and packaged so that it can be controlled and properly expressed once inside the host plant.

3. Third step

- Modified gene will then be mass produced in a host cell in order to make thousands of copies.
- When the gene package is ready, it can then be introduced into the cells of the plant being modified through a process called transformation.

4. Fourth step

- Most common methods used to introduce the gene package into plant cells are biolistic transformation (using a gene gun) or Agrobacterium-mediated transformation.
- Once the inserted gene inserted stable, inherited, and expressed in subsequent generations, then the plant is considered a transgenic.

5. Fifth step

- Backcross breeding is the final step in the genetic engineering process, where the transgenic crop is bred and selected in order to obtain high quality plants that express the inserted gene in a desired manner.
- Length of time in developing transgenic plant depends upon the gene, crop species, available resources, and regulatory approval.
- It may take 6-15 years before a new transgenic hybrid is ready for commercial release.

3.4. Potential benefits of GM plants

- Higher crop yields
- Reduced farm costs
- Increased farm profit
- Improvement in health and the environment

UNIT-VI

SEED SCIENCE AND TECHNOLOGY

1

Flowers, Pollination and Fruits

1.1. Floral biology

1.1. 1. Types of inflorescence

- Group or cluster of flowers arranged on a stem that is composed of a main branch or a complicated arrangement of branches.

1. Simple

- **Indeterminate:** Main branches continues to grow
- Raceme: Radish, mustard
- Spike: Achyranthus, amaranthus
- Corymb: Candy tuft
- Umbel: Coriander
- Spadix: Maize, palms, Anthurium
- Capitulum: Sunflower
- Catkin: Mulberry, oak, *Acalypa indica*
- **Determinate:** Terminal bud forms a terminal flower and other flowers grow from lateral buds
- Cyme
- Monochasium
- Helicoids: Begonia
- Drepanium
- Scorpioid: Cotton
- Cincinnus
- Rhipidium
- Dichasial: Jasmine, teak, ixora
- Pleiochasium

2. Compound

- Compound raceme*: Delonex regi*a, neem, *Cassia fistula* (panicle), paddy, sorghum
- Thyrse
- Anthela
- Thyrsoid
- Compound umbel: Carrot, coriander

1.1.2. Flowers

1. Parts of flower

- **Calyx**: Green covering formed out of the sepals.
- **Calyx lobes**: Sepals
- **Corolla**: Tube of petals
- **Corolla lobes**: Petals
- **Peduncle**: Stalk of the flower
- **Bract**: Any modified leaf or scale on a flower stalk or at the base of a flower
- **Sepal**: A structure that cannot be distinguished as a sepal or a petal (Lillies and tulips)
- **Stamen**: Pollen producing part of a flower, usually with a slender filament supporting the anther.
- **Anther**: Part of the stamen where pollen is produced.
- **Pistil**: Ovule producing part of a flower. The ovary often supports a long style, topped by a stigma. The mature ovary is a fruit, and the mature ovule is a seed.
- **Stigma**: Part of the pistil where pollen germinates.
- **Ovary**: Enlarged basal portion of the pistil where ovules are produced

1.1.3. Types of flower

1. **Hermaphrodite** (Bisexual or perfect flowers)

- Have both male (androecium) and female (gynoecium) reproductive structures, including stamens, carpels and an ovary.

2. **Unisexual (Imperfect or incomplete)**

- Either functionally male or female.

- **Staminate or male**: Flowers with an androecium, but no gynoecium
- **Pistillate, carpellate or female**: Flowers with a gynoecium, but no androecium.

1.1.4. Sex expression

1. Monoecious

- An individual that has both male and female reproductive units.
- Radish, cucurbits, maize, castor, jack, coconut, arecanut
- Protandrous: Maize
- Protogynous : Pearl millet

2. Dioecious

- Population having separate male and female plants
- Papaya, grapes, palmyra, nutmug , date palm

3. Subdioecious

- Population produces normally male or female plants, but some are hermaphrodite
- **Gynomonoecious:** Has both hermaphrodite and female structures; Ailanthus
- **Andromonoecious:** Has both hermaphrodite and male structures; Sorghum, sunflower
- **Subandroecious:** Has mostly male flowers, with a few female or hermaphrodite flowers
- **Subgynoecious:** Has mostly female flowers, with a few male or hermaphrodite flowers

4. Polygamy

- Plants with male, female, and perfect (hermaphrodite) flowers on the same plant
- **Trimonoecious:** Banana, castor

1.1.5. Inflorescence and flower types in agricultural crops

Crops	Inflorescence	Flower	
1. Cereals			
	Rice	Panicle	Perfect
	Wheat	Panicle	Perfect
2. Millets			
	Pearl millet	Head	Perfect
	Maize	Male: Tassel	
		Female: Silk	Monoecious
	Ragi	Panicle	Perfect
	Thinai	Panicle	Perfect
	Samai	Panicle	Perfect
	Varagu	Pancile	Perfect
	Panivaragu	Panicle	Perfect
	Sorghum	Panicle	Perfect
	Bajra	Panicle	Perfect
3. Pulses			
	Red gram	Axillary receme	Bisexual
	Black gram	Axillary raceme	Bisexual
	Green gram	Racemose	Bisexual
	Bengal gram	Solitary	Bisexual
	Soybean	Solitary	Perfect
4. Oilseeds			
	Groundnut	Solitary	Bisexual
	Sesame	Receme	Bisexual
	Sunflower	Capitulum	Perfect
	Castor	Raceme	Monoecious
5. Fibre crops			
	Cotton	Solitary	Perfect

1.2. Pollination

1.2.1. Types of pollination

1. Self pollination (Autogamy)

- Transfer of pollen from anthers to stigma of same flower; Rice, wheat, green gram, black gram, groundnut, tomato.

Group	Crops
Legumes	Pea, redgram, cowpea, soybean, blackgram, greengram, lentil, khesari, rajma, sunhemp, guar
Cereals and millets	Wheat, rice, barley, oat, ragi
Vegetables	Potato, tomato, brinjal, chilli, okra, lettuce
Forage crops	Wheat grass, burr clover, subterranean clover, velvet bean
Oilseeds	Groundnut, sesamum, linseed
Fruit trees	Apricot, peach, citrus

2. **Cross pollination (Allogamy)**: Transfer of pollen from male reproductive organ of one plant to female reproductive organ of another plant; Maize, bajra, sunflower, cucurbits, cabbage, cauliflower

- Hydrophily (By water): Hydrilla, water hyacinth
- Anemophily (By wind): Maize, sugarcane, bajra, coconut, date plam
- Entomophily (By insects): Sunflower, maize, bajra
- Ornithiophily (By birds): Bombax
- Canthophily (By beetles): Cucurbits, mango
- Symphytophily (By sawfly): Lucerne, carrot
- Waspophily (By wasp): Coriander
- Mitophily (By bees): Sunflower
- Myophily (By flies): Onion
- Malacophily (By slugs): Aroids
- Psycophily (By butterflies): Castor, clover
- Palenophily (By moths): Dianthus, Hesperogucca
- Chirotophily (By bats): Kapok

3. **Often cross pollination:** In many self pollinating species, cross pollination may occur 5 to 30%; Sorghum, redgram, sesame, cotton, brinjal, chilli, bhendi etc.

- Geotenogamy: Pollination of different flowers in same plant
- Xenogamy: Pollination between two different plants; Varalaxmi cotton: Laxmi x SB 289 E.

1.2.2. Mechanism facilitating self pollination

- **Bisexuality or hermaphroditism:** Presence of both reproductive organs in a flower.
- Self pollination is the rule
- **Homogamy**: Both the sex organ mature at the same time; Mirabilis.
- **Cleistogamy:** Pollens shed within the closed flowers so that self pollination is obligatory; *Commelina benghalensis*, oxalis, peas, groundnut, *Viola tricolor,* wheat, barley, oat, some grasses, bengal gram, horse gram.

- **Chasmogamy:** Opens and exposes stamens and styles to environment; Rice, wheat, barley, oat.
- **Hidden stamen and stigma:** Some floral organs (As keel petal in legumes) hide or cover stamens and stigmas to avoid cross pollination; Pea, red gram, black gram, green gram, soybean.
- **Anther position:** Stigma remains densely and closely surrounded by anthers; Tomato, brinjal.

1.2.3. Aids of cross pollination

- **Dichogamy:** Maturation of anthers and stigma at different times
- **Protandry**: When anthers mature earlier than stigma of same flower; Maize, Helianthus, leucas.
- **Protogyny**: When stigma matures earlier than anthers of same flower; Pearl millet, Ficus, mirabilis.
- **Unisexuality or dicliny:** If plant becomes dioecious, cross pollination becomes indispensible; Mulberry.
- **Monoecious:** Maize, cucurbits
- **Dioecious:** Papaya, grapes, palmyra, nutmug, date palm
- **Self sterility:** Plants which fail to set self seed, when grown in isolation or self pollinated.
- **Heterostyly:** In one type, stamens are long and style is short, but in second type, stamens are short and style is long; Brinjal
- **Herkogamy:** Barrier is formed between stigma and stamens of same flower; Vinca, Iris, *Clerodendrum thompsonii*, silk cotton etc.
- **Self incompatability:** In some plant, self pollination prevents fertilization; Brassica, sunflower.

1.2.4. Inflorescence and flower types in agricultural crops

Crop	Pollination behavior	Anthesis time	Stigma receptivity
Cereals			
Rice	Self	8-10 am	3 days from the time of anthesis
Wheat	Self		
Millets			
Ragi/finger millet	Self	1-5 am	
Barley	Self		
Oats	Self		
Maize	Cross/protoandry	Extends 2-14 days	For 24 hours
Bajra/cumbu	Cross /protoandry		2-3 days after emergence
Sorghum	Often cross	8 pm -2 am	Up to 2 days before blooming
Pulses			
Black gram	Self	1-4 am	
Green gram	Self		
Bengal gram	Self	3 - 9 am	
Soybean	Self		
Red gram	Often cross		
Oilseeds			
Groundnut	Self		Commences at 6 am and continues up to 8 am
Sesame	Often cross	5 -7 am	
Sunflower	Cross	5 -8 am	2-3 days
Castor	Cross		5-6 days
Fibre crops			
Cotton	Often cross		
Jute	Self		

1.3. Fruits and their structures

1.3.1. Fruit types

1. **Fleshy fruits**

- **Simple fleshy fruits**
- **Drupe**: Fruit with a single seed enclosed by a hard, stony endocarp or pit; Mango, peach, plume, cherry, cocoa, coconut and coffee.
- **Pomes**: Bulk of flesh comes from enlarged floral tube or receptacle that grows around the ovary; Apple, pear, quince, loquat.

- **Compound fleshy fruits**
- **Berry**: Develops from a compound ovary and contains more than one seed. Having thin skin and soft pericarp; Banana, grapes, papaya, sapota, arecanut, avocado.
- **Pepos**: Relatively thick rinds; Pumpkins, watermelon, cucumber.
- **Hesperidium**: Leathery skin containing oils; Citrus, aonla, carambola.

2. **Dry fruits**

- **Dehiscent**
- **Follicle**: Splits along one side or seam; Larkspur
- **Legume**: Splits along two sides or seams; Grams
- **Silique**: Splits along two sides or seams, but seeds borne on central partition are exposed when two halves separate; Sesamum, crucifers.
- **Capsule:** Consist of at least two carpels, and split in a various ways; Okra or bhendi, jimson weed.
- **Indehiscent**
- **Achene**: Sunflower
- **Nut**: A large, indehiscent achene, commonly one seeded;Almond
- **Grain/caryopsis:** An achene in which fruit wall and seed coat are inseparable; Wheat, corn, rice, sorghum, pearl millet etc.
- **Samara**: A winged achene or nut; Maple
- **Schizocarp:** Carrot

3. **Aggregate fruit**

- Derived from a single flower with several to many pistils. Individual pistils mature as a clustered unit on a single receptacle; Raspberries, Strawberries.

4. **Multiple fruit**

- Derived from several individual flowers in a single inflorescence; Pineapples, figs.

1.3.2. Inflorescence and fruit types in agricultural crops.

Crops	Fruit type	Seed type
Cereals		
Rice	Caryopsis	Monocot
Wheat	Kernel/Caryopsis	Monocot
Millets		
Sorghum	Caryopsis	Monocot
Barley	Caryopsis	Monocot
Bajra	Caryopsis	Monocot
Maize	Caryopsis	Monocot
Fibre crops		
Cotton	Boll	Dicot
Jute	Pod	Dicot
Pulses		
Bengal gram	Pod	Dicot
Green gram	Pod	Dicot
Soybean	Pod	Dicot
Red gram	Pod	Dicot
Black gram	Pod	Dicot
Oilseeds		
Groundnut	Pod	Dicot
Sesame	Silique	Dicot
Sunflower	Achene	Dicot
Castor	Schizocarp	Dicot

2

Seed and Its Importance

2.1. Seed

- Defined as a mature ovule or a reproductive unit formed from fertilized ovule, consisting of an embryo, reserve food and a protective cover.

2.1.1. Differentiate between seed and grain

Seed	Grain
Should be a viable and vigorous one	Need not be a viable one
Should be physically and genetically pure	Not needed
Should satisfy seed standards such as maximum germination and pure seed and minimum seed moisture	No such requirements
Should satisfy minimum seed certification standards	No such requirements
Treated with pesticide /fungicide to protect seed against storage pests and fungi	Never be treated with any chemicals, since used for consumption
Respiration rate and other physiological and biological processes should be kept at low level during storage	No such specifications
Should be certified / truthfully labelled	No such condition
Should satisfy all the quality norms	Not considered
Should never be converted into grain, unless warranted	Converted as seed

2.1.2. Types of seed

- Based on storage organ, seeds are classified into monocot and dicot seeds.

1. **Monocot seeds/Monocotyledons**

- Also known as albuminous or endospermous seeds.
- Single seeds embedded in a fruit called caryopsis/utricle.
- Seed coat and fruit coat are fused to form pericarp
- Embryo is minute and confined to one end of the seed and is viable with 2n number of chromosome.

- Storage tissue is known as endosperm and is non-viable with 3n number of chromosome and occupies the major portion of the seed.
- Embryo consists of radicle and plumule which are enclosed in coleorhiza and coleoptiles, respectively.
- Single cotyledon of the monocot is known as scutellum.
- Aleurone layer is a protein body which remains at the periphery of the endosperm.

2. **Dicot seeds/Dicotyledons**

- Embryo consists of a primary axis and a living storage tissue, cotyledons; enclosed in the seed coat
- Do not have endosperm which is fully utilized for the development of embryo.
- Sometimes, they have well developed endosperm (Castor, fenugreek)
- They are ex-endospermous or ex-albuminous in nature (Legumes).
- In some cases perisperm (Remaining unfertilised portion of nucellus) is also seen (Sugar beet, coffee, black pepper).

2.1.3. Components of seed

1. **Seed coat**

- Outer covering of seed.
- Develops from the 2 integuments of ovule.
- Outer layer of seed coat is smooth and rough known as testa and formed from outer integument.
- Inner layer of the seed coat is called the tegumen and formed from inner integument.

2. **Embryo (2n)**

- Mature ovule consisting of an embryonic plant.
- Miniature plant consisting of plumule, radicle and cotyleldon.
- Plumule and radicle without the cotyledon is known as primary axis.

3. **Radicle and coleorhiza**

- Rudimentary root of a plant compressed in embryo is radicle, which forms primary root of young seedling.
- Enclosed in a protective cover known as coleorhiza.

4. Plumule and coleoptiles

- First terminal bud of plant compressed in embryo that gives rise to first vegetative shoot of seedling.
- Enclosed in a protective cover known as coleoptile.

5. Cotyledon

- Compressed seed leaves.
- Single cotyledon present in monocots is called scutellum, while two cotyledons are present in dicots.

6. Scutellum

- Modified seed leaf.

7. Endosperm (3n)

- Develops from primary endosperm nuclei formed by two polar nuclei and one sperm nuclei.
- Stores food for developing embryo.

2.1.4. Appendages of seeds

1. Awn

- Thorn like projection in the tip of the seeds.
- In paddy, bract tip was elongated in to the awn.

2. Hilum

- Scar mostly white in colour present on the lateral side of the seed.
- Represents the attachment of seed stalk to the placenta of the fruit to mother plant. Eg. Pulses

3. Caruncle

- White spongy outgrowth of the micropyle end in some species. Eg. Castor, tapioca.

4. Aril

- Coloured fleshy mass present on the outside of seed. Eg. Nutmeg.

5. Hairs

- Minute thread like appendages present on the surface of the seed. Eg. cotton

6. Wings

- Papery structure attachment to the side of the seed coat either to a specific side of the seed coat or to all sides. Eg. Moringa

2.1.5. Parts of normal seedling

- Normal seedling should have all essential parts on germination after a certain period of growth.

1. **Root (Primary)**

- Part of the plant growing underground.
- Radicle develops into root.
- Grows straight and must be free from damages and decays.

2. **Shoot**

- Part of the plant growing above the ground
- Develops from the plumule.

3. **Epicotyl**

   Portion of the shoot above the cotyledons

4. **Hypocotyl**

   Region of shoot below the cotyledon and above the root.

5. **Coleoptile**

- Leathery covering protecting the plumule and helps in the emergence.
- Also helps in elongation of the cotyledons in monocots.

6. **Coleorhiza**

- Sheath which covers the radicle.
- Elongation of hypocotyl region.

2.2. Seed quality

2.2.1. Characteristics of good seed quality

1. **Improved variety**

- Must be truly superior to existing ones.
- Improved seed should be carrier of all latest technologies, which favours assured high yield.

2. **Genetic quality**

- Trueness to type.
- Breeder /nucleus seed: 100 % genetic purity
- Foundation seed: 99.5 % genetic purity
- Certified seed: 99.0 % genetic purity

3. **Physical quality**

- Physical composition of seed lots.

- Should be free from inert matter and other impurities.

All crops	:	98 %
Carrot	:	95 %
Bhendi	:	99 %
Other crops	:	98 %
Sesame, soybean, jute	:	97 %
Groundnut	:	96 %

4. Seed germination and vigor (Physiological quality)

- Germination refers to ability of a seed to give to a normal seedling under normal conditions.
- Vigour refers to sum total of all attributes that give effective plant stand in the field.
- Should have high physiological vigour and stamina.

5. Planting value

- Planting value is the real worth of a seed lot for raising the crop.
- Pure live seed = Pure seed % x Germination % x 100

6. Health quality

- Refers to the presence or absence of disease organisms insect pests.
- Should be free from designated pests and diseases.

7. Seed moisture

- Most critical factor in the maintenance of seed germination and viability during storage
- Dried to safe moisture content.

8. Should be free from other crop seeds

- Expressed in number/kg

Crop	Designated inseparable other crop seeds
Barley	Wheat, Oats and Gram
Oats	Wheat, Gram and Barley
Wheat	Oats, Gram and Barley

9. Should be free from objectionable weed seeds
10. Should have good shape, size, colour, etc., according to specifications of variety
11. Should possess high longevity and shelf life.

2.2.2. Advantages of quality seed

- Saves your money as you need less seed.
- Quality seed means a higher percentage of seed emergence.

- Minimum of replanting
- Seedlings will be strong or vigorous.
- Plants stand will be uniform
- Plants will grow more quickly.
- Greater resistance to stress and diseases.
- Uniform in ripening which means more uniformity in the harvested grain.
- An yield increase by 15-20 %
- Less admixtures and fetch higher price in the market.

2.3. Role of seed in agriculture

- Seeds acts as a carrier of a new technology
- Evolving new high yielding varieties and hybrids and their suitable seed management practices will result in increased yield from unit area of land.
- Increasing food grain production.
- Seed is a basic tool for secured food supply
- Because of the introduction of the high yielding varieties and hybrids, sufficient requirements of seeds and food grains can be produced
- Import of food grains from foreign countries is curtailed.
- Seed acts as a principal means to secure crop yield under less favourable production areas.
- Should have ability to tide over drought and submerged conditions and give normal yield.
- Seed acts as a medium for rapid rehabilitation of agriculture in case of natural disaster
- Sufficient carryover stock of seed can be preserved to resow during floods or severe drought.

2.4. Goals of Seed Technology

- Rapid multiplication
- To increase agricultural production through supply of good quality seeds; high yielding varieties and hybrids will be developed by the plant breeders.
- Timely supply
- To ensure the supply of high yielding varieties and hybrids to the farmers in time, so that the planting schedule of the farmer is not disturbed and they are able to use the good quality seeds in planting.
- Assured high quality seeds to obtain increased food production.
- Reasonable price within the reach of average farmer.

3

Seed Generation System

3.1. Seed generation system

- Seed multiplication is nothing but production of a particular class of seed from specific class of seed up to certified seed stage.
- Success of a seed program depends on
    - Rate of genetic deterioration
    - Seed multiplication ratio
    - Total seed demand

3.2. Seed multiplication

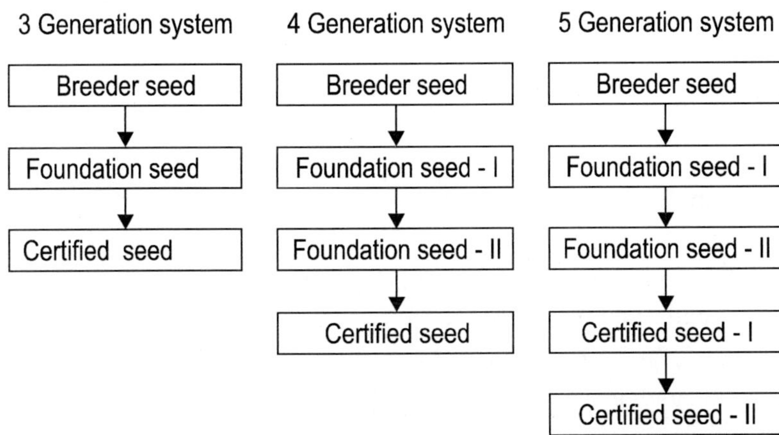

- Five generation system is followed only when multiplication is very poor
- There should be heavy demand
- Should be of self - pollinated crop

- For most of often cross pollinated and cross pollinated crops, 3 and 4 generation models are suggested
- Eg. Castor, red gram, jute, green gram, rape, mustard, sesame, sunflower and vegetables

3.3. Class of seed

3.3.1. Nucleus seed

- No specified tag
- Maintained by concerned breeder of sponsoring institution or breeder himself
- Maintenance breeding
- Genetic purity: 100%

3.3.2. Breeder seed

- Progeny of nucleus seed
- Golden yellow tag
- Produced by concerned breeder or sponsoring institution or breeder himself
- Field inspection by breeder seed monitoring team to check genetic purity (Grow out test)
- Sometimes breeder stages are I and II; for stage II, source of seed is stage I.
- Genetic purity 99.9%; physical purity 98%; moisture content less than 12%

3.3.3. Foundation seed (stage I and stage II)

- Progeny of breeder seed
- White tag
- Genetic purity is 99.5% or more
- Produced by State Agricultural Universities, State Seed Farms, other Government Farms, National Seed Corporation, Cooperative Agencies and Central and State Seed Corporations, Private sectors.
- Field inspection by Certification agency to check minimum required standards of genetic and physical purity and other quality standards.
- Can be produced in two stages viz. stage I and stage II.

3.3.4. Certified seed

- Progeny of foundation seed.
- Genetic purity is 99% or more.
- Produced by the State and National Seeds Corporations, Private seed Companies and Farms of Progressive Growers.
- Azhur blue colour tags.

- Certified by Certification agency to check minimum required standards of genetic and physical purity and other quality standards
- For paddy and wheat, two generations from foundation seed is permitted
  - Certified seed I - Certified seed II
- Barley, garden pea, groundnut, soyabean, are eligible up to 3 generations from foundation seed
  - Certified seed I-Certified seed II-Certified seed III

3.3.5. Truthful labeled seed

- Seed produced by private seed companies and sold under truthful labels.
- Companies should maintain field and seed standards as per Aeed Act.
- Opal green colour tags.
- Truthful labelling is compulsory, while seed certification is voluntary.

3.3.6. Differences between certified seed and truthful labeled seed

Certified seed	Truthful labeled seed
Certification is voluntary	Truthful labeling is compulsory for notified kind of varieties
Applicable to notified kinds only	Applicable to both notified and released varieties
Should satisfy both minimum field and seed standards	Tested for physical purity and germination
Seed certification officer, seed inspectors can take samples for inspection	Seed inspectors alone can take samples for checking the seed quality.

3.4. Flow chart

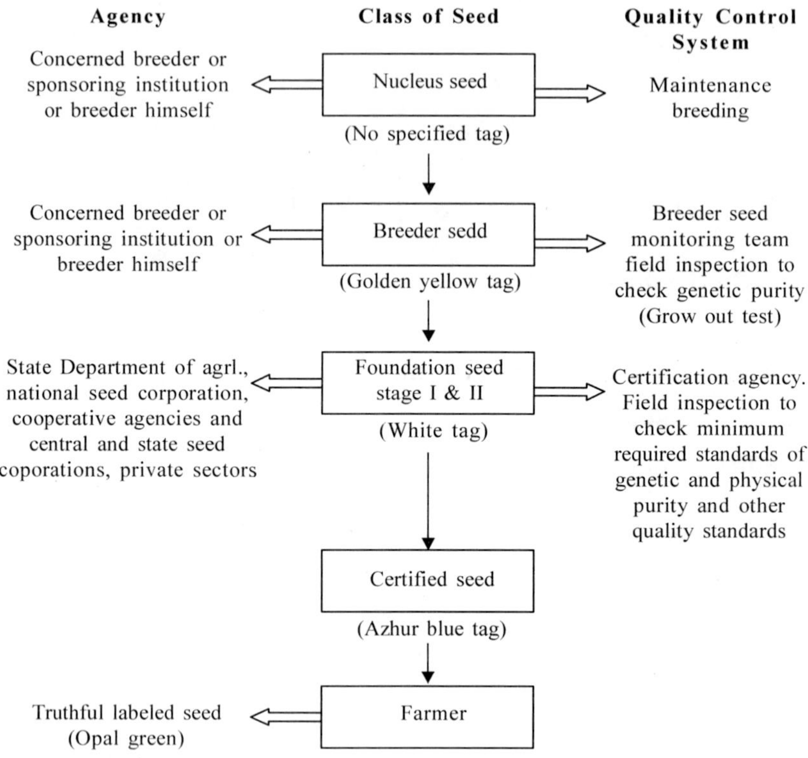

4

Seed Policy and Planning

4.1. Seed policy

4.1.1. Seed policies formulated for governing seed quality control system

Year	Events
1966	Seeds Act passed (came into force in 1968).
1968	Seed Rules.
1972	Seed Certification Agencies and Central Seed Certificate Board formed.
1983	Seed Control Order.
1986	Essential Commodity Act, Consumer Protection Act.Breeder Seed Supply Policy. Before 1986, Breeder seed was supplied by NSC only for production of foundation seed, which were done by private sector.
1988	A New Seed Policy to encourage import of seed and planting material was announced.
1993	Plant Variety Act
1999	Protection of Plant Varieties and Farmers' Right.
2001	Protection of Plant Varieties and Farmers Act 2001.
2001-02	National Seed Policy 2002.
2003	PPV&FRA implemented in India
2004	New seed bill submitted

4.1.2. Organizations involved in implementation of seed policy

1. Seed production

- **National Seeds Corporation (NSC - 1963)**
- Produces and distributes breeder seed, foundation seed and certified seed.
- Regional office at state level to perform its function.
- Also takes up processing, storage, training programs, publicity of varieties, distribution of vegetable seeds etc.
- **State Farm Corporation of India (SFCI)**
- Production and distribution of breeder and foundation seeds at national level.

- **State Seed Corporation**
- Formed by respective states and producing foundation and certified seeds.
- **Department of Agriculture**
- Seed requirement is met out by department of Agricultural through seed production in state seed farms and seed farms organized in the farmers holding.
- **Private sector**
- Multinational seed companies in India

Rank	Company name	Country
1.	Monsanto (US) + Seminis	USA
2.	Dupont/Pioneer	USA
3.	Syngenta	Switzerland
4.	Groupe Limagrain	France
5.	KWS AG	Germany
6.	Land O' Lakes	USA
7.	Sakata	Japan
8.	Bayer Crop Science	Germany
9.	Takii	Japan
10.	DLF-Trifolium	Denmark

- **Co-operative sector**
- Produces seeds either through contract farmers or share holders.

2. Seed quality control system

- Indian seed industry is based on Seed Act 1966
- Seed Rules was enacted in 1969.
- Government Agencies /organizations involved in seed quality control system are.
    - Seed certification agency
    - Seed law enforcement wing
    - Seed testing laboratory
    - State variety release and notification committee

3. Seed research

- ICAR - NSP started in 1974, renamed as NSP (Crops) in 2003

- Phase I : 1976 - 1981
- Phase II : 1981 - 1986
- Phase III : 1987 - till date

4.2. Seed planning

4.2.1. Seed Multiplication Ratio (SMR)

$$SMR = \frac{\text{Seed yield}}{\text{Seed rate}}$$

4.2.2. Seed Renewal Period (SRP)

- Seeds undergo genetic deterioration due to
- Developmental variation
- Mechanical mixture
- Mutation
- Natural crossing
- Minor genetic variation
- Selective influence of pests and diseases
- Techniques of plant breeder.
- Hence, seed should be renewed after certain generations adopting generation systems.

4.2.3. Seed Replacement Rate (SRR)

SRR = x / y x 100

Where

X = Quantity of actual quality seed sown / used in an area /location

Y = Quantity of quality seed (certified) required for the entire production area/location

- Seed replacement rate in India is around 15–20%
- 100% for hybrid seeds.

4.2.4. Multiplication ratio, seed renewal period and seed multiplication stages.

S. No.	Seed crop	Multiplication ratio	Seed renewal period (times)	Seed multiplication stages		
				BS	FS	CS
1.	Paddy	152	4	1	1	2
2.	Wheat	49	4	1	1	2
3.	Barley	26	4	1	1	2
4.	Maize hybrid	248	1	1	1	1
5.	Maize variety	115	3	1	1	1
6.	Sorghum hybrid	179	1	1	1	1
7.	Sorghum variety	94	3	1	1	1
8.	Bajra hybrid	380	1	1	1	1
9.	Bajra variety	175	3	1	1	1
10.	Ragi/finger millet	420	4	1	1	2
11.	Gram and peas	24	3	1	1	2
12.	Pigeon peas	150	3	1	1	1
13.	Other pulses	125	3	1	1	1
14.	Groundnut	18	5	1	2	2
15.	Brassicas	200	3	1	1	1
16.	Sesamum	200	3	1	1	1
17.	Linseed	42	4	1	1	2
18.	Other oil crops	73-100	3	1	1	1
19.	Cotton	46	3	1	1	1
20.	Jute	120	3	1	1	1
21.	Fodder	75	3	1	1	1

4.2.5. Planning for seed production

- Basic information needed on planning for seed production for different classes of seeds in varieties and hybrid are:

1. Varieties

- Seed rate
- Seed yield
- Multiplication ratio
- Extent of reduction in yield
- Seed requirement of certified seed
- Area to be covered with certified seed

- Processing losses
- Fruit yield in case of vegetables
- Seed recovery (%)

2. Hybrids

- Planting ratio
- Seed rate of A, B and R lines
- Seed yield of A, B and R lines
- Multiplication ratio

5

Seed Village Concept

5.1. Seed village

- Compact area approach in which the whole village is involved in seed production.
- Concept is a win-win model (Deepak Mullick)

5.2. History

- Started in 1964 in Jounti village, Khanjawala block of Delhi state
- 15 jounti farmers cultivated 70 acres with high yielding varieties of wheat. Formed the Jawahar Jounti Seed Co-operative Society Ltd
- Dr. M.S. Swaminathan has guided in wheat
- Milestone in green revolution
- Empowerment of farm women
- Women have been successfully involved in the seed production program
- Livelihood security
- Ranebennur block of Karnataka observed 4-5 times higher labour engaged in seed production than crop production.

5.3. Organizations involved in seed village approach

5.3.1. Advanta India

- British-American Tobacco, formed a joint venture in 1994 with Zeneca (Formerly part of Britain's ICI) to research, develop, produce and market hybrid seeds for Indian farmers.
- ITC Zeneca has recently changed its name to Advanta India
- Deepack Mullick, a 25 year ICI veteran, was Managing Director
- Over 5,000 village farming families took part in this scheme

- Supplied parental seed to the village farmers
- Offers technical guidance and procures seed at pre-determined price much higher than commercial crop.

5.3.2. AID - Association of Indian Development

- AID people worked directly at the village level.
- Networking with other NGO's/groups in the region, and elsewhere in India
- Empower villages and outsiders through Community Service Hours and Village Gram Sabhas

5.3.3. MSSRF - M.S. Swaminathan Research Foundation

- To train unskilled women in producing quality seed and planting materials
- To improve the capacity of farm women in seed processing and packaging
- To link farm families with private and public sector seed companies so that small farm families get the benefit of scale in seed production and marketing.
- Launched in 1996 with financial support from Council for Advancement of People's Action and Rural Technology involving Indo-American Hybrid Seeds (IAHS) and many other small companies.
- Implemented in Kannivadi, Srirengapuram and Vadugapatty of Dindigul district of Tamil Nadu.
- Two projects
  - Seed village
  - Farm level biopesticide production
- 2,455 men and women were trained for 19,712 days in seed production and management
- Bhendi, cotton and cluster beans were the most important crops
- In seed processing and seed quality management, around 552.5 women labour days/ha was engaged.
- Average wage for hybrid seed production is Rs. 35/6 hr work day and for tomato, it was Rs. 40 compared to Rs. 25 for an 8 hr day for regular agricultural operation.
- Average net income in hybrid seed was Rs. 15,698/acre and in open pollinated seed production, it was Rs. 4,936/acre.
- Reddiar Chatiram Seed Growers Association was formed and registered during 1998-99

- Centre for Sustainable Agriculture and Rural Development Research and Action (CENSAR), NGO with farmers and agricultural labourers as its members, was formed

5.4. Advantages of seed village

- Educate the farmers requiring seed production
- Infrastructure development is easy
- Problems of isolation distance does not arise/solved
- Mechanization is easier
- Labour force is trained (Cotton-emasculation/dusting/roguing etc.,)
- Easier for Government agencies for inspection
- Ensure production of good quality seed

5.5. Disadvantages

- Co-operation of farmers is lacking
- If any disease or pest occurs, whole area is affected
- Operational difficulties may arise in particular time

6

Principles of Seed Production

6.1. Genetic principles

6.1.1. Maintenance of genetic purity during seed production

1. **Steps suggested by Home (1953)**

- Use of approved seed.
- Inspection and approval of fields prior to planting.
- Field inspection and approval of growing crops at critical stages for verification of genetic purity, detection of mixtures, weeds, and for freedom from noxious weeds and seed borne diseases, etc.
- Sampling and sealing of cleaned lots.
- Growing of samples of potentially approved stocks for comparison with authentic stocks.

2. **Steps suggested by Hartmann and Kester (1968)**

- Providing adequate isolation to prevent contamination by natural crossing or mechanical mixtures.
- Roguing of seed fields prior to stage at which they could contaminate seed crop.
- Periodical testing of varieties for genetic purity.
- Avoiding genetic shifts by growing crops in areas of their adaptation only.
- Certification of seed crops to maintain genetic purity and quality of seed.
- Adopting the generation system.
- Grow out tests

6.2. Agronomic principles

6.2.1. Varietal seed production

1. Selection of suitable agro-climatic region

- A crop variety should be grown in the adapted areas in terms of photoperiod and temperature.

- Sensitive varieties should be grown in their specific localities
- Regions with moderate rainfall, humidity and temperature more suited
- For effective pollination, prevalence of moderate rainfall and wind is much suitable
- Flowering should coincide with dry sunny period

2. Season

- Select season much suitable for seed production and without isolation problem.
- Off season is much suited for raising seed crop as it avoids isolation problem.
- Prevalence of cool and dry weather is preferable
- Excessive rain leads to a higher incidence of diseases results in mould attack and seed discoloration (Eg. Bleached seed in peas).
- Heavy rains or unusual rains may result in delayed maturity and precocious germination (*in situ* germination in rice and wheat).
- High temperatures cause desiccation of pollen resulting in poor seed set and quality (Lab-lab due to contabascence of anthers; flower shedding in tomato and chilli).
- Temperature between 24-38°C is most favorable.
- High humidity and temperature encourages production of diseased seeds.

Disease symptom	Crop
Blonded seed	Peas
Dimpled seed	Peas, french bean, cluster bean
Off-coloured seed	Pulses

3. Land selection

- Free from volunteer plants, weed plants and other crop plants.
- In the preceding season, same crop should have not been grown.
- Avoid areas where isolation is a problem.
- Avoid coastal belts and marshy places.
- Avoid areas of endemic diseases and pests.
- Select compact areas for hybrid seed production.
- Availability of skilled labourers especially for hybrid seed production.
- Fields near the processing operations with transporting and marketing facilities are most suitable.

4. Isolation distance

- Distance maintained between the seed crop and source of contaminant
- **Time isolation:** Adjusting time of sowing of seed crop and contaminating crop to prevent synchronized flowering. Eg. Rice, maize
- **Distance isolation:** Space maintained between seed crop and source of contaminant in terms of distance. Eg. For cotton 50 m.
- **Barrier isolation:** Topographic features like hills, woodlot, vegetative barrier like maize, sesbania, sugarcane etc., a distance over 30 m
- Artificial barrier (Plastic sheets above 2 m height)

5. Land selection

- Select well leveled field for uniform maturity.
- Select well drained fields nearest to irrigation sources.
- Select fertile fields.
- Soil pH should be around 7
- Avoid soil moisture stress
- Poor drained soils cause wilt diseases in tomato and chilli.
- Soil should have adequate macro and micro nutrients for producing vigorous and viable seeds.

Nutrients	Deficiency	Crop
Potassium	Blossom end rot	Tomato
Calcium	Blotching	Tomato
Boron	Black rot	Cabbage and cauliflower
Molybdenum	Whip tail	Crucifers
Manganese	Marsh spot	Peas
Boron	Hollow heart	Garden pea
Boron	Darkened or blackened plume	Groundnut

6. Selection of seed

- Seed must be from authenticated seed source
- Appropriate class of seed must be used for further multiplication.
- Eg. For production of foundation seed, breeder seed must be used.
- Proof of purchase (Bill) is necessary to identify the source along with details on the tag *viz.,* name of the agency, purity, germination, validity period.

7. Sowing

- Sow seeds in equal distance & sowing depth should be 2.5 times of seed size.
- Line sowing is advisable for seed crop

8. Fertilizer management

- Lack of N delays field emergence,
- Lack of P delays flowering,
- K is responsible for filling and lusture of crops
- Zn is essential for fertilization.
- Selenium is responsible for germination in onion.
- Cu deficiency leads to poor embryo growth.
- Mn deficiency causes marsh spot in peas.
- Boron deficiency results in poor seed development and hollow heart in peas and black rot in cole crops particularly cabbage and cauliflower, blackened or darkened plumule in groundnut
- Iron deficiency causes sterility,
- Molybdenum deficiency causes bleaching in peas and whiptail physiological disorder in crucifers

9. Irrigation

- Immediately after sowing, irrigate the field if it is a direct sown crop or if transplanted, first irrigate the field and transplant the seedlings.
- After sowing, irrigate the field at 3rd day as life irrigation and thereafter, once in a week or 10 days interval depends on the water requirement of the seed crop.
- Over irrigation can promote vegetative growth, lodging, nutrient imbalance, while under irrigation delays flowering, stunted growth, reduced filling and immature drying.
- Irrigation at critical stages like flowering, fruit and pod formation, seed filling and maturation is must.

10. Foliar spray

- 2 sprays of 40 ppm of NAA at flowering in tomato and chilli will arrest the flower drop.
- In cotton, spraying of 2% DAP or KCl at boll formation stage improves seed quality.
- In pulses, spraying of 2% DAP at flowering stage produces bolder seeds

11. Roguing

- Very important operation to be carried out for the seed crop by the grower in order to maintain the purity of seed crop

- The physical contaminant rogues are
  - Weeds/noxious weeds
  - Objectionable weed plants
  - Volunteer plants
  - Designated diseased plants
  - Other crop plants

12. Method of harvest

- Once over harvest: Rice, sorghum
- Multi picking: Cotton, brinjal, chillies etc.

13. Drying, grading, treatment and bagging

- Avoid sun drying between 12 noon and 2 pm.
- Artificial drying is done at 70-85°F (30-32°C), but never exceeding 110°F (40°C).
- Grading is done with various types of grader and upgrading machines.
- In cotton, delinting should be done
- Seed treatment with recommended fungicides.

6.2.2. Hybrid seed production

1. Planting ratio

- Ratio to be maintained between female and male lines in order to have good amount of pollen grains for proper pollination and fertilization.
- Eg. Planting ratio for sunflower hybrid seed production is 4:1.
- In cotton hybrid seed production, it is called block system, female and male are raised in 8:1 ratio in separate blocks not more than 5 m distance.

2. Border rows

- Seed field should be planted with border rows all the four sides with male parent seeds to have more pollen source and avoid contamination with other crops.
- Number of border rows may be from 4 to 12 depends upon the crop.

3. Synchronized flowering

- Uniformity of flowering between parental lines.
- To achieve uniformity, any one of techniques can be done

- Staggered sowing
- Withholding irrigation
- Application of urea or DAP

- **Staggered sowing**: Adjusting sowing time of parental lines in order to have synchronized flowering between parental line for proper seed set.
- Early flowering line should be sown later.
- Eg. Sunflower KBSH -1 hybrid seed production; male parent is late flowered by 6 days than female.
- So, sowing of male line 6-D-1, 6 days earlier than female line ensures synchronized flowering
- **GA_3 application rice**: Exertion of female panicle from boot leaf is a problem. To have complete exertion, spray GA_3 at 70-75 g/ha twice during flowering time

4. Special techniques for hybrid seed production

- **Detasseling**: Removal of tassel from the female row in maize
- **Emasculation and dusting:** In cotton.

5. Supplementary pollination

- To have good seed set
- **Rice:** Rope pulling or rod driving
- **Sunflower:** Rubbing or installation of bee hives

6. Genetic contaminants

- **Off types:** Deviant of seed plant. Common for varieties and hybrids
- **Pollen shedders:** Presence of B line in A line in hybrid seed production field
- **Shedding tassel:** Presence of tassel in the female row of maize hybrid seed production plot
- **Selfed bolls:** In cotton hybrid seed production plot, selfed bolls without emasculation and dusting
- **Selfed fruits:** In the case of other vegetables

7. Method of harvest

- In hybrid seed production, harvest the R line (Male) first, remove from the field and do one final roguing in the female row and do the harvest.

7*

Varietal Seed Production Technology

*ChapterStart from next page

Crop	Isolation (m)	Season	Seed rate (kg/ha)/Seed treatment	Spacing(cm)/ Seedling age	Fertilizer (NPK kg/ha)	Foliar spray/ Seed treatment	Harvesting	Grading (Sieve size)	Seed yield (kg/ha)
Rice ***(Oryza sativa)***			Seed soaking in 0.5% KNO_3 for 16h			2%DAP 1st spray-60 days 2nd spray-80 days		7/64"	3000-6000
Short duration (105-110 days)		November-December	60	20x30 13-22 days old seedlings	120:40:540 $ZnSO_4$ at 25		28 DAA		
Medium duration(110-125 days)	3	November	40	20x15 25-30 days old seedlings	150:50:60		35 DAA		
Long duration (125-145 days)		August	30	20x20 30-35 days old seedlings	150:50:80		35 DAA		
Maize ***(Zea mays)***	200	November-December June-July Sep-Oct	10 Seed soaking in 2 % KH_2PO_4 for 8h	60 x 20 45 x 15	150:75:75	-	Sheath turns yellow & dry (Moisture 25 %)	18/64"	2500
Wheat ***(Triticum aestivum)***	150	November	85-100	20-22.5 (Row spacing)	120:60:40	-	At 30-35% Moisture	6mm (TS) 1.8/2.1mm (BS)	3000-4000
Sorghum ***(Sorghum bicolor)***	FS: 200 CS: 100	June-July October-November	7.5-10 Seed soaking in 2% KH_2PO_4 for 10h	45 x 15	50:50:50		40-45days after 50% flowering, 20-22% moisture, black layer at base of seed	9/64"	2000
Pearl millet ***(Pennesitum glaucum)***	FS: 400 CS: 200	November-December	3.75 Seed hardening with 2 % KCl for 16 h	45 x 20 18 days old seedling	100:50:50		20 to 25 DAF, small dark layer at seed bottom	4/64"	2000

Crop	Isolation (m)	Season	Seed rate (kg/ha)/Seed treatment	Spacing(cm)/ Seedling age	Fertilizer (NPK kg/ha)	Foliar spray/ Seed treatment	Harvesting	Grading (Sieve size)	Seed yield (kg/ha)
Red gram (*Cajanus cajan*)	100	June-September February-March	20-25 Seed soaking in 100 ppm $ZnSO_4$ for 3h	45 x 30 (Short duration) 90 x 30 (Long duration)	25:50:25	2 % DAP at first flowering and 15 days after	75-80 DAS	3.35 mm (Larger seed) 2.8 mm (Medium seed)	2000-2500
Blackgram (*Vigna mungo*)	5	June-September February-March	20-25 Seed soaking in 100 ppm $ZnSO_4$ for 3h	30 x 10	25:50:25	1 % DAP at first flowering and 15 days after	55-65 DAS/ 30 days after 50% flowering	2.36 mm	1200-1300
Greengram (*Vigna radiata*)	5	June - September February-March	8-10 Seed soaking in 100 ppm $ZnSO_4$ for 3h	30 x 10	25:50:25	1 % DAP at first flowering and 15 days later	55-65DAS/ 30 days after 50% flowering	2.36 mm	1000-1200
Cowpea (*Vigna unguiculata*)	FS: 10 CS: 5	February-March	20-25	45 x 20	25:50:25	2 % DAP on 25-35 DAS	70-80 DAS/ 30 days after 50% flowering	3.50 mm (Small seed) 4.00 mm (Bold seed)	1300-1400
Field lablab (*Lablab purpureus* var. *lignosus*)	5	March-April September-October	20-25	45 × 15	25:25:25	2 % DAP on 45 DAS	90-140 DAS/ 70% pods turn straw colour	18/64"	700
Soybean (*Glycine max*)	3	June-July September-October	40	30 x 10	25:50:25	2 % DAP on 25-35 DAS	30days after 50% flowering / 70% pods turn yellow	14/64"– 12/64"	1800-2100
Bengalgram	3	November	75-90 Seed soaking in 1% KH_2PO_4 or 3-4 h	30 x 10	25:50:25	-	30-40 DAA	13/64" or 18/64"	2000-2500

Crop	Isolation (m)	Season	Seed rate (kg/ha)/Seed treatment	Spacing(cm)/ Seedling age	Fertilizer (NPK kg/ha)	Foliar spray/ Seed treatment	Harvesting	Grading (Sieve size)	Seed yield (kg/ha)
Cotton (*Gossypium sp*)	FS: 50 CS: 30	July-August	FS: 15-20 DS: 7.5-10	75 × 30/ 90 × 45/ 120 ×60	120:60:60 / 80:40:40 / 60:30:30	1 % DAP on 70,80 and 90 DAS	45-55 DAA	18/64" (FS) / 10/64" (DS)	
Acid delinting: Conc. H_2SO_4 at 100 ml/kg of seed									
Jute (*Corchorus spp*)	50	Late February to May	Line sowing: 5 - 7 Broadcasting : 7 - 10	30 × 10-15	20:20:20	-	-	-	1300
Groundnut (*Arachis hypogea*)	5-10	June – July Sep – Oct	BT: 100-130 ST: 80-100	30×15	40:40:60 +10 kg borax Gypsum at 400 kg/ha on 40-45 DAS	1 % FeSO4 and 50 DAS	Drying and falling down of lower leaves, inner side of the pod turns black	9/64"	1200-1400 (Pod yield of ST) 800-1000 (Pod yield of BT)
Soaking in 0.5 % $CaCl_2$ for 6h, incubate soaked seed in between two moist gunny bag for 12h, separate radicle sprouted seeds at 2h interval, shade dry and sow									
Sunflower (*Helianthus annus)*	CS: 200 FS:400	April-August	15 Soaking in 0.5 % KH_2PO_4 for 16h/2 % $ZnSO_4$ for 12h	45 × 30	60:45:45	-	Back of head turns to yellow, bracts turn brown,	9/64"	1500
Gingelly (*Sesamum indicum*)	CS: 50 FS:100	Jan-Feb May-June	2 - 4	30 × 30	20:10:10 2kg $MnSO_4$	-	Seeds turn chocolate brown, stem turns brown, most pods turn yellow	4/64"	300-350
Castor (*Ricinus communis*)	FS:200 Cs:150	Jun-Sep Feb-May	10	90-120 x 45-60 40-50 x 20-25	40:40:20		90-120 DAS	8/64"	3000
Safflower (*Carthamus tinctorius*)	FS:400 CS:200	Sep-Oct	10-15		20:30:20		150-180 DAS	BSS 6 x 6	1500-2000

Crop	Isolation (m)	Season	Seed rate (kg/ha)/Seed treatment	Spacing(cm)/ Seedling age	Fertilizer (NPK kg/ha)	Foliar spray/ Seed treatment	Harvesting	Grading (Sieve size)	Seed yield (kg/ha)
Safflower (*Carthamus tinctorius*)	FS:400 CS:200	Sep-Oct	10-15	40-50 × 20-25	20:30:20		150-180 DAS	BSS 6 × 6	1500-2000
Tomato (*Lycopersicon esculentum*)	25	May-June December	350g/ha	75×60 80×75 45×45 25-30 days old seedling	150:60:60 120:60:60 100:100:100	NAA 20-40ppm Borax spray:2.5kg/ha	40 DAA	BSS 10 × 10	100-150
Seed soaking in 100ppm IA A for 16 or 2 % KNO_3 for 48 h; Seed extraction using HCl at 2ml/kg of pulp for 30 minutes									
Brinjal (*Solanum melongena*)	FS:200 CS:100	May – June	300-400g/ha	75 × 60 60 × 45 30 - 35 days old seedling	100:100:60	NAA 20ppm,	50 DAA	5/64"	300-350
Seed soaking in GA_3 at 200ppm for 12h ; Seed extraction using HCl at 10ml/kg of pulp for 45 minutes									
Chilli (*Capsicum annuum*)	FS:400 CS:200	Round the year June - July February-March	1kg/ha	30 x 30 45 × 45 4-6 weeks old seedling	140:70:35	NAA 50ppm Ethrel 400ppm	35-41 DAA	8/64"	400-600
Seed treatment with *Trichoderma viridi* at 4 g and *Pseudomonas fluorescens* at 5g/kg of seed									
Sweet pepper/ bell pepper (*Capsicum* sp.)	FS: 400 CS:200	June – July Nov-Feb Sep-Oct	1	90 × 45	140:75:35	2ppm 2,4D spray	50 DAF Reddening of fruit	-	80-90
Bhendi (*Abelmoschus esculentus*)	FS:400 CS:200	Jun-July Nov-Dec	10-12	45 × 30 60 × 20 60 × 30	20:50:30	DAP 0.5% NAA 75mg/l Ethrel 150 mg/l 2,4-D 10 mg/l.	35 DAA Dry fruits with hair line cracks	10/64"	1000-1400
Onion(*Allium cepa*) Bulb production	FS: 1000 CS:500	June - September	8-10 100 ppm GA_3 soaking for 3 h	20 × 20	60:60:30	-	110-115 DAT	-	1200-1500

Crop	Isolation (m)	Season	Seed rate (kg/ha)/Seed treatment	Spacing(cm)/ Seedling age	Fertilizer (NPK kg/ha)	Foliar spray/ Seed treatment	Harvesting	Grading (Sieve size)	Seed yield (kg/ha)
Seed production	FS: 1000 CS:500	October-March	3000-3500 kg bulb/ha Select 40-50 g bulb and cut 1/3rd portion of top	50 x 10	60:60:30	-	A week after 75% plants show neck fall 3-4 pickings	10×10 mm sieve	800
Pumpkin *(Cucurbita moschata)*	FS:800 CS:400	July-August February-March	2	2-2.5 ×1-1.5m	100 g of NPK 6:12:12 mixture as basal+10 g of N (Top dressing) per pit	Ethrel 250ppm four times from 10-15 DAS	Fruits turn orange or yellow and drying of stalk	16/64"	500-1000
Ashgourd *(Benincasa hispida)*	FS:800 CS:400	June-July Jan-Feb	2.5 Seed soaking in 0.5 % KH_2PO_4	2.5 × 2 m	Urea 30g SSP 72g Potash 19 g per pit	Ethephon 250 ppm four times starting from four leaves stage	Seeds have complete ash coating	16/64"	80-100
Seed extraction using HCl at 1:1 ratio for 30min. (Diluted 6 times with water)									
Bittergourd *(Momordica charantia)*	FS:800 CS:400	Late October June-July	1.8 - 2	2.5 x 2m	Urea 30g SSP 72 g MP 19 g per pit		Fruits turn to bright yellow	16/64"	100-150
Ribbedgourd *(Luffa acutangula)*	FS:1000 CS:500	June- July January-February	2.5 - 3.0	2.5 x 2m	Urea 20g SSP 90 g MP 15 g (basal) Urea 22 g (top dressing) per pit	Ethrel 100-200 ppm from first flowering at weekly intervals	60 DAA Fruits dry and turn straw yellow	16/64"	300
Snakegourd *(Trichosanthes cucumerina)*	FS:1000 CS:500	June- July January-February	1.5	2.5 x 2m	100 g 6:12:12 NPK mixture + 10 g N (top dressing) per pit	Ethrel 100-200 ppm from first flowering at weekly intervals	Skin colour changes from yellow to orange	16/64"	200

ST: Spreading type; BT: Bunch type; FS: Fuzzy seed; DS: Delinted seed; DAS: Days after sowing; DAT: Days after transplanting; DAA: Days after anthesis; DAF: Days after flowering; SSP: Single super phosphate; MP: Muriate of potash; PS: Potassium sulphate; TS: Top screen; BS: Bottom screen; BSS: British Standard Sieve.

* For Nursery and other practices of rice, refer Unit II Chapter 4

8*

Hybrid Seed Production Technology

*ChapterStart from next page

Crop	Isolation (m)	Season	Seed rate (kg/ha)		Planting ratio	Border rows	Spacing (cm)	Fertilizer (kg/ ha)	Method of hybrid seed production/Special techniques	Seed yield (kg/ha)
			M	F						
***Rice**	100	Kharif: May- June Rabi: December – January	10	20	8:2	4	F × F-10 × 15 F × M-20 M × M- 30 × 15	150:60:60	CMS Rod driving/Rope pulling for supplementary pollination	100
For synchronous flowering: Spray 2 % DAP for late flowering parent; spray 2 % urea for early flowering parent; spray 75 g/ha GA_3, first on 15-20 % ear head emergence and second on next day to enhance seed set. Time isolation: 25 days; Barrier isolation: Vegetative barrier: Sesbania, maize, sugarcane -30m distance, Artificial barrier: Plastic sheet at 2m height										
Maize	FS:400 CS:200		10	15	Single cross - 4:2 Double cross 6:2 3 way cross 6:2	4	60 x 20	200: 100: 100	Detasseling in female parent	1500
Sorghum	FS: 200 CS:200	October - December	5	7.5	A:B-4:2 A:R-5:2	4	F: 45 x 30 M: 45 x 10	100: 50:50	CMS	1500
Synchronisation of flowering: Staggered sowing: Late flowering parent should be sown early and late sowing for early flowering parent; Spray 500 ppm malic hydrazide /300 ppm CCC/skip one irrigation to advancing parent for delay flowering; 1 % urea spray to late parent.										
Pearl millet	FS: 1000 CS: 200	November - December	0.75	3.5	6:1or 4:2	4	45 x 20	100: 60: 40	CMS	1500
Redgram	200	June - September February - March	5	30	4:2	4	45 x 30	25: 50: 25	GMS	

Crop	Isolation (m)	Season	Seed rate (kg/ha)		Planting ratio	Border rows	Spacing (cm)	Fertilizer (kg/ ha)	Method of hybrid seed production/Special techniques	Seed yield (kg/ha)
			M	F						
Cotton	FS:50 CS:30	June - July September - October	3-3.5	8-10	8:2 (Block system)	4	Male: 60 × 45 (or) 90 × 60 Female: 120 × 60	100: 50: 25	Emasculation and hand pollination	
Sunflower	600	April - August	1-1.5	4-4.5	3:1	4	45x30	60: 45: 45	CMS, Rubbing with muslin cloth for pollination	
Castor	FS: 200 CS: 150	June - September February - May	4	6	3:1 4-6:1	4	90-120 x 45-60	40: 40: 20	Temperature based sex expression < 32℃- Female expression >32℃-Male expression	1000 -1200
Tomato	FS: 200 CS: 100	May - June	25g	100g	5:1	4	60 x 60 60 x 45	150: 60: 60	Emasculation and pollination Stalking of female parent	100 - 150
Brinjal	FS:200 CS:100	May - June	50g	200g	5:1	4	75 x 60 60 x 45	100: 100: 60	Emasculation and pollination	600 - 800
Chilli	FS:400 CS:200	Jan - Feb, June -July Sep - Oct, Oct - Nov.	0.25	1.00	5:1	4	90 x 45	120:60:30	Emasculation and pollination	100 - 150
Bhendi	FS:400 CS:200	January - March June - August	1	4	5:1	4	75 x 30 60 x 45	100:60:50	Emasculation and pollination	1000 -1200

Pumpkin	FS:1500 CS:1000	August - October	2	6	3:1	4	2 x 1.5 m	100:75:50	Emasculation and pollination	500
Bittergourd	FS:1500 CS:1000	August - October	1	4	4:1	4	2.5 x 2 m	100:75:75	Emasculation and pollination	500
Ribbedgourd	FS:1500 CS:1000	June - July February - March	0.75	3.25	4:1	4	2.5 x 2 m	50:40:30	Pinching of male flowers/ Chemical sex expression/ Use of gynoecious lines/ Male sterility	250 - 300
Snakegourd	FS:1500 CS:1000	June - July Dec - Jan.	0.5	1.5	4:1		2.5 x 2m	60:50:40	Emasculation and pollination	220 - 250

F: Female parent; M: Male parent; CMS: Cytoplasmic male sterility; GMS: Genetic male sterility.

Genetic contaminants:

Pollen shedders and partials: Rice, sorghum, pearl millet

Shedding tassel: Maize

Fertile plants: Redgram

Selfed bolls: Cotton

Selfed fruits: Tomato, brinjal, chilli

* For nursery and other prachces, refer Unit II Capter 4

9

Seed Processing

9.1. Purpose

- To separate and dry seeds to safe moisture, remove various undesirable materials, weed seeds, other crop seeds, damaged seeds, uniform size grading and seed treatment to upgrade overall seed quality.

9.2. Harvesting

Types of harvests	Crops
Once over harvest	Paddy, sorghum, maize, , sunflower, groundnut, gingelly
2-3 harvests	All pulses
4-5 harvests	Bhendi, gourds
6-8 harvests	Tomato, brinjal
Harvest methods	Manual (Pulses, vegetables)
	Mechanical using reaper, binder, thresher, combine, corn picker (All cereals and millets)
Traditional methods	Rubbing and squeezing (Rice, pearl millet, finger millet, pulses) Beating (Pulses) Stripping (Groundnut) Treading by animals or tractor (Cereals, pulses) Failing and pounding (Rice) Animal threshing (Sorghum, pearl millet, finger millet, pulses) Tractor/ power tiller threshing (Cereals, pulses)
Threshers	Pedal thresher (Rice - 8-12h/ha) Drum thresher (Cereals, pulses) Multicrop thresher (Cereals, pulses) Maize thresher Paddy thresher Sunflower thresher Groundnut decorticator Vegetable seed extractor

9.3. Seed extraction

- Seed removal in dry fruits is known as threshing, while that of wet fruits is known as extraction.

Seed extraction	Separation of seed from wet fruits
Dry extraction	Beating and separation (All field crops)
Wet extraction	Pulping / soaking / fermenting and seed separation (Tomato, brinjal, gourds)
	Fermentation: Ferment fruit pulp for 24-48h and separate seed (Tomato, brinjal)
	Acid method: Add HCl at 10 ml/kg of pulp for 30min., wash 3-4 times and separate seeds. Seed recovery (0.5 to 1.0 %), bright coloured with high storability (Tomato, brinjal)
	Alkali method: Washing soda at 900g/litre of boiling water
	Mechanical method: Tomato seed extractor
	Ashgourd seed extraction: Cut open, scoop out seeds and soak in 1:0.6 con. HCl for 30 min. Seed recovery 0.5 to 0.75 %.

9.4. Seed drying

Seed drying	Elimination of moisture
	Two stages – Removal of surface moisture which caused imbalance between moisture of surface and inner portion of seed leading to integration of outer layer.
Equilibrium moisture	Rate of moisture loss is equal to rate of gain
Factors	Initial moisture
	Size and capacity of bin
	Rate of air blow
	Temperature and RH of atmosphere
	Static pressure
	Drying temperature
Methods	1. Natural drying - Avoid drying between 12 noon and 2 pm.2. Mechanical drying using forced natural air drying or forced heated air drying not exceeding 42°C.
Driers	Rectangular bin drier, continuous flow drier (LSU drier, non mixing column drier)

9.5. Seed processing

9.5.1. Physical characteristics that aid in grading and upgrading

- **Seed size:** Air screen cleaner
- **Seed weight:** Specific gravity separator
- **Seed length:** Indented cylinder separator
- **Surface texture:** Magnetic separator
- **Seed colour:** Colour separator
- **Seed shape:** Spiral separator

9.5.2. Flow chart for processing sequence

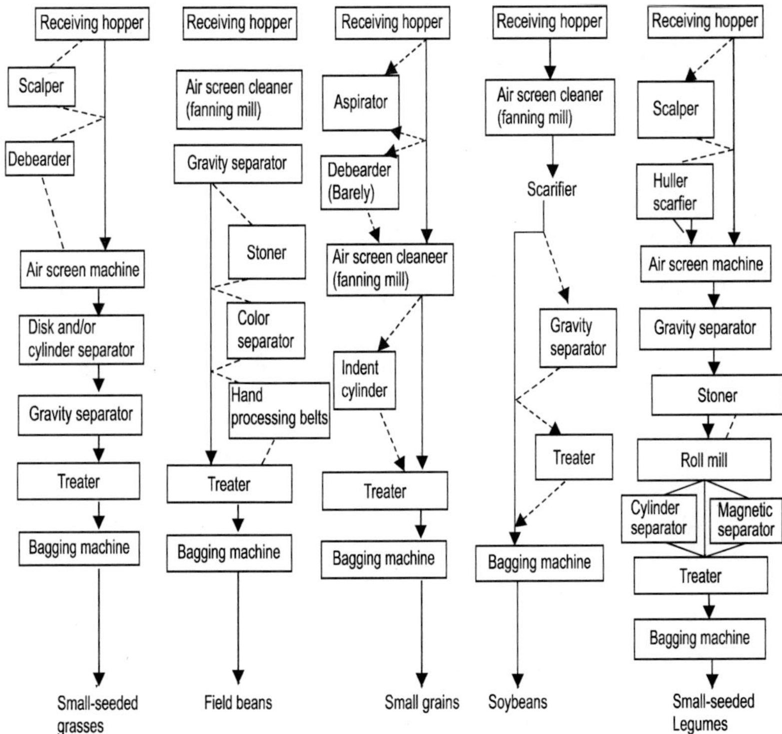

9.5.3. Pre cleaning

1. Pre cleaner

- Removes trash, stones, clods, plants remnants, chaff, other inert materials which are larger or lighter than seeds
- **Scalper:** Removes large materials and allow seed to pass through
- Unit consists of a vibrating or rotating screen or sieve having perforation large enough to allow the rough seeds pass through readily
- **Debeader:** Removes awns, beads, dehull, polish
- Rubbing the seeds and clip the seeds of oats, debeard barley, thresh white cap in wheat, remove awns in paddy, dehull some grass seeds and polish the seed
- **Huller - Scarifier**: Removal of husk and scratching of seed coat
- Abrade the seeds between two rubber faced surface or impel seeds against roughened surfaces, such as sandpaper.

9.5.4. Cleaning and grading

- Second stage of cleaning is carried out with air blasts and vibrating.

1. Air-screen cleaner cum grader

- Machine size varies from small, two-screen farm models to large industrial cleaners with 7-8 screens.
- Three cleaning elements:
- **Aspiration**: Light material is removed from the seed mass.
- **Scalping**: Good seed are dropped through screen openings, but larger material is carried over the screen into a separate spout.
- **Grading**: Good crop seed ride over screen openings, while smaller particles drop through.

9.5.5. Upgrading

- In final stage of processing, removal of seeds larger or smaller than required size and removal of cracked, damaged or defective seeds is accomplished.

1. Specific gravity separator

- Makes use of a combination of weight and surface characteristics of seed to be separated.
- A mixture of seeds is fed onto the lower end of a sloping perforated table.
- Air is forced up through the porous deck surface and the bed of seeds by a fan, which stratifies the seeds in layers according to density with the lightest seeds and particles of inert matter at the top and the heaviest at the bottom.
- An oscillating movement of the table causes the seeds to move at different rates across the deck.
- Lightest seeds float down under gravity and are discharged at the lower end, while the heaviest ones are kicked up the slope by contact with the oscillating deck and are discharged at the upper end.

2. Indented cylinder separator

- Separate seeds according to the length.
- Equipment consists of a slightly inclined horizontal rotating cylinder and a movable separating trough.
- Inside surface has small closely spaced hemispherical indentations.
- Small seeds are pressed into the indents by centrifugal force and can be

removed. The larger seeds flows in the centre of the cylinder and is discharged by gravity.

3. Magnetic separator

- Separates seed according to its surface texture or related seed characteristics.
- Seed is treated with iron filings, which adhere to rough surface alone.
- They are passed over a revolving magnetic drum and separated from smooth, uncoated seed.

4. Electronic colour separator

- Separation based on colour is necessary because the density and dimensions of discoloured seed are the same as those of sound seed.
- Seed that differs in colour is detected by the photo cells, which generate an electric impulse.
- Impulse activates an air jet to blow away the discoloured seed

5. Spiral separator

- Separates according to its shape and rolling ability.
- Consists of sheet metal strips fitted around a central axis in the form of a spiral.
- Seed is introduced at the top of the inner spiral.
- Round seeds roll faster down the incline than flat and irregularly shaped seeds, which tend to slide or tumble. The orbit of round seed increases with speed on its flight around the axis, until it rolls over the edge of the inner flight into the outer flight where it is collected separately.

6. Liquid/water flotation

- Used to separate paddy seeds.
- Liquids with a density or specific gravity between that of the full and empty seed are used.
- Specific gravity of the liquids used is such that the full seed sinks and the empty seed and light debris float.
- Ten litres of water taken in plastic bucket of 15 litres capacity
- A good quality fresh egg is dropped in the water and the egg sinks in water and reaches the bottom.

- Then, commercial grade common salt (Sodium chloride) is added little by little and dissolved in water.
- As the density of the salt solution increases, the egg rises up.
- Addition of salt is stopped when surface of egg to size of a 25 paise coin is visible above solution.
- Now, the density of water is suitable for separation of quality seeds.
- To this salt solution, add 10 kg of paddy seeds.
- Seeds with low density float on surface of solution, whereas seeds with high density sink.
- Seeds sunk in solution are washed for 2-3 times in water and then, used.
- By this method, 20 kg of seeds required for one acre can be upgraded.

10

Seed Act and Rules

10.1. Seed Act (1966) and Seed Rules

- To ensure the availability of quality seeds, GOI has enacted Seed Act in October 1966 and Seed Rules in September 1968.
- Seed Act passed by the Indian Parliament in 1966.
- Implemented in October, 1969.
- Objective is for regulating the quality of certain notified kind/varieties of seeds for sale and for matters connected therewith.
- Have 25 sections.
- Seed (Control) order 1983 promulgated under Essential Commodities Act, 1955 to ensure the production, marketing and distribution of the seeds.

10.2. Main features of Seed Act, 1966

10.2.1. Sanctioning legislation

- Sanctioning legislation authorizes formation of advisory bodies, Seed Certification Agencies, Seed Testing Laboratories, Foundation and Certified Seed Programs, Recognition of Seed Certification Agencies of Foreign Countries, Appellate Authorities, etc.

1. Central Seed Committee

- To advise Central and State Governments on all matters related to seeds
- To advise Government regarding notification of such kinds/varieties and to regulate the quality of seeds.
- To advise Government on minimum seed standards for kinds/varieties brought under purview of Act.
- To recommend procedure and standards for certification, grow out tests and analysis of seeds.
- To recommend to Central Government the suitability of any seed certification agency established in any foreign country for the purposes of this Act.

- To recommend the rate of fees to be charged for analysis of samples by the Central and State Seed Testing Laboratories and for certification by Certification Agencies.
- To advise the Central and State Governments regarding suitability of seed testing laboratories.
- To send its recommendations and other pertinent proposals related to the Act to the Central Government.
- To carry out such other functions as are supplemental, incidental or consequential to any of the functions conferred by the Act or Rules.

2. Central Seed Certification Board

- To deal with all problems related to seed certification and to co-ordinate the work of State Seed Certification Agencies.

3. State Seed Certification Agencies

- Certify seeds of any notified kinds of varieties.
- Outline procedure for submission of applications for growing, harvesting, processing, storage and labeling of seeds intended for certification till the end so as to ensure that seed lots finally approved for certification to meet out prescribed standards under the Act, or these rules.
- Maintain a list of recognised breeder seeds
- Verify for variety eligible for certification, seed source used for planting for authenticated and record of purchase in accordance with these rules and that fees have been paid.
- Take sample and inspect seed lots produced under the procedure laid down by the certification agency and have such samples tested to ensure that the seed conforms to the prescribed standards.
- Inspect seed processing plants to see that the admixtures of other kinds and varieties are not introduced.
- Ensure that action at all stages, *e.g.* field inspection, seed processing plant inspection, analysis of samples taken and issue of certificates (Tags, marks, labels and seals) is taken expeditiously.
- Undertake educational programs designed to promote the use of certified seed, including a publication of listing on certified seed growers and sources of certified seed.
- Grant certificates (Tags, labels, seals, etc.) in accordance with the provisions of the Act and these rules

- Maintain such records as may be necessary to verify that seed fields for the production of certified seed are eligible under the rules
- Inspect fields to ensure minimum standards for isolation, roguing, use of male sterility and similar factors that are maintained at all times, as well as to ensure that seed borne diseases are not present in the field as per the standards for certification.

4. Central Seed Testing Laboratory

- Initiate testing program in collaboration with the State Seed Laboratories designed to promote uniformity in test results between all seed testing laboratories in India.
- Collect data on quality of seeds found in market and make this data available to Committee.
- Carry out such other functions as may be assigned by the Central Government from time to time.
- Act as referee laboratory in testing seed sample for achieving uniformity in seed testing.
- State STLs have to send 5 % samples to CSTL along with their analysis results.

  Now, it is renamed as National Seed Research and Training Centre (Locaked at Varanasi, UP) Since October, 2005.

5. State Seed Testing Laboratories

- To carry out the seed analysis work of the State in a prescribed manner.

6. Appellate Authority

- Appellate authority appointed through an official notification in the Gazette by the State Governments is to look into grievances of certified seed producers against seed certification officials.

7. Recognition of seed certification agencies of foreign countries

- Central Government, on recommendation of Central Seed Committee and by notification in the official Gazettee, may recognise any seed certification agency established in any foreign country, for purpose of Seeds Act, 1966.

10.2.2. Regulatory legislation

- To control quality of seeds sold in the market including suitable agencies for regulating seed quality.

10.3. Minimum limit and labeling of the notified kind/varieties of seed

- Quality control as envisaged in the Act is to be achieved through pre and post marketing control, **voluntary certification** and **compulsory labeling** of seeds of notified kind / varieties.

10.3.1. Power to notify the kind/varieties

- New varieties evolved by the SAUs and ICAR institutes are notified and released under **section 5** in consultation with central seed committee and its sub committees constituted under **section 3 and 3(5)**.

10.3.2. Labeling provision

- Minimum limits for germination, physical purity and genetic purity of varieties/ hybrids have been prescribed and notified for labeling seeds of notified kind/varieties under **section 6(a)**.

10.3.3. Seed testing

- Provision to set up a CSTL (NSRTC) and STLs to discharge functions under **section 4(1) and 4(2).**
- In 1968, there were 23 STLs in the country. At present, there are 102 STLs.

10.3.4. Seed Analyst

- State Government could appoint the Seed Analysts through notification in the Official Gazette under **Section 12** defining his area and his jurisdiction.
- Seed Analyst should posses certain minimum qualification as prescribed under **clause 20 part IX** of Seed Rules.

1. Duties of seed analyst

- On receipt of a sample for analysis, the seed analyst shall first ascertain that the mark and the seal of fastening as provided in **clause (b) of sub-section (1) of Section 15** are intact and shall not the condition of the seals ther on. It can be opened by any other officer authorised in writing in that behalf by the seed analyst, who shall record the condition of the seal on the packet.
- Shall analyse the same according to the provision of the Act and Seed Rules.
- Shall deliver the report copy of the analysis to persons specified in **sub-section (1) of Section 16** of Act.

- Shall, from time to time, forward to State Government reports on result of analytical work done by him.

10.3.5. Seed Inspectors

- State Government, under **section 13** may appoint such a person as it thinks fit, having prescribed qualification (**Clause 22 part IX of Seed Rules**) through notification, as a Seed Inspector
- Define the areas within which he shall exercise jurisdiction for enforcing the seed law.
- Every Seed Inspector shall be deemed to be a public server within the meaning of **section 21** of the **Indian Penal Code (45 of 1860)** and shall be official subordinate to such authority as the State Government may specify in this behalf.
- He has power to examine records, register and document of the seed dealer.
- **Duties** are defined in **clause 23 of part IX of Seed Rules**.
- Issue stop sale order in case the seed in question contravenes the provision of relevant Act and Rules for which he can use form no. III. When he seizes any record, register documents or any other material, he should inform a magistrate and take his order for which he can use form no. IV.

1. Powers of seed inspectors

- To take samples of seed of any notified kind/variety from any person selling such seed or purchaser of consignee and send such samples for analysis to the seed analyst notified for the area
- To enter and search, at all reasonable time, with such assistance, if any, as he considers necessary, any place in which he has reason to believe that an offence under this Act has been or is being committed and order in writing the person in possession of any seed in respect of which the offence has been or is being committed, not to dispose of any stock of such seed for a specific period not exceeding thirty days or unless the alleged offence is such that the defect may be removed by the possessor of the seed, seize the stock of such seed.
- To examine any record, register, document or any other material object found in any place mentioned in **clause (2)** and seize the same, if he has reason to believe that it may furnish evidence of the commission of an offence punishable under this Act and exercise such other powers as may be necessary for carrying out the purposes of this Act or any Rule made there under.

- On demand to pay the cost of seed, calculated at the rate at which such seed is usually sold to the public, to the person from whom the same is taken.
- To break open any container in which any seed of any notified kind or variety may be contained, or to break open the door of any premises where any such seed may be kept for sale.
- Provided that the power to break open the door shall be exercised only after the owner or any other person in occupation of the premises, if he is present therein, refuses to open the door on being called upon to do so.
- Where the seed inspector takes any action under **clause (a) of sub-section (1)**, he shall, as far as possible, call not less than two persons to be present at the time when such action is taken and take their signatures on a memorandum to be prepared in the prescribed form and manner.
- The provisions of the **Code of Criminal Procedure, 1898,** shall, so as may be, apply to any search or seizure made under the authority of a warrant issued under **section 98 of the said Code.**

10.3.6. Penalty

- If any person, contravenes any provision of the Act or Rule or prevents a seed inspector from taking sample under this Act or prevents a seed inspector from exercising any other power conferred on him could be punished under **section 19** of the Act with a fine of five hundred rupees for the first offence.
- In the event of such person having been previously convicted of an offence under this section with imprisonment for a term, which may extend to six months or with fine, which may extent to one thousand rupees or with both.

10.4. Seed certification

- To maintain and make available to the public through certification of high quality propagating material of notified kind/varieties so grown and distributed as to ensure genetic identify and genetic purity.
- Certified standards in force are Indian Minimum Seed Certification Standards and seed certification procedures form together for the seed certification regulations.
- Seed of notified varieties under **Section 5** of the Seed Act shall be eligible for certification.

10.5. Restriction of export and import of seeds - Seeds control order, 1983

- **Section 17** define as under
- No person shall for the purpose of sowing or planting by any person (Including himself) export or import or cause to be exported or imported any seed of any notified kind or variety unless, a) it conforms to the minimum limits of germination and purity specified for that seed under **clause (a) of Section 6** and, b) its container bears in the prescribed manner the mark or label with the correct particular thereof specified for that seed under **clause (b) of section 6.**

10.5.1. Gist of the Seed control order 1983

1. Issue of licence to dealers

- All persons carrying on the business of selling, exporting and importing seeds will be required to carry on the business in accordance with terms and conditions of licence granted to him for which dealer make an application in duplicate in Form 'A' together with a fee of Rs. 50/- for licence to licencing authority unless the State Government by notification exempts such class of dealers in such areas and subject to such conditions as may be specified in the notification.

2. Renewal of licence

- A holder of licence shall be eligible for renewal upon and application being made in the prescribed form 'C' (in duplicate) together with a fee of rupees twenty before the expiry of licence or at the most within a month of date of expiry of license for which additional fee of Rs. 25/- is required to be paid.

3. Appointing of licencing authority

- State government may appoint such number of persons as it thinks necessary to be inspector and define the area of such inspector jurisdiction through notification in the official gazette.

4. Time limit for analysis of samples by STL

- Time limit for analysis of samples and suspension/cancellation of licence may be done by licencing authority after giving an opportunity of being heard to the holder of license, suspend or cancel the licence on grounds of mis-representation of a material particular or contravention in provision of the order.

6. Suspension/cancellation of licence

- Licensing authority may after giving an opportunity of being head to the holder of licence, suspend or cancel the licence on grounds of mis-representation of material particular or contravention in provision of the order.

7. Appeal

- State government may specify authority for hearing the appeals against suspension/cancellation under this order and the decision of such authority shall be final. Any person aggrieved by an order of refusal to grant or amend or renew the licence for sale, export/import of seed may within 60 days from the date of order appeal to the designated authority in the manner prescribed in the order.

8. Miscellaneous

- Licencing authority may on receipt of request in writing together with Rs. 10/- form amend the licence of such dealer. Every seed dealer are expected to maintain such books, accounts and records to this business in order and submit monthly return of his business for the preceding months in Form 'D' to the licencing authority by 5th day of every month.

10.6. New seed policy

- Implemented from October 1, 1988.
- Policy laid a special emphasis on
  - Import of high quality of seeds
  - A time bound program to modernize plant quarantine facilities
  - Effective implementation of procedures for quarantine/post entry quarantine
  - Incentives to encourage the domestic industry

10.6.1. Import of quality seeds

- Bulk import of seeds of coarse cereals, pulses and oil seeds may replace local production.
- Instead of bulk import of seeds or technology, it is better to import germplasm of superior variety and could be developed locally to meet the demand i.e., incorporate the advantages of exotic variety to the local types or even direct multiplications after adaptive trials.

- Import of flower seeds could be encouraged in order to earn foreign exchange through export of flowers and it can be imported under (OGL) open general licence.
- But, there is a fear of introduction of new pests and diseases as they are coming without post entry quarantine check up.

10.7. Key to the Seeds Act, 1966 and the Seeds Rules, 1968

S. No.	Subject	Related section(s) of the Seed Act	Related rules of Seeds Rules	Related forms of Seeds Rules
1	**Minimum limits and labeling (Seed Quality control)**			
	General	1,2,23,25	Part I-1, 2 Part XI-38	
	Central Seed Committee	3	Part II-3, 4	
	Seed Testing Laboratory	4, 16(2), 16(3), 16(4)	Part III-5 Part X-36	
	Notification of kinds and varieties	5		
	Labeling provisions	6, 7, 17	Part V-7, 8, 9, 10, 11, 12Part VI-13	
	Seed Analyst	12, 16(1)	Part IX-20, 21, 33, 35	VII
	Seed Inspector	13, 14, 15	Part IX-22,23 Part X-24, 25, 26, 27, 28, 29, 30, 31, 32, 34, 37, 39	III, IV, V, VI, VIII
	Penalty	19, 20, 21 22		
	Exemption	24		
2	**Seed Certification**			
	Certification agency	8, 18	Part IV-6	
	Certification	9, 10	Part VI-14Part VII-15, 16, 17	I, II
	Appeal	11	Part VIII-18, 19	

11

Seed Certification Techniques

11.1. Seed certification

- Legally sanctioned system for quality control on seed multiplication and production.
- Involves field inspection, pre and post control tests and seed quality tests.

11.2. Concepts of seed certification (AOSCA - Association of Official Seed Certifying Agencies)

- Pedigree of all certified crops must be essential.
- Integrity of certified seed growers must be recognized.
- Field inspection must be made by qualified field inspectors.
- Verification trials to establish and maintain satisfactory pedigree of seed stock.
- For keeping proper records, establish and maintain satisfactory pedigree of seed stock.
- Standard should be maintained for purity and germination.
- Principles of sealing seeds to protect both grower and purchaser must be approved.

11.3. Phases of seed certification

- Six broad phases
  - Receipt and scrutiny of application.
  - Verification of seed source, class and other requirements of the seed used for raising the seed crop.
  - Field inspections to verify conformity of the standing crop to the prescribed field standards.
  - Supervision at post harvest stages including processing and packing.

- Seed sampling and analysis, including genetic purity test and / or seed health test, if any, in order to verify conformity to the prescribed seed standards.
- Grant of certificate and certification tags, bagging and sealing.

11.4. Seed certification procedures

11.4.1. Seed source verification

- Seed producer should submit relevant evidence such as certification tags, seals, labels, seed containers, purchase records, sales records etc., during scrutiny of the application and/or during the first inspection of the seed crop to confirm that the seed used for raising the crop has been obtained from a source approved by the agency and conforms to the class of seed required for seed production.
- A source verification register should be maintained by seed producer for verification by agency staff.

11.4.2. Submission of application and sowing report

- Application for certification (Sowing report) should be submitted in the prescribed **FORM-1 in triplicate within 30 days** from the date of sowing or **15 days from the date of transplanting** to the concerned Assistant Director of Seed Certification.
- Seed certification charges such as registration fee, inspection fee, grow out test charges, seed testing charges, etc. should be remitted along with FORM-1.
- Separate FORM-1 should be submitted for certification of each variety.
- FORM-1 should contain complete details of the name and address of the seed producer; season of production; name and address of the grower; location of the seed plot; crop/variety and class of seed to be produced; area under seed production; details of parental seed materials used with lot number; date of sowing and the particulars of seed certification charges remitted.
- In a single application, the maximum area that can be offered for certification is 25 acres. Additional area will require separate application.
- No refund of inspection and registration fee will be made once the seed plot has been visited/inspected by the Seed Certification Officer.
- Seed producer should assist the agency staff in locating the seed plots during the first inspection.

- Seed producers should guide their growers in agronomic practices, pest/ disease control etc.,
- Only seed from plots meeting all the prescribed field standards for certification is accepted for processing at the recognised seed processing plants.
- All the necessary care should be taken to avoid admixture during harvesting, threshing and transportation.
- Harvested seed produce from the approved fields should be brought to the seed processing unit and seeds are certified within 2 1/2 months from the date of harvest.
- Certification of a seed lot will be taken up only, if the seed lots have met the prescribed field and seed standards.

11.4.3. Field inspection

- Essential step in verifying conformity of seed crop to prescribed minimum seed certification standards
- Field standards consist of
- Minimum preceding crop requirement have been specified to minimize genetic contamination from the disease and volunteer plants.
- Minimum isolation requirement has been specified to minimize genetic contamination.
- Number of field inspections and specified stage of crop have been described to ensure verification of genetic purity and other quality factors.

1. Field inspection stages

Stages of crop	The plant characters based on which rogues / offtypes are to be identified and removed
Seedling	Other crop seedling (Volunteers), diseased seedling, early types with rapid growth.
Vegetative	Plant stature (Weak / healthy / bushy), leaf colour, stem colour, hairyness, early / late type.
Flowering	Early / late type, flower colour, anther colour, petal spot (Cotton), pollen shedder, pollen partials, (Pearl millet, rice, sorghum hybrids) and shedding tassel (maize hybrid)
Maturation	Early / fruit shape (Conical / cylindrical/oval), fruit colour (Mottled / pure), shape and size of fruits, fruit characters (Ridged non-ridged), selfed bolls (Cotton hybrid)
Harvesting	Early maturing plants, late maturing plants, re-checking of all above characters.

2. Verification of the seed used for sowing purpose

Factors for verification	Class of seed used for sowing propose		
	Breeder	**Foundation**	**Certified I**
Source	Seed Corporation	Seed Corporation	Seed Corporation
Colour of certificate/tag	Yellow	White	Azure blue
Seal	ICAR/Institute	Seed Certification Agency	Seed Certification Agency
Purchase receipt	Seed Corporation	Seed Corporation	Seed Corporation

* State Agriculture Universities *(SAU's)

3. Cropping history of the field

Year/season	Crop Last present	Variety
	Last year/season crop was different	
	Seed is separable	Permitted
	Seed is not separable	Not permitted
	Last year/season crop was the same	
	Variety was different	Not permitted
	Same variety but class was lower	Not permitted
	Last year/season crop was same with same variety and same or superior class	
	But, field /seed was rejected due to seed-borne diseases	Not permitted
	But, field/seed was rejected due to objectionable weeds	Not permitted
	Passed during seed certification	Permitted

4. Minimum gap required for seed production of different varieties of same crop in selected field

Season	Crops
One season	Rice, wheat, pearl millet, sorghum, soybean, sesame, mustard
Two seasons	Groundnut, berseem

5. Isolation distance verification

- **Varietal seed production**
- 3 m: Rice, soybean
- 5 m: Black gram, green gram, cowpea, field lab-lab
- 5-10 m: Groundnut
- 30 m: Cotton CS
- 50 m: Cotton FS, jute, gingelly CS
- 100 m: Sorghum variety, red gram, gingelly FS

- 150 m: Castor CS
- 200 m: Maize, sorghum FS, pearl millet CS, sunflower CS, castor FS, safflower CS,
- 400 m: Sudan grass FS and CS, fodder sorghum FS and CS, pearl millet FS, sunflower FS, safflower FS,
- **Hybrid seed production**
- 30 m: Cotton CS
- 50 m: Cotton FS
- 100 m: Rice
- 200 m: Maize CS, sorghum, pearl millet CS, red gram
- 400 m: Maize FS
- 600 m: Sunflower
- 100 m: Pearl millet FS,
- **Varieties and hybrids**
- 1600m for FS and 1000m for CS of all crucifers except carrot (1000m for FS; 800m for CS)

  FS: Foundation seed ; CS: Certified seed.
- Minimum isolation distance to avoid transmission of loose smut

Crop	Field of crops infected with loose smut	Isolation distance (m)	
		FS	CS
Wheat	Wheat, triticale, rye	150	150
Barely	Barely	150	150
Triticale	Triticale, wheat, rye	150	150
Oat	Oat	150	150

6. Field counts

- **Number of inspections**
- Minimum: 2
- Maximum: 4
- Compulsory: 1 inspection at flowering

Area of the field crops (acres)	No. of counts to be taken
0–5	5
6–10	6
11–15	7
16–20	8
21–25	9

Crop	No. of plants/head per count
Bhendi, brinjal, bulb crops, capsicum, castor, root crops, chilli, cotton, cucurbits, groundnut, maize, redgram, tomato.	100
Beans, cowpea, gram, leaf crops, moong (Greengram), sunflower, mustard, peas, sesame, sunhemp, safflower	500
Bajra, paddy, sorghum, wheat, millets, jute	1000

- **Field inspection method**
- No. of steps required for one field count =

$$\frac{\text{No. of plant / earheads to be observed in one field count}}{\text{Number of steps required for one crop}}$$

- If 1000 plants to be observed
  - For one step = 25 plants
  - So, for 1000 plants = 40 steps
  - i.e., 40 (steps) x 25 plants = 1000 plants
- **Minimum number of field inspections and recommended crop stages for seed certification**

Crop	Number	Stage
Field and fodder crops		
Maize synthetic and composite	2	1st before and 2nd during flowering
Greengram, Blackgram, chickpea, rajmash, lentil, kidney bean, Indian bean, horse gram	2	1st before and 2nd during flowering and fruiting
Soybean	2	During flowering and second at maturity with shedding of leaves
Napier grass		45 and 100 days after planting
Groundnut, linseed, minor millets, cotton, castor, berseem, lucerne, Indian clover	2	1st during flowering and 2nd before harvesting
Rice, wheat, oat, barley, triticale	2	Between ear emergence and harvesting
Sorghum, pearl millet, rapeseed, mustard, niger, sesame, sunflower, jute, field pea, buffel, dinanath, forage sorghum, guinea, marvel, setaria, stylo, teosinte	3	Before flowering, during flowering and prior to harvesting (Pod stage for fieldpea)

7. Purity maintenance

- **Roguing**
- Removal of unwanted plants (Rogues, off types and volunteer plants) which can interfere with genetic, physical and health purity of the crop.

- **Maximum permitted limits for off-types (%)**

Crop	Foundation seed	Certified seed
Napier grass (*Pennisetum purpureum*)	0.01	0.3
Sorghum, pearl millet, barely, linseed, niger, safflower, minor millets	0.05	0.1
Paddy, wheat, triticale, oat	0.05	0.1
Buffel, dharaf, dinanath, guinea grass, setaria and stylo	0.10	0.1
Blackgram, Greengram, chickpea, fieldpea, lentil, pigeonpea rajmash, lathyrus, cowpea, Indian bean, rice bean, cluster bean, kidney bean, horse gram, groundnut, sesame, cotton, castor	0.10	0.2
Sunflower *	0.10	0.2
Soybean, rape seed and mustard, taramira, teosnite	0.10	0.5
Berseem, senji (*Meliotus* sp.), marvel grass	0.20	1.0
Jute	0.50	1.0
Maize** (Composite, synthetic and open pollinated varieties)	1.00	1.0

- * Sunflower: Sterile plants of the variety under seed production program are not considered as offtype.
- ** Maize: This is the MSCS for off type, pollen shedding plants when 5 % or more of the plants in seed field have receptive silk.

8. Special types of rogues in hybrid seed production

- **Pearl millet, sorghum, paddy, sunflower**
- **Pollen shedders:** Presence of 'B' line in 'A' line.
- **Pollen partials:** A part of the earhead of the female parent (A line) shedding pollen here and there.
- **Maize**
- **Shedding tassels:** Part of male inflorescence shedding pollen.
- **Cotton**
- **Selfed bolls:** Bolls produced by self fertilization in female parent of cotton.
- **Redgram**
- **Fertile plants:** In GMS system, 50% of plants will be sterile and 50 % will be fertile.
- On the basis of viable pollen (Plumpy), the plants are to be removed at flowering.

9. Objectionable weeds

Maximum permitted limit (%) of objectionable weeds at field level

Crop	Weed	Foundation seed	Certified seed
Field crops			
Rice	Wild rice (*O. sativa var. fatua)*	0.01	0.02
Rapeseed, mustard, Taramira	Satyanashi (*Argemone mexicana*)	0.05	0.1
Safflower	Wild safflower (*Catharanthus oxyacantha*)	None	None
Sunflower	Wild *Helianthus*	None	None
Oat	Wild oat (*Avena fatua)*	0.01	0.02
Fodder crops			
Berseem	Chicory (Kasni) (*Cichorium intybus*)	None	0.05
Lucerne	Dodder (*Cuscuta* spp.)	None	0.05
Napier grass	Canada thistle (*Cirsium arvense)* dodder (*Cuscuta* spp.), Johnson grass (Saru) (*Sorghum halepense),* quack grass (Thitch grass, couch grass) *Agropyron repens*), *C. arvensis*	None	None,

11.4.4. Field rejection

- If seed field do not meet minimum field standards, field is liable for rejection,
- Action taken should be communicated immediately to the seed grower.

11.4.5. Inspection of seed processing

- Make as many inspections of seed lots, records and the processing plant as may be required.
- To ensure that the admixtures of other kinds and varieties are not introduced during the seed processing,

11.4.6. Sampling and sending to STL

- Take seed samples of all seed lots and send to STL for evaluation of purity, germination and moisture.
- If seed lots fail to meet the standards, resampling and retesting should be encouraged.

11.4.7. Tagging and sealing

- Upon receiving a satisfactory report, tagging and sealing should be done under supervision of agency.
- Affixing of tags and seals on the containers complete the process of certification of seeds.

11.4.8. Control plot testing

- Should arrange for a post season grow-out test of all the hybrid seed lots as prescribed in the standards

11.4.9. Extension of the validity period

- Extension of validity period of certified seed (9 months) shall be for a period of six months at each subsequent validation as long as the seed conforms to the prescribed standards.

12

Seed Testing

12.1. Seed standards

- Minimum seed certification standards can be grouped into two groups.
  - General seed certification standards
  - Specific crop standards
- Consists of field standards and seed standards.
- Minimum percentage of pure seed and germination.
- Maximum permissible limits for inert matter and other crop seeds.
- Maximum permissible limits for objectionable weeds and seeds infected by seed borne diseases.
- Maximum permissible limits for moisture content have been prescribed for the safe storage of seeds.

12.1.1. Minimum germination for foundation seed and certified seed

- 60% - Bottlegourd, pumpkin, bittergourd, watermelon
- 65% - Cotton varieties, bhendi
- 70% - Castor, groundnut, sunflower, soyabean, tomato, brinjal, cluster bean
- 75% - Sorghum, ragi, pearl millet, redgram, greengram, blackgram, cowpea, cotton hybrids, peas, french beans
- 80% - Rice, maize inbreds and single cross hybrid, horsegram, safflower, sesamum, niger, jute
- 85% - Wheat, bengalgram
- 90% - Maize hybrids, composites, synthetics

12.1.2. Seed standards for different crops

Particulars	Germination (Min) (%)		Moisture (Min) (%)		Pure seed (Min) (%)		Inert matter (Max) (%)		O.C.S.(No /kg) (Max)		O.D.V. (No /kg) (Max)		O.W.S. (No /kg) (Max)	
	F	C	F	C	F	C	F	C	F	C	F	C	F	C
Paddy	80	80	13	13	98	98	2	2	10	20	10	20	10	20
Finger millet	75	75	12	12	97	97	3	3	10	20	-	-	10	20
Sorghum	75	75	12	12	97	97	3	3	10	20	10	20	5	10
Wheat	85	85	12	12	98	98	2	2	10	20	-	-	10	20
Maize														None
Hybrids	-	90	-	12	-	98	-	2	-	10	-	10	-	
Inbreds and single crosses	80	-	12	-	98	-	2	-	5	-	5	-	None	
Composites and synthetics	90	90	12	12	98	98	2	2	10	20	-	None		
Pearl millet	75	75	12	12	98	98	2	2	10	20	-	None	None	
Red gram	75**	75**	9	9	98	98	2	2	5	10	10	20	5	10
Green gram and black gram	75**	75**	9	9	98	98	2	2	5	10	10	20	5	10
Horse gram	80	80	9	9	98	98	2	2	None	10	5	20	None	None
Bengal gram	85**	85**	9	9	98	98	2	2	Non	10	5	20	None	None
Cowpea	75**	75**	9	9	98	98	2	2	None	10	5	20	None	10
Castor	70	70	8	9	98	98	2	2	None	10	5	20	None	None
Groundnut	70	70	8	9	98	98	4	4	None	-	-	10	None	None
Sunflower	70	70	9	9	98	98	2	2	None	None	None	None	5	10
Safflower	80	80	9	9	98	98	2	2	None	None	None	None	5	10
Seasamum	80	80	9	9	97	97	3	3	10	20	10	20	10	20
Niger	80	80	9	9	98	98	2	2	10	20	10	20	10	20
Soybean	70**	70**	12	12	98	98	2	2	None	10	5	10	5	10
Cotton hybrids	75	75	10	10	98	98	2	2	5	10	-	-	5	10
Cotton varieties	65	65	10	10	98	98	2	2	5	10	-		5	10
Jute	80	80	9	9	97	97	3	3	10	20	10	20	10	20

F: Foundation seed; C: Certified seed; O.C.S: Other crop seeds; O.D.V: Other distinguishable varieties;

O.W.S: Objectionable weed seeds; ** Including hard seeds.

12.2. Seed sampling

12.2.1. Seed lot

- Specified quantity of seed of a variety

12.2.2. Sample

- A representative sample of a seed lot

1. Types of samples

- **Primary sample:** A small portion taken from different representative part of the seed lot.
- **Composite sample:** A mixture of all the primary samples taken from the seed lot.
- **Submitted sample:** Samples submitted to the seed testing laboratory.
- Submitted sample size is ten times higher than the working sample size, but restricted to 1000 g
- **Working sample:** A reduced quantity of sample taken from submitted sample in the laboratory for use in a given quality test.

2. Working sample size of different crops

Working sample (g)	Cereals and Millets	Pulses	Oilseeds	Fibres	Forages	Green manures & other crop	Vegetables
1					Doob	Poppy	Celery
2					Blue panic, guinea grass, setaria grass, shaflal		
3					Bird grass, buffel grass, marvel grass, paragrass		Lettuce, carrot
4			Raket sated		Fenugreek		Parsley, Chinese cabbage
5					Lucerne, Venezuela grass		

Working sample (g)	Cereals and Millets	Pulses	Oilseeds	Fibres	Forages	Green manures & other crop	Vegetables
6	Finger millet				Egyptian clover		
7	Little millet		Sesame		Stylo		Tomato variety & hybrid, amaranth, turnip
8	Barnyard millet, Kodo millet						Onion
9	Italian millet						
10				Jute	Indian clover		Cabbage, cauliflower, broccoli, knol-khol
15	Common millet, pearl millet		Niger		Napier grass		Brinjal, chilli, sweet pepper, muskmelon, cucumber
25				Cotton hybrid	Sudan grass		Chow-chow, spinach, sweet potato, tapioca
30			Linseed				Radish
35				Cotton variety			
40	Paddy						Coriander
50		Horsegram			Velvet bean	Sugar beet	
60		Lentil			Hemp, indigo		Goa bean, jack bean, sword bean
70				Roselle, sunnhemp			Ashgourd, bottlegourd
75		Kidney bean					
90	Sorghum				Teosinite	Daincha	
100							Asparagus cluster bean
120	Wheat, barley, triticale, oat	Greengram					

Working sample (g)	Cereals and Millets	Pulses	Oilseeds	Fibres	Forages	Green manures & other crop	Vegetables
125			Sunflower				
140							Okra
150		Blackgram					
180			Safflower				Pumpkin
250							Snakegourd, Spongegourd, Indian squash, water melon
300			Pigeonpea				
400			Cowpea				Ridgegourd
450		Chickling vetch					Bittergourd
500		Indian bean	Soybean				
600			Groundnut				Field bean
700			French bean				Winter squash
900	Maize	Pea					
1000		Bengalgram		Castor, groundnut			
Broad bean,				(pods)			lima bean, scarlet summer bean

3. Sample for moisture estimation

- Should be packed in moisture vapour proof container and submitted to STL separately.

4. Sealed sample

- Container should be sealed and cannot be opened or closed again without destroying.

12.2.3. Sampling intensity

1. Container sampling

Upto 5 containers	Sample each container and always take atleast 5 primary samples
6 to 30 containers	Sample 5 containers or one in every 3 containers, whichever is greater
31 to 400 containers	Sample 10 containers or atleast one in every 5 containers, whichever is greater
401 or more containers	Sample 80 containers or atleast one in every 7 containers, whichever is greater

2. Bag sampling

Upto 500 kg	Atleast 5 individual samples
501 to 3000 kg	One individual sample for each 300 kg, but not less than 5 samples
3001 to 20,000 kg	One individual sample for each 500 kg, but not less than 10 samples
20001 and above	One individual sample for each 700 kg, but not less than 40 samples

12.2.4. Sampling devices

1. Sleeve or stick type trier

- For cereals (Except maize) in bags, a trier around 76 cm long and 2.5 cm outer diameter, with six slots is suitable.

2. Bin sampler

- Larger and up to 160 cm in length and 3.8 cm diameter with 6–9 slots.

3. Nobbe trier

- Suitable for sampling of seed lot in a bag.
- About 50 cm long including a handle of 10 cm and a point of about 6 cm

12.2.5. Sampling procedure

1. Bag sampling

- Probing the horizontal bag diagonally from corner to corner.
- Inserted with slots closed and facing down.
- Upon full insertion into the bag, the slots should be rotated upwards to dislodge any seed which may have been carried along with the trier sections and the slots opened

2. Bulk sampling

- Five to seven probes must be made.

3. Hand sampling

- Non free flowing seeds should be sampled by hand
- Hand is inserted into the bag
- Upon reaching the appropriate location, hand is closed and a sample is taken.

12.3. Mixing and dividing of seeds

12.3.1. Mixing and dividing

- To obtain homogenous representative working sample

12.3.2. Dividers

- **Boerner divider:** Works on gravitational force
- **Soil divider:** Mix and divide large and chaffy seed based on gravitational force
- **Gamet divider:** Based on centrifugal force

12.3.3. Methods of dividing

- **Mechanical divider:** Not suitable for chaffy seeds
- **Random cup method:** Suitable for seeds that need working sample upto 10 g
- **Modified halving method:** Alternate cups have no bottom
- **Spoon method:** Suitable for single small seeded species
- **Hand halving method:** Suitable for chaffy seeds

12.4. Purity analysis

- To estimate biotic and abiotic impurities

12.4.1. Purity components

1. Pure seed

- Seeds of kind/species stated by the sender.
- Includes all botanical varieties of that kind/species.
- Immature, undersized, shrivelled, diseased or germinated seeds are also pure seeds.
- Also includes broken seeds, if the size is > ½ of the original size except in leguminaceae, and cruciferae where the seed coat entirely removed are regarded as inert matter.

2. Other crop seed

- Refers to the seeds of crops other than the kind being examined.

3. Weed seed

- Includes seeds of those species normally recognized as weeds or specified under Seed Act as a noxious weed.

4. Inert matter

- Includes seeds like structures, stem pieces, leaves, sand particles, stone particles, empty glumes, lemmas, paleas, chaff, awns, stalks longer than florets and spikelets.

12.4.2. Purity test equipments

- Purity work board
- Magnifying glass of 3x to 15x magnification
- An analytical balance (Accuracy 0.1 mg) with a capacity of 200 mg and a precision balance (Accuracy with about 1 kg capacity would serve the purpose)
- Small hand sieves with different standard sizes of perforations
- A small seed blower
- Spoon, spatula, forceps, needles, shallow trays, and watch glasses

12.4.3. Procedure

- Required quantity of the working sample is spread on the work table.
- Identify and isolate these seeds individually in the lot which have different shape, size, colour and surface texture as distinguishable under transmitted light.
- Such separation results in splitting the seed sample into three distinct components.
- True seed belonging to the kind declared
- Other seeds including weed seeds and those not belonging to the variety or type
- Inert matter/particles.
- Percentages of the components are determined as follows

$$\% \text{ of components} = \frac{\text{Weight of individual component}}{\text{Total weight of all components}}$$

- Reported in percentage ; 4 decimals, if sample weight is 1 g.

- If results is in border line, duplicate test can be done

Seed sample weight	Decimal
< 1 g	4
> 1 g < 10 g	3
> 10 g < 100 g	2
> 100 g < 1000 g	1

- Groundnut: Purity analysis is done on pods
- Paddy and sunflower: Huskless seeds are counted

12.5. Seed moisture determination

- Weight loss due to drying is calculated in percentage
- Sample should be packed in vapour proof container
- Estimate moisture in 2 min. after repacking for seeds which need grinding; 30 min. for seed that does not

12.5.1. Method

1. Indirect method

- Moisture meter and NMR
- Moisture meter: Moisture is directly proportionate to EC

2. Direct method: Hot air oven method

- Hot air oven: Complete evaporation / removal of moisture
- 4 - 5 g of sample for seeds having < 8 mm size
- 10 g for seeds having >8 mm
- Grind the larger seeds E.g. Paddy, wheat, oat, barley, sorghum, maize, blackgram, greengram, chickpea, pigeonpea, peas, beans, soybean, groundnut, castor, cotton
- Low constant method : 103 ± 2^0C for 17 ± 1 h ; for oily seeds
- High constant method : $130 + 3^0C$; maize 4 h, cereals 2 h, others 1 h

12.5.2. Estimation

$$\textbf{Moisture content (\%)} = \frac{M_1 - M_3}{M_2 - M_1} \times 100$$

$$= \frac{\text{Loss in weight}}{\text{Weight of seed}} \times 100$$

Where,

M_1 = Weight of the empty container and cover

M_2 = Weight of container with seed before drying

M = Weight of container with seed after drying

12.5.3. Report

- Expressed in percentage in one decimal
- Difference between two tests should not exceed 0.2 %
- If, repeat the test

12.6. Germination test

12.6.1. Media

1. Germination paper

- Must be free from chemical residues.
- Most common papers used are germination paper, filter paper or rice straw paper.
- Thickness of paper used should be more than 2 mm when moist.
- Water used must be reasonably free from acid, alkali, organic material or other impurities.
- Use either tap, distilled or de-ionized water

2. Sand

- Do not use excess water.
- Fine sand should be used.
- Make sure that the sand is clean by sterilizing before use.
- Quarry or river sand is better than shore sand, which must be washed thoroughly to remove all salt.

12.6.2. Method

- **TP:** Small seeds, light requiring seed
- **BP:** Medium sized seeds, unwind the rolled paper for observation
- **Sand:** Larger seeds, use river sand, holes distance should be 4 - 5 times the seed diameter

12.6.3. Seedling evaluation

1. Normal seedling

- A well developed root system with primary root except in certain species of poaceae which normally producing seminal root or secondary root
- A well developed shoot axis consists of elongated hypocotyl in seedlings of epigeal germination.
- A well developed epicotyl in seedlings of hypogeal germination
- One cotyledon in monocotyledon and two in dicotyledon
- A well developed coleoptile in poaceae containing a green leaf.
- A well developed plumule in dicotyledon.
- Seedlings with following slight defects are also taken as normal seedlings. Primary root with limited damage, but well developed secondary roots in leguminosae (*Phaseolus*, *Pisum*), poaceae (maize), cucurbitaceae (*Cucumis*) and malvaceae (cotton).
- Seedlings with limited damage or decay to essential structures, but no damage to conducting tissue.
- Seedlings which are decayed by pathogen, but it is clearly evident that the parent seed is not the source of infection

2. Abnormal seedling

- Seedling with loss of essential structures; diseased seedling source from nearby seedling; damaged, diseased and decayed seedling

3. Hard seed

- Remain as such at final count

4. Fresh ungerminated

- Imbibed seed, but no germination seed

5. Dead seed

- Seeds on pressing yield white substances

12.7. Quick viability test

- Developed by Lakon (1942) in Germany

12.7.1. Principle

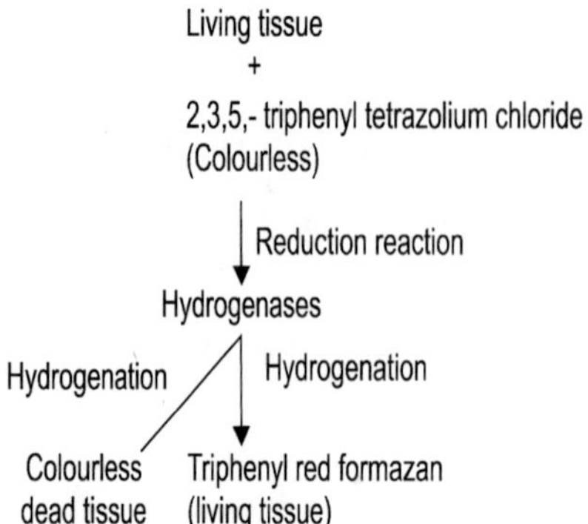

12.7.2. Method

1. Preparation of TZ solution

- 0.25 to 1.0% concentration
- Optimum pH 7 for good staining

2. Pre-conditioning of seed

- Soak in water 3 - 4 h

3. Preparation of seed

- Cut longitudinally (Maize, sorghum etc)
- Cut laterally (Small seeded grass)
- Pierce with needle (Small seeded grass)
- Remove seed coat (Dicots with hard coats)
- Conditioning only (Large seeded legumes)
- No preparation (Small seeded species)

4. Staining

- Immerse in TZ solution and kept in dark

5. Evaluation

- Germinable and non-germinal seed based on staining pattern

12.8. Grow out test

- To measure genetic purity
- Pre-requisite for all hybrids
- In addition, to check efficacy of Certification Officer

12.8.1. Sample size

- Minimum of 400 plants
- Also depend on the maximum permissible off-type plants prescribed for the species under consideration in the Indian Minimum Seed Certification Standards in accordance to the following table.

Maximum permissible off - types (%)	Minimum genetic purity (%)	Number of plants required per sample
0.1	99.9	4,000
0.2	99.8	2,000
0.3	99.7	1,350
0.5	99.5	800
1.00 and above	99.0 and above	400

12.8.2. Procedure

- Grow crop as per standard package of practices, but no thinning
- Grow one standard sample for every 10 sample under test

12.8.3. Reporting

- Other crop per cent
- Off type per cent

Reject number for prescribed standards and sample size

	Reject numbers for sample size of	
	800	400
99.5 (1 in 200)	8	*
99.0 (1 in 100)	16	8
95.0 (5 in 100)	48	24
90.0 (10 in 100)	88	44
85.0 (15 in 100)	128	64

* Indicates that the sample size is too small for a valid test

12.9. Seed viability tests

12.9.1. Vital colouring

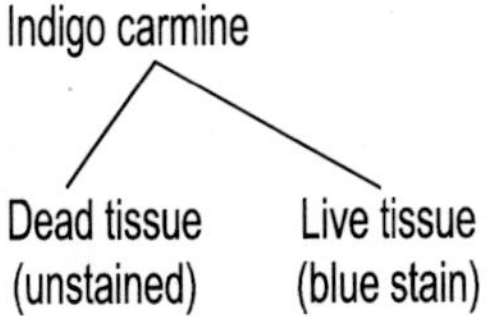

12.9.2. Catalase content

- High content - more viable

12.9.3. Peroxidase

H_2O_2 + Guaiacol → Blue, tetraguaiacoquinone

12.9.4. Dehydrogenase

$$\text{Colourless TZ solution} \xrightarrow[\text{Raduction}]{\text{Dehydrogenase}} \text{Re d form azan}$$

12.9.5. Ferri chloride test

- Soak seeds in 20% $FeCl_3$ for 15 min Black coloured dead tissue (Damaged seed)

12.9.6. Conductivity test

- Deteriorated seed leachate had high EC

12.9.7. Excised embryo test

- Quick assessment of germination than normal test

12.9.8. X-ray analysis

- **Nondestructive:** Empty, damaged and ill filled seeds are assessed
- **Contrast method:** Treat with aqueous $BaCl_2$ - Impregnated dead tissue show lighter on the film than unimpregnated living tissue

12.10. Seed vigour

- Sum total of all properties of seed leading to production of vigourous and healthy seedling under a range of stressful environments.

12.11.1. Vigour tests

Physical tests	Performance tests	Stress tests	Biochemical tests
Seed size, physical Soundness, seed density	First count, speed of germination, seedling length, seedling dry weight, strong and weak seedling, vigour index, seed metabolic efficiency, mobilization efficiency	Cold test, cool germination test, brick gravel test, paper piercing test, compacted soil test, accelerated aging test, paper-piercing test, exhaustion test, water sensitivity test, osmotic stress.	Gluiamic acid decarboxylase activity (GADA test). tetrazolium test, respiration and RQ, Mitochondrial activity, ATP level, membrane integrity

1. Seed size

- 100/1000 seeds weight (Random samples)

$$\text{Speed of germination} = \frac{\text{Number of seeding removed daily}}{\text{Day after planting}}$$

2. Vigour index

- Germination percent x Seedling length on the day of final count (Root length and shoot length in cm).

3. Seed metabolic efficiency (SME)

- Amount of efficiency determined by respired for producing 1 g of dry root and shoot
- Higher the value of SME, lower the efficiency of seed as more seed reserve would be used for producing root and shoot.
- Amount of food material respired (RESP) is calculated as

  RESP = SDW – (SHW + RTW + RSW)

  Where,

  SDW - Dry weight of seed before germination

  SHW – Dry weigth of shoot

  RTW – Dry weight of root

  RSW – Dry weight of seed after germination

  SME – SHW + RTW/RESP

3. Mobilization efficiency (ME)

- Seed with higher mobilization efficiency is considered as vigorous because of its capacity to supply maximum food material to seedling.
- Weighed seeds are placed on paper towel and kept for germination in a germinator for required period of time at optimum temperature.
- On the day of final count, seedlings and cotyledons are dried separately at 100°C for 24 h.

$$ME = \frac{\text{Increase in dry weight of embryonic axis (Dry wieght of seedling)}}{\text{Decrease in weight of cot yledons}}$$

4. Brick gravel test (Hiltner test)

- Hiltner (1912) to show pathogenic infection, porous brick gravel of 2-3 mm diameter, 2.5 cm thick layer of moist gravel.

5. Accelerated aging test

- Expose seed to 40±1°C temperature and 99 ± 1% RH.

6. Cool germination test

- Expose seed to 18°C constant temperature . Eg. Cotton

7. Cold test

- For maize; 70% water holding capacity soil spread on sand for 2 cm thickness; exposed to 10°C for one week and then, at 25°C temperature; seedling count on 4th day after 45°C exposure.

8. Exhaustion test

- Cereals and legumes; keep germination roll towel @ 10°C in complete dark and count at 20 days.

9. Water sensitivity test

- 70 g of dry sterilized sand; low moisture level - 9 ml; high moisture level of 18 ml distilled water at 25°C for 7 days.

$$WST = \frac{I - II}{I} x 100$$

I – Germination per cent at 9 ml

II – Germination per cent at 18 ml

10. Osmotic stress

- 75 g mannitol / lit of water

12. E.C. test

- Immerse 2 - 5 g seeds in 100 ml of water at 25 ± 1°C temperature for 10-12 h; expressed as dSm^{-1}.

12. GADA test

- Corn, wheat

13. Methyl cellosolve test (TZ test)

- Extract red formazon in 100 ml methyl cellosolve; read at 480 nm in spectronic 20.

14. RQ test

- Ratio of volume of CO_2 evolved / unit time to the volume of O_2 consumed unit time.

12.12. Varietal identification

12.12.1. Approaches

Physical tests	Physiological tests	Chemical tests	Molecular marker tests	Plant characters
Seed size, shape, weight, texture and colour, endospermic characters, appendages	Coleoptile colour, seedling illumination, hypocotyl colour, cotyledon colour, disease resistance, sensitivity to DOT, GA_3 soak, 2,4-D	Phenol, modified phenol, alternate phenol, alkaloid, seed coat peroxidase, sodium hydroxide, potassium dichromate, KOH bleach, ferrous sulphate colour, hydrochloric acid and iodine tests	RFLP, RAPD, SSR, ISSR, SCAR, AFLP, SNP	For example in rice, leaf blade length, width, leaf colour, colour of ligules, leaf colour, colour of lower crown area, venation pattern of leaf in sheath, length of intemode, pubescence & leaf shape or blade with auricle and flowers, booting stage, leaf and leaf sheath anthocyanin colouration, presence or absence of auricles, collar ligules and altitude of culm

12.12.2. Seed morphological characters

- To observe physical test, seed size, shape, colour, weight, texture etc: 4 x 100 seeds

1. Seed size

Size	Length	Breadth
Short	< 7.5 mm	<2 mm
Medium	7.5 – 1 mm	2 – 2.4 mm
Long	9 – 12 mm	2.4 – 3 mm
Very long	> 12 mm	> 3 mm

2. Seed shape

- Example: Long slender – IR 50
- Medium bold – IR 20
- Short bold – ADT 37

3. Seed colour

- Use day light or UV light
- **Maize:** CO 1 - Orange
- COH 1 - Deep yellow
- COH 2 - Yellow
- **Cowpea:** CO2 - Molted colour
- C152 - Light brown
- CO 4 - Greyish black
- **Lab lab:** CO 9 - Marron
- CO 10 - Black
- **Groundnut:** Off white, white rose, tan, red, dark red

4. Endospermic character

- Pearl spot in the center and small
- Pearl spot in the center and extended
- Pearl spot in the center and side extended
- Pearl spot in the center and side slightly extended

5. Appendages/special characters

- **Groundnut :** Beak or reticulation – absent or present (slight, medium, prominent, very prominent).
- **Blackgram:** Prominent hilum – Vamban 1, T9; rudimentary hilum – CO 5, TMV 1.

12.12.3. Physiological test

1. Coleoptile colour

- Colour is due to anthocynin content
- **Sorghum:** CO 28 – Reddish stem; CO 27 and CO 25 - Green stem
- Grain amaranthus: Swarna – Reddish stem; CO 4 – Green stem

2. Seedling illumination test

- Expose seedling to fluorescent light in dark room
- **Oats:** White cultivars - Fluorescent type; Yellow cultivars - Non fluorescent type
- **Festuca:** Red fescue (*Festuca rubra*) - Yellow root; Hard fescue (*Festuca lenefolia*) - Green root
- **Loliluim:** *L. multiflorum* - Fluorescent root; *L. perenna* - No fluorescent root
- **Alluim:** *A. cepa* - Yellow; *A. porrum*, *A. fistulosum* - Bluish

3. Hypocotyl colour

- Beet root as white, yellow, light red, red

4. Cotyledon colour

- Brassica's; immerse seedlings in 85 – 90% ethanol; lemon or orange colour

5. Disease resistant test

- Wheat cv. Kalyansona – Rust resistant
- Sharmila – Rust susceptible

6. GA soak test

- Rice, maize
- 25 ppm, GA soaking @ 30°C for 4 days

Coleoptiles length increase over control	Categorization as response under GA soak test
Less than 25%	Very slow response
26-50%	Low response type
51-75%	Medium response type
76-100 %	High response type
More than 76%	very high response type

7. 2, 4-D test

- Rice, maize
- 5 ppm 2, 4 – D at 30°C for 4 days

Coleoptile length reduction over control	Categorization as response under 2,4-D test
Less than 30%	High resistant type
30-60%	Medium resistant type
More than 60%	Low resistant type

12.12.4. Chemical test

1. Phenol test

- Rice, wheat, barley, oats, rye grass, kenkucky blue grass
- 1% phenol at 30°C for 4h
- Colour intensity: 0-Negative, 1-9 – light brown to deep black.

2. Modified phenol test

- Soak seeds in 0.45% $CuSO_4$ or 1% NaOH or 0.6% Na_2CO_3 for 18 h before phenol soak

Cu	Fe	Na
No reaction	Light brown, brown greyish	Bright yellow, yellowish brown deep wine

- For pearl millet hybrids, 1% phenol solution with a drop of 0.25% ammonia solution at pH 12 for 24 h.
- All male shows black colour
- All female shows intermediate colour

3. Alternate phenol test

- Soak in warm water for 10 minutes and then in 1% phenol

4. Lugols solution test/Alkaloid test

- Lupinous varieties
- Soak in lugols solution (Iodine + potassium iodide + water)
- Observe for presence or absence or alkaloids

5. Seed coat peroxide test

- Soybean
- Place seeds in 10 drops of 0.5% fuaicol + 1 drop of 0.1% hydrogen peroxidase
- Reddish brown – High activity
- Colour less – Low activity

6. Sodium hydroxide test

- Paddy, wheat, garden pea, fodder peas
- 5% Na_2OH

7. Potassium dichromate

- 1% for garden pea

8. KOH bleach test

- Sorghum
- 1:5 (w/v) potassium hydroxide plus 5-25% NaOCl for 5-10 minutes
- Presence or absence of tannic acid

9. Ferrous sulphate colour test

- Paddy
- 1.5% $FeSO_4$ for 4h

10. Hydrochloric acid test

- *Cucumis sativa, C.melo*
- 1N HCl

12. Iodine test

- To distinguish grain and weedy type millets
- 5% iodine
- Seeds of weedy type turns black brown in 5-7 minutes.

12.13. International Seed Testing Association (ISTA)

- Founded in 1924, with the aim to develop and publish standard procedures in the field of seed testing, with member laboratories in over 70 countries world wide.
- ISTA membership is truly a global network.
- ISTA secretariat is located at Zurich, Switzerland

12.13.1. Major achievements

- International Rules for Seed Testing, guaranteeing worldwide annually updated, harmonized, uniform seed testing methods

- Accreditation Programme including Accreditation Standard, Proficiency Testing Programme and Auditing Programme guaranteeing worldwide harmonized, uniform seed testing
- Issuing of the ISTA International Seed Lot Certificates by officially independent ISTA accredited and authorized laboratories
- Promoting of research, training, publishing and information in all areas of Seed Science and Technology and cooperation with related organizations such as ISF, OECD, UPOV and many others.

12.13.2. ISTA certificates

1. Seed lot certificate (Orange or green certificate)

- **Orange certificate** is issued when sample is drawn officially from the lot under the authority of a member station.
- **Green certificate** is issued when sample is drawn officially from the lot under the authority of a member station and testing for seed quality attributes is done in another country by a member station.
- A member station in the country where the lot is located shall be responsible for sampling, sealing and labeling and for despatech of the sample to the testing station in another country for analysis of the desired seed quality and issuance of the certificate.

2. Seed sample certificate (Blue certificate)

- Refers to the sample submitted for testing and shall be printed on blue paper.

12.13.3. Publications

- Seed Science and Technology (SST) is one of the leading international journals featuring original papers and review articles on seed quality and physiology as related to seed production, harvest, processing, sampling, storage, genetic conservation, habitat regeneration, distribution and testing.

13

Seed Health

13.1. Seed pathology

- Study of relationships between pathogens and seeds

13.1.1. Types of pathogen

- Fungi, bacteria and virus

13.1.2. Nature of infection

1. Soil borne diseases

- Caused by fungi only
- Wilting, yellowing, stunting and plant death
- Root rot: *Rhizoctonia*, *Sclerotina, Pythium, Phytophthora, Sclerotium, Fusarium, Cylindrocladium* and *Armillaria*
- Wilt: *Fusarium*, *Verticillium*
- Seedling blight and damping off: *Pythium* , *Phytophthora, Rhizoctonia, Sclerotium* and *Fusarium*

2. Seed borne diseases

- **Internally seed borne diseases:** Pathogen attacks seed, endosperm and embryo
- **Externally seed borne diseases**: Pathogens externally carryover on the seeds.

13.1.3. Mechanism

1. Systemic infection through seed

- Infection through vascular system: *Xanthomonas*

- **Examples of some infections that occur through the vascular system.**

Pathogens	Crops
Fusarium oxysporum	Pumpkin, pea, tomato
Plasmopara halstedii	Sunflower
Septoria glycines	Soybean
Verticillium dahliae	Spinach and sugar beet
Xanthomonas campestris	Bean, cabbage, rice, sweet pepper
Pseudomonas syringae pv. *lachrymans*	Cucumber

- Systemic infection through flower, fruits (*Fusarium*, *Pseudomonas*)
- Penetration through stigma: *Ustilago* spp
- Penetration through ovary wall: *Colletotrichum*
- Penetration through wounds and natural opening: *Xanthomonas campestris* pv. *phaseoli* in bean and *Pseudomonas syringae* pv. *lachrymans* in cucumber

2. Seed contamination or infestation

- Pathogen sticks on the seed surface (*Alternaria* spp)
- Accompanying contamination

Pathogens	Crops
Alternaria brassicae, *A. brassicicola*	Crucifers
A. longipes	Tobacco
A. radicina	Carrot
Ascochyta pinodella	Pea
Drechslera sorokiniana, *D. oryzae*	Rice
D. avenae	Oats
Fusarium oxysporum f. sp. *callistephi*	China aster
Sclerotia of *Rhizoctonia solani*	Eggplant, pepper, tomato
Tilletia caries, T. foetida, T. controversa, and *Urocystis agropyri*	Wheat

- **A number of bacteria contaminates seed surfaces.**

Pathogens	Crops
Corynebacterium flaccumfacjens pv. *flaccumfacjens*	Bean
P. syringae pv. *lachrymans*	Cucumber
P. *syringae* pv. Tomato, Tobacco mosaic virus, Tomato mosaic virus	Tomato
C. rnichiganense pv. *michiganense*	Pepper
Xanthomonas campestris pv. *campestris*	Cabbage
Pepper Mosaic virus	Pepper

3. Seed transmission

- **Systemic transmission**
- Infection of embryo: *X. campestris*
- Non embryo infection: *Corynebacterium michiganense* pv. *michiganense* in tomato and *Xanthomonas campestris* pv. *campestris* in cabbage,
- Episperm contamination: *Ustilago, Puccinia*
- **Non systemic transmission**
- Embryo infection: *Ascochyta pisi*
- Episperm infection: *Septoria nodorum*
- Episperm contamination: *Neovossia* spp
- Accompanying contamination:*Sclerotina*

13.1.4. Designated diseases

Crop	Common name	Causal organism
Paddy	Bunt	*Nepyossia horrida*
Wheat	Loose smut	*Ustilago tritici*
Sorghum	Kernel smut or grain smut	*Sphacelotheca sorghi*
Bajra	Head smut	*Sphacelotheca reiliana*
	Downy mildew/Green ear	*Sclerospora graminicola*
	Ergot	*Claviceps microcephala*
	Grain smut	*Tolyposporium pencillariae* *T. senegalense*
Greengram	Ashy stem blight	*Macrophomina phaseoli*
	Anthracnose	*Colletotrichum lindemuthlanum*
Cowpea	Ashy stem bright	*Macrophomina phaseoli*
	Anthracnose	*Colletotrichum lindemuthianum*
	Ascochyta blight	*Ascochyta* spp.
Cotton	Black arm,	*Xanthomonas malvacearum*
	Alternaria leaf spot,	*Alternaria macrospora*
	Helminthosporium leaf spot.	*Helminthosporium spicifemm*
Sunflower	Leaf spot	*Cercospora phaseolorum*
	Downy mildew	*Plasmopora halstedii*
	Vericillium wilt	*Verticillium dahliae*
Sesamum	Ascochyta blight	*Ascochyut phaseolorum*
Toria (Indian rape)	Alternaria blight	*Alternaria* sp.
Rai (Mustard)	Alternaria blight	*Alternaria sp.*
Yellow sarson	Alternaria blight	*Alternaria* sp.
Brown sarson	Alternaria blight	*Alternaria* sp.
Indian bean	Halo blight	*Pseudomonas phasiolicola*
	Bacterial blight	*Xanthomonas* spp.
	Ascochyta blight	*Ascochyta* spp.
French bean	Antliracnose	*Colletotrichum lindemuthianum*
	Bacterial blight	*Xanthomonas* spp.

	Antliracnose	*Colletotrichum lindemuthianum*
	Bean mosaic	*Macrosiphum pisi*
Dry bean	Bacterial blight	*Xanthomonas phaseoli*
Rajma	Bean mosaic	*Phaseolus* virus I
Sugarcane	Red rot	*Glomerella tucumanensis*
	Smut	*Ustilago scitaminea*
	Wilt	*Cephalosporium sacchari*
	Grassy shoot disease	Mycopplasma - like - organism
	Leaf scald	*Xanthomonas alhilmeans*
Tomato	Early blisht	*Alternaria solani*
	Leaf blight	*Xanthomonas vesicatpria*
	Mosaic	Tobacco Mosaic virus
	Leaf spot	*Stemphylium solani*
Brinjal	Blight	*Phomopsis vexans*
Capsicum and chilli	Late blight	*Alternaria solani*
	Antliracnose	*Gloeosporium piperitum*
Okra	Yellow vein mosaic viru	Hibiscus virus 1
Cucurbits	Mosaic	Cucumis mosaic virus
Potato	Brown rot	*Pseudomonmas solunaceantm*
	Root knot nematode	*Meloidogyne incognita*
Cauliflower, cabbage,	Black leg	*Leptosphaeria maculans*
knol-kohl	Black rot	*Xanthomonas campesiris* pv. *Campestris*
	Soft rot	*Erwinia carolovora*
Lettuce	Mosaic	Lettuce mosaic virus
Celery	Leaf blight	*Seploria apiicola*
Petunia	Leaf blotch	*Cercospora petunia*
	Leaf spot	*Ascochyiu peiuniae*
	Phyllostica leaf spot	*Phyllosticta petuniae*
	Leaf bright	*Alternaria alternate*
	Crown rot	*Phytophthora parasitica*
	Mosaic	Tabacco Mosaic virus
	Cucumber mosaic	Cucumber mosaic virus

13.2. Seed health testing

13.2.1. Field fungi

- Pathogen infection in the field while seed on the plant itself at 90-100 % RH and 24 - 25% moisture: *Alternaria*, *Curvularia* spp. *Cephalosporium* sp. *Fusarium* sp. *Drechslera* sp. *Phoma* sp. and *Verticillium* sp.

13.2.2. Storage fungi

- Invade seeds after harvest and in storage at 70-90% RH. Storage fungi are represented by 10-15 group species of *Aspergillus* and several species of *Penicillium*.

13.2.3. Methods of testing

1. Seed borne fungi detection

- **Dry seed examination**: Presence of *Acervuli*, *Pycnidia*, *Pericthecia*, *Sclerotia* on seed surface and mixed with seed.
- Seed coat discolouration, reduction in size and weight, shrivelled seed.
- **Blotter method**: Incubate in water soaked filter paper at 22°C and 12h alternate light and dark for 7 days and observe for fungi
- **Agar plate method**: Incubate in agar medium and procedure as above.
- **Seedling symptoms test**: Observe fungi in the germinated seedling
- **Wash test**: For smut and bunt fungi except smut of wheat and barley
- **NaOH test**: Soak seeds in 0.2% NaOH for 24h and separate black colour seeds infected by karnal bunt of wheat and kernel smut of rice
- **PCR**: Used to detect DNA of microorganisms

2. Seed borne bacteria detection

- **Dry seed examination**: As in fungi
- **Growing on test**: To identify *Xanthomonas* and *Pseudomonas*
- **Agar medium**: As in fungi
- **Indicator test**: Inoculate the water suspension of seed sample into the infiltration of primary leaf node of 10 days old bean seedling and observe for appearance of lesions
- **Serological test**: Test the supernatant of a washing of seed sample with antiserum of suspected pathogen
- **ELISA test**: Reaction between antibodies and antigen is amplified by sandwich method

3. Seed borne viruses detection

- Dry seed examination
- Growing on test
- Indicator test
- Serological test

13.3. Storage pests

13.3.1. Storage loss

- Weight loss upto 30 per cent
- Loss of market value due to discolouration

- Mould attack and caking
- Reduce germination

Damage due to	% of loss
Pests	36-43
Rodents	35-72
Fungus	4-8
Birds	12-34
Moisture	9-21
Others	1-7

13.3.2. Storage pests

- Rice weevil (*Sitophilus oryzae*) – Make round holes on seeds of rice, maize etc.
- Angoumois grain moth (*Sitotroga cerealella*) – Larva cause damage on seeds of rice, maize, sorghum and wheat
- Lesser grain beetle (*Rhyzopertha dominica*) – Insect damaged seeds of sorghum, maize
- Larger grain beetle (*Prostephanus truncates*) - Higher weight loss than others
- Bruchids (*Callosobruchus maculatus*) – Infected pulse seeds in the field itself and eats embryo only by making round holes
- Indian meal moth (*Plodia interpunetella*) – Damages seed only
- Rice moth (*Corcyra cephalonica*) – Major primary pests of cereals, groundnut
- Red flour beetle (*Tribolium castaneum*) – Feeds on already damaged seed
- Khapra beetle (*Trogoderma granaria*) – Attacks seeds of groundnut, cereals and legumes
- Saw-tooth beetle (*Oryzaephilus surinamensis*) – Feeds on damaged seeds of cereals, cowpea and cocoa
- Granary weevil (*Sitophilus granaries*) – Larva and adult damage seed
- Confused flour beetle (*Tribolium surinamensis*) – Larva and adult damage seed
- Long headed flour beetle (*Latheticus oryzae*) – Larva damages seed

13.3.3. Management

1. Sanitation

- Remove infested material.

- Do not mix new seed with old material that should be thoroughly fumigated.
- Clean the storage structure
- Brush away all traces of spilled seed, dust, etc.
- Remove dust from handling equipment and machinery
- Disinfect sacks and baskets by sunning or chemical treatment
- Take control measures early to prevent infestation of crops maturing in the field.
- Spray malathion 50EC at 5 litres per 100 m^2 area
- Fumigate with aluminium phosphide at 1 tablet (3g) per m^2 space thrice at 40-60 days interval

2. Assessment of damage

- Visual inspection - Examine for discolouration and nature of damage for different pests
- X-rays radiography - Examine to x-rays at 22KV, 3MA for 10 seconds at 30 cm distance
- Transparency test - Seed coat becomes transparent, if boil with lactophenol for 10 - 20 min.

- **Symptoms of damage caused by some storage pests**

Symptoms	Crop	Pest
Brown/damaged seed	Maize, rice, sorghum, wheat, oat, barley	Red flour, saw toothed and flat grain beetle
Circular holes	Pulses, wheat, rice, maize, sorghum and grasses	Pulse beetle, rice weevil, angumois moth
Irregular holes	Spices	Lesser grain borer, cigarette beetle
Seeds reduced to husk	Wheat	Khapra beetle
Webbing	Maize, rice, wheat	Indian meal moth
Musty odor and dust	All seeds	Mites

13.4. Seed borne nematodes

- May be internal or external
- Seed gall - *Anguina* sp.
- Discolouration in groundnut - *Ditylenchus destructor*

13.4.1. Seed transmitted nematodes

- *Anguina, Aphelenchoides, Ditylenchus, Heterodora, Rhadinaphelenchus*

13.4.2. Important seed borne nematodes

Crop	Seed borne nematodes	Scientific name	Symptoms	Management
Rice	White tip	*Aphelenchoides besseyi*	Infectes seeds. The nematodes lie in a quiescent stage beneath the half of the seeds. Whitening of the top 3-5 cm of the leaf leading to necrosis. Distortion of the flag leaf that encloses the panicle.	Seed treatment with hot water at 55°C for 10-15 minutes. Apply cartaphydrochloride @ 1.0 kg a.i/ha at 45 OOP. Apply carbofuran 3 G in nursery @ 1.0 kg a.i/ha 7 days before transplanting.
Cereals	Ufra	*Ditylenchus angustus*	Externally attached to seeds.	Delaying sowing or planting to revive the nematodes with first showers in the season and killing them by starvation rotation with jute or sesame. Application of carbofuran or mocap or monocrotophos in between planting and flooding
Wheat, triticale, rye	Ear cockle	*D. dipsaci* stem eelworm	Distorted leaves and stems are evident prior to heading. As diseased plants approach maturity, galls are formed in the florets, replacing the kernals. Seed galls in cereals are known as ear cockle. The galls are similar in shape to the seed, they replace hut are dark brown in colour.	Seed cleaning and crop rotation. Galls may be also removed by submersion of the seed in 20% brine solution (galls float to the surface), followed by thorough washing in water. Hot water treatment at 54°C for 10 min.
Pigeon pea	Cyst ncmatode	*Heterodera cajani*	Infected plants show yellowing, stunting, poor vigour and pod formation. At seedling stage of plant, pearl-like or lemon-shaped white female can be found attached with roots.	Summer ploughing of fields during hot months. Crop rotation with non host crops for 2-3 years. Seed treatment or soil application with *Pseitdomonas fluorescent* and *Trichoderma viride* @ 5+5 g/kg seed
Tomato, brinjal, chilli,	Root Knot nematode	*Meloidogyne incognita*	Presence of galls on the roots. Plants wilt rapidly.	Application of *Pseudomonas fluorescens* @ 10 g/rrr in nursery.

bhendi, snake gourd		Growth may be retarded and leaves may be chlorotic. Seedlings die in the seed bed and seedlings do not survive transplanting.	Crop rotation with non-hosts or resistant crops Use of resistant cultivars. Application of carboruran 3 G @ 1 kg a.i/ha.
Aster	*Aphelenchoides ritzemabosi*	Infestations usually start at the base of the stem and work upwards in the direction of the flower head. In severe attacks, most of the leaf area is killed and the flower head itself may become infested.	

13.4.3. Management

- White tip (*Aphelenchoides besseyi*) in rice – Hot water treatment @ 55°C for 10-15 min. or carbofuran
- Ufra (*Ditylenchus angustus*) – Carbofuran
- Ear cockle (*D. dipsaci*) – Hot water @ 54°C for 10 min.
- Cyst nematode (*Heterodera cajani*) in pigeonpea – *Pseudomonas* + *Trichoderma* @ 5+5 g/kg of seed

13.4.4. Detection

- Dry seed examination
- Desection technique
- Disection and staining
- Washing and extraction

13.5. Seed treatment

- Application of pesticides to seed to control seed pathogen and insects for safe storage

13.5.1. Methods of seed dressing

- Dust slurry, liquid, mist and gaseous form

13.5.2. Equipment

- Treatment drum, slurry treater, fumigation chamber

13.5.3. Types of pre storage treatment

- **Halogenation:** Expose the seeds to bleaching powder (calcium oxychloride) to dry or vapour form
- **Antioxidant:** Treat seeds with vitamin A, C, E to quench free radicals
- **Seed protectants:** Treat seeds with insecticides and fungicides to control seed and soil borne diseases and insects. Eg. Bavisin @ 2-4 g/kg
- **Seed disinfection:** Control of deep seated internally borne pathogens
- **Seed disinfestation:** Control of surface and externally borne pathogens
- **Fumigation:** Fumigate the seeds at 14% moisture with aluminium phosphide at 2-3 tablets/cu.m. Each tablet weighs about 3 g.

14

Seed Storage

14.1. Seed storage

- Maintenance of high germination and vigour until sowing

14.1.1. Principles of storage

- Seed storage conditions should be dry and cool
- Effective storage pest control
- Proper sanitation in seed stores
- Before placing seeds into storage, they should be dried to safe moisture limits
- Storing of high quality seed only, *i.e.*, well cleaned treated as well as of high germination and vigour.

14.1.2. Stages of seed storage

- Storage on plants (Physiological maturity until harvest)
- Harvest, until processed and stored in a warehouse
- In storages (Ware houses)
- In transit (Rail wagons, trucks, carts, railway shed etc)
- In retail stores
- On the user's farm

14.1.3. Classification

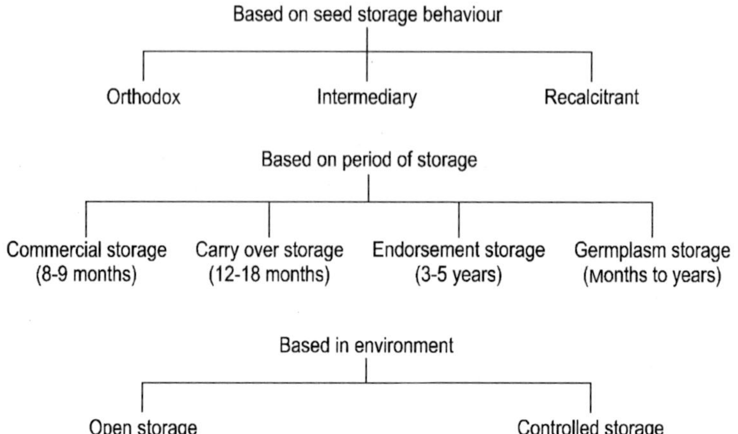

14.1.4. Differentiate between orthodox seed and recalcitrant seed with examples

Orthodox seed	Recalcitrant seed
Can be dried to 6-10% moisture depending upon the species	Cannot be dried below 30-35% moisture and should be maintained at this moisture
No decline in germination, if dried	A decling in germination, if dried below this moisture
Tolerant to desiccation (5-7% moisture) and low temperatures (0-5°C)	Intolerant to desiccation (30-35% moisture) and low temperature (12-15°C)
Examples. Cereals, Pulses, Oilseeds etc	Example. Rubber, Coffee, Coconut, Mango, Citrus, Swietenia, Jamun
Prevalent in arid and semi arid environments	Prevalent in warm humid climates
Can be stored for several years	Can be stored for few days to few months
Small to medium size seed often with hard seed coat	Usually medium to large size and heavy seeds
Dormancy often occurs	Dormancy absent or weak
Not metabolically active when shed	Metabolically active when shed
Accumulation of dry weight ceases before maturation	Accumulation of dry weight upto the time of seed dispersal

14.1.5. Types of storage

1. Based on storage behaviour

- Orthodox: cereals, millets, pulses, oilseeds
- Recalcitrant: Rubber, coffee, mango, citrus
- Intermediary: Neem

2. Based on environment

- Open storage
- Controlled storage

3. Based on genetic make up

- **Microbiotic:** Short lived (Onion, soybean, groundnut)
- **Mesobiotic:** Medium lived
- **Macrobiotic:** Long lived

14.1.6. Factors

1. Internal factors

- Genetic
- Initial seed quality
- Provenance
- Weather during seed formation
- Pre-harvest infestation
- Seed moisture

2. External factors

- RH
- Temperature
- **Harrington rules**
- For every decrease in 1 % moisture, life doubles; This rule is applicable between moisture content of 5 -14%.
- For every decrease in 10^0F (5.6^0C) temperature, life doubles; This rule applies between O°C to 50°C.
- Sum of RH and temperature in F should not exceed 100. i.e. 50% RH + 50^0F = 100.
- Gases
- Microflora, insects and mites

3. Packaging container

- Moisture pervious container (Cloth bag, paper bag)

- Moisture impervious and vapour pervious container (Polyethylene bags of 100 - 300 gauge thickness)
- Moisture impervious and vapour proof container (Polyethylene bags of 700 gauge thickness, aluminium foil pouches)

14.1.7. Seed store sanitation

- Pre-cleaning of seed store thoroughly and disinfect it
- Store should be dry and cool
- If use old bags, disinfect / fumigate them
- Use wooden pallets for stacking upto 6-8 bags height in criss-cross
- Spray malathion 50 EC @ 5 lit /100m^2 for every 3 weeks
- Aluminium phosphate @ 3 g /cu. m for fumigation
- In pulses, insect infestation comes from field (e.g. bruchids).

14.2. Seed deterioration

- Inexorable and irreversible event in seed

14.2.1. Basis of seed deterioration

1. Morphological changes

- Discolouration (Darkening in groundnut)

2. Ultrastructural changes

- Loss of membrane integrity leading to increased leakage of solutes
- Swollen mitochondria results in decreased respiration
- Dissociation of ribosomes results in retardation of protein synthesis
- Degeneration of ER

3. Biochemical changes

- Loss of enzymes activity (Amylases, proteinases, cytochrome oxidases, glyceraldehyde phosphate dehydrogenases, catalase, peroxidase)
- Reduced germination
- High RQ value

14.3. Germplasam storage and cryopreservation

14.3.1. Germplasm

- Collection and conservation of plant genetic resources

1. Conservation methods

- ***In situ* conservation**: Conservation in natural habitat
- ***Ex situ* conservation**: Preservation outside natural habitat
- **Seed bank**: Collection and preservation of seeds of food and other crops
- Not for exchange
- Meet out the requirements due to natural disasters
- Millennium seed bank lab @ Royal Botanical Gardens, Kew, London
- Svalbard Global Seed Vault, Norway
- ***In vitro* conservation**: Conserve at low/non greezing temp. of 1 to 9°C
- Paddy - IRRI, Philippines
- Wheat and maize - CIMMYT, Mexico
- Cassava - CIAT, Columbia; IITA, Nigeria
- Patato - CIP, Peru
- Sweet potato - CIP, Peru; AVRDC, Taiwan; IITA, Nigeria

14.3.2. Cryopreservation

- Preserve in liquid nitrogen (-196°C)
- Proposed in 1968 for cell culture maintenance
- Conservation of shoot tips in pear, plum, olive etc.

15

Seed Industry and PPV & FR Act

15.1. Seed industry

15.1.1. Seed industry overview

Year	
1905	IARI
1908	AOSA: Association of Official Seed Analysts
1919	ICIA: International Crop Improvement Association, USA
1921	ESTA: European Seed Testing Association (Denmark)
1922	SCST: Society of Commercial Seed Technologist
1922	CSAAC: Commercial Seed Analyst Association of Canada
1924	ISTA: International Seed Testing Association (Switzerland)
1928	Royal Commission on Agriculture submitted report citing that facilities for increasing the supply of breeder seeds were inadequate and that varietal purity of seeds was not maintained.
1929	ICAR: New Delhi; CRISAT: International Crop Research Instt., for Semi-Arid Tropics, Hydrabad
1952	Grow more seed enquiry committee (First 5 year plan, 1951-56)
1956	Seed health testing was established in IARI
1957	All India Co-ordinated Maize Improvement Project began
1960	Similar projects on Sorghum and Pearl millet began
1961	Maize hybrids released
1961	Seed multiplication team review and World seed year – First STL (CSTL)
1963	National Seeds Corporation established
1964	First sorghum hybrid released
1965	First pearl millet hybrid released; 250 t of dwarf wheat was imported from CIMMYT, Mexico
1966	Seed Act passed (Came into force in Oct, 1969).
1968	Seed Review Team report was submitted; Seed Rules framed
1969	AOSCA: Association of Official Seed Certifying Agency, USA
1969	SFCI (State Farm Corporation of India) framed
1969	Tarai Development Corporation established with World Bank Assistance; concept of Compact Area Approach for seed production
1970	STL was started in Tamil Nadu

1971	1st cotton hybrid (C.T. Patel), MSCS (Minimum Seed Certification Standard)
1971	National Commission on Agriculture submitted report; Indian Society of Seed Technologist framed
1972	TNAU: Seed Technology was started
1972	ISST organized first All India Seed Seminar
1973	First seed technology course in India at TNAU
1975	Project report on NSP was submitted
1976	NSP I
1981	1st workshop on Seed Technology under NSP was held.
1981	NSP II
1983	Seed Control Order (Under Essential Commodities Act, 1956)
1986	Seed Certification was started in India
1987	An expert group on seed was constituted and the report was submitted in June, 1989
1988	A New seed policy to encourage import of seed and planting material was announced in September.
1990	NSP III
1994	GOI signed: GATT approval, ISTA rules were revised
2001	National Seed Policy
2002	PVPFR
2004	New Seed Bill

15.2. PPV&FR Act

- 2001 - Enacted
- 2003 - Regulations formulated
- 2005 - PPV&FR Authority came into force

15.2.1. Features

- Protects plant varieties and plant breeders
- Recognizes and protects farmers variety and develops farmers varieties
- Protects public interest
- Promotes food security

15.2.2. Registration of varieties

- Covers all plant expect microbes
- Started registration of 12 plants species in May 2007 (Rice, bread wheat, maize, sorghum, pearl millet, chickpea, pigeonpea, greengram, blackgram, lentil, field pea and kidney bean)

15.2.3. Farmers right

- Farmers varieties is registered, if fulfill DUS requirements

15.2.4. Breeders right

- Exclusive right to produce, sell, market, distribute, export and import of registered variety

15.2.5. Essentially derived variety

- EDV derived from initial variety, but should differ atleast for one character

15.2.6. Duration of protection

- 18 years for trees and vines
- 15 years for other crops

15.2.7. Researchers right

- Right to use registered variety for research purpose

15.2.8. National Gene Fund

- Benefit sharing, compensation, on-farm conservation of traditional varieties and land races.